The Latin American Studies Book Series

The Latin American Studies Book Series promotes quality scientific research focusing on Latin American countries. The series accepts disciplinary and interdisciplinary titles related to geographical, environmental, cultural, economic, political, urban and health-related research dedicated to Latin America. The series publishes comprehensive monographs, edited volumes and textbooks refereed by a region or country expert specialized in Latin American studies.

The series aims to raise the profile of Latin American studies, showcasing important works developed focusing on the region. It is aimed at researchers, students, and everyone interested in Latin American topics.

Submit a proposal: Proposals for the series will be considered by the Series Advisory Board. A book proposal form can be obtained from the Publisher, Andrea Brody-Barre (andrea.brody-barre@springer.com).

Lucía Angélica Magnin

The Landscape and Mobility of Hunter-Gatherer Societies

Archaeological Distributions in the Central Plateau of Santa Cruz, Argentina

Springer

Lucía Angélica Magnin
Consejo Nacional de Investigaciones
Científicas y Técnicas (CONICET)
La Plata, Argentina

División Arqueología
Facultad de Ciencias Naturales y Museo -
Universidad Nacional de La Plata
(FCNyM-UNLP)
La Plata, Buenos Aires, Argentina

ISSN 2366-3421 ISSN 2366-343X (electronic)
The Latin American Studies Book Series
ISBN 978-3-032-08126-1 ISBN 978-3-032-08127-8 (eBook)
https://doi.org/10.1007/978-3-032-08127-8

This Springer imprint is published by the registered company Springer Nature Switzerland AG
The registered company address is: Gewerbestrasse 11, 6330 Cham, Switzerland

With this book, I hope to make a contribution to the knowledge and respect for the diversity of cultures that inhabited the Patagonian region in the past and still do so today. I dedicate this book to the indigenous people of Patagonia, past and present, who have courageously defended their history and indigenous ancestry.

I also dedicate it to the National System of Public Education and Science of my country, and to my family, the most important source of support for this work.

Foreword

Opening a New Door Fifteen years ago, when the doctoral thesis that gave rise to this book was written, the use of Geographic Information Systems (GIS) in archaeology was still incipient in Argentina. Although there were international precedents in spatial analysis, landscape archaeology and distributional studies applied to nomadic societies, in local contexts such as Patagonia and the Pampa—historically occupied by hunter-gatherer groups—the incorporation of GIS tools was still a distant horizon. In this panorama, Lucía Magnin's work represents a milestone. Her research opened a line of study that, over time, consolidated as a robust approach to explore issues related to human mobility, settlement configuration, raw material provisioning, and the circulation of goods, ideas, and people, at different temporal and spatial scales. The development of this proposal had the valuable support of teachers from the Aerial Photointerpretation Department of the Faculty of Natural Sciences and Museum (UNLP), where Lucía began her training in innovative methodologies and analytical techniques applied to the complex archaeological record of the Deseado Massif.

This pioneering GIS work in Argentina opens a new analytical perspective to apply to archaeological statements of local and regional scale proposed for that time. This approach enriches the theoretical and practical contents of the archaeological experience in the same region of the Deseado massif and multiplied the questions about territoriality, identity, mobility, and access to resources. In this context, the open-air sites took on an unprecedented significance for the construction of communication and social circulation networks among hunter-gatherers. The revaluation of the documentary and interpretative potential of these archaeological manifestations is possible with the correct methodological application presented in this book.

I want to emphasize that prior to this work, for the Deseado Massif, the geographic approximation detail was inappropriate for fine-grained archaeological spatiotemporal analyses and the only thing available to archaeologists working in the region was until 2010 a cartographic accuracy of scales 1:500,000 and only for certain places of the massif in 1:250,000. Therefore, this work is a methodological advance because it makes the GIS data for the locality available for regional and extra-regional comparison.

This work adds spatial resolution to the archaeology of this area and provides a corpus of data from numerous opn-air sites and isolated finds recorded in La Primavera. This data complements the existing archaeological knowledge of the area, which, except for the local flint qarries, had not been analyzed until now. Previous research had focused on the in-pepth analyses of the stratigraphic sequences and rock art in the Maripe cave and several rock shelters in the canyon. The research shows the great importance contributed by the study of surface sites and subsurface stratigraphies, from which the author obtained radiocarbon dates as well as archaeological contexts corresponding to the late Holocene. This systematic documentation shows the informative potential of this abundant evidence of Patagonian contexts. The application of GIS allows the author to consistently integrate empirical statements with theoretical ones, without "leaping into the void" of the heterogeneous information derived from caves with reliable stratigraphies and rock art and open-air sites. This methodological link results in models about human habitation and landscape transformations. In this context, and taking the La Primavera valley with the adjacent sierras and plateaus as a landscape unit, the work captures landscape changes throughout the entire hunter-gatherer occupation involving some 9500 years. Thus, the proposed model characterizes three different forms of the use of these spaces, giving a reliable picture of the important moments of socio-environmental change, comprising the early Holocene, the middle Holocene, and the late Holocene. The archaeological evidence in the different microenvironments (open-air caves, mallines, plateaus, etc.) is very varied over time. Its accurate methodology allowed us to provide new information about the effective occupation of the high plateaus toward the later Holocene, while during the exploration and colonization phases, toward the early and middle Holocene, caves and valley bottoms were the preferred places for human groups to set up residential camps. The various frames of reference used here are adequate to cover the different questions generated by the approach to the study of a space, which has become a place and a human history, in a broad temporal perspective of thousands of years. In this way, caves are no longer seen as the only places for the settlement of residential bases. This interpretative category is taken to archaeological sites on the surface and enriches the proposals of the settlement patterns of Patagonian hunter-gatherers.

The use and the deep methodological development of the GIS tool is a valuable contribution to the regional study of the area, but it can also be applied to different regions and archaeological situations.

Off-Site Archaeology, Views, and Perspectives I want to emphasize that the local work was built on the basis of a systematic use of theoretical–methodological tools and innovative techniques that aim to explore the use of space on a broad temporal scale and in all its socio-cultural dimensions. All variables are represented on a georeferenced basis, thus we can assure that the enormous quantity and variety of archaeological and environmental data compiled were considered with an equitable treatment, which increases the quality of comparisons and interpretations and results in models of space use trends in the long term.

In this line, Lucía advanced in taking me out of the caves and showing me the fabulous knowledge that we can find "out in the open" and then relating it to the extensive stratigraphic sequences that we had also worked on together. This study is a true reference for hunter-gatherer archaeology, and we are pleased to see it successfully multiplied today in other archaeological localities of the Patagonian plateau region. Other important concepts and details provided by this work have been left out of this review, but I leave to the reader the stimulating task of discovering little by little, in the precision of her text and the excellent and unmissable figures, the subtle relationship made by the author between the present and the past, hunter-gatherers and landscapes.

July 2025

Dr. Laura Lucía Miotti
Consejo Nacional de Investigaciones
Científicas y Técnicas (CONICET)
La Plata, Argentina

División Arqueología
Facultad de Ciencias Naturales y
Museo - Universidad Nacional de La
Plata (FCNyM-UNLP)
La Plata, Buenos Aires, Argentina

Preface

This work is an adaptation of my doctoral thesis. It was defended in 2010 at the Facultad de Ciencias Naturales y Museo de la Universidad Nacional de La Plata in La Plata, Buenos Aires, Argentina, to earn my Ph.D. in natural sciences. The original work is in Spanish and is titled *Distribuciones arqueológicas en la Meseta Central de Santa Cruz. Implicancias para los estudios de uso del espacio y movilidad de sociedades cazadoras recolectoras*. The thesis director was Dr. Laura Lucía Miotti, and the thesis is hosted at http://sedici.unlp.edu.ar/handle/10915/24773.

The time elapsed between the creation and publication of this study is due to the demands of scientific careers in Argentina, as well as the publishing opportunities that occurred throughout my academic and personal history. Nevertheless, the book remains relevant to current Patagonian archaeology. It is sufficient to say that no other location studies exist for continental South Patagonia today.

As part of the "Latin American Studies" book series, I am confident that this work will increase its visibility and generate renewed interest in this line of analysis.

The book is structured into nine chapters derived from the original thesis. The original study problem, methodology, results, and discussion constitute an advance in the field of Patagonian hunter-gatherers, with the study of landscape use forming the core of this work. This translation from the Spanish includes updated contextual information on recent advances and perspectives in the region. Chapter 10 is a new addition, presenting a restudy of locational patterns through the application of novel, multimodel selection statistical techniques for Patagonian hunter-gatherer archaeology.

Readers should be aware that except for Chap. 10, this work is a translation of a 15-year-old study. Although the most relevant bibliography was updated in the process, some absences or omissions may be apparent if the study was conducted today. The time gap between the doctoral thesis and book publication is also evident in the manuscript's focus on the importance of Geographic Information Systems (GIS) for archaeological spatial modeling, a subject that is well-established in academic research today. On the other hand, the use of GIS in archaeological spatial modeling is a highly relevant practice in the field of archaeology today, but the systematic approach applied here is novel in Patagonian archaeology. GIS tools are essential for managing, analyzing, and visualizing spatial data in archaeological research. These tools aid in interpreting past human behavior and landscapes. The valuable information about spatial patterns obtained for the La Primavera locality awaits similar research in other sectors of the Deseado Massif and other regions to study mobility and the configuration of seasonal and annual territories. I hope the detailed theoretical–methodological framework serves as excellent material for how to proceed.

I would like to dedicate my final words to the indigenous peoples of Patagonia. When I began working in this area, I had no connection to the urban and rural Tehuelche, Mapuche, and Mapuche-Tehuelche communities of Santa Cruz Province, Argentina. These communities are known today for their struggles to defend their rights. Today, we have embarked on a path of conducting archaeology in territories where indigenous peoples continually contest materialities, the past, and the present. We actively seek encounters to foster the participation of indigenous peoples and the various social actors involved.

La Plata, Argentina Lucía Angélica Magnin

Acknowledgments

After defending my thesis, I looked for different ways to publish it as a book. The color maps and photographs were an important part of the work, but they implied high publishing costs. Some years later, Dr. Jorge Rabassa, the Springer Nature Regional Editor for Latin America of the Earth System Sciences and Latin American Studies book series, presented me with an opportunity. When he extended an open invitation to researchers willing to publish their original work at the book presentation ceremony for the archaeological site Piedra Museo at the La Plata Museum, I realized this could be an opportunity to publish my doctoral thesis as a book. The kind assistance of Ms. Juliana Pitanguy, editor of Latin American Studies; Ms. Banu Dhayalan, production editor; and especially Dr. Germán Gasparini was fundamental to achieving my goal of writing a scientific book in a foreign language for a broad audience.

I thank Dr. Facundo Palacio for his guidance on the GLM application in the final chapter of the book.

I would like to express my gratitude to Dr. Juan Bautista Belardi and Dr. César Parcero Oubiña, who kindly served as evaluators for this work, for their valuable contributions and observations, which were intended to improve this book.

I am always thankful to my thesis director for my original doctoral thesis work. Laura Miotti for her generosity and confidence. I would also like to thank Juan Carlos Gómez, the co-director of the CONICET scholarship program, for sharing his knowledge and opening his home, which also served as a high-tech remote imaging lab. Finally, I thank the many interns, collaborators, partners, and colleagues who participated in the field and laboratory work, especially Santiago Medel, Enrique Terranova, Bruno Mosquera, Franca Griffiths, Raúl González Dubox, Juan Pablo Pérez, Gisela Becerra, and Sebastián Carreño. Finally, my family Santiago, Jujlián, and Isabela were very patient and understanding of the extra time I had to dedicate to this work. I am deeply grateful to them.

The provincial authorities of Santa Cruz permitted the research. The Municipality of Pico Truncado and the rural inhabitants provided invaluable logistical support and other assistance during the challenging fieldwork.

This work was made possible by postgraduate scholarships from Argentina's National Council for Scientific and Technical Research (CONICET). Financial support was provided by various grant institutions that subsidized projects directed by Dr. Laura Lucía Miotti, including the National University of La Plata (UNLP) and the National Agency for Promotion of Science and Technology (ANPCyT). I also received significant support from the Marie Curie Host Fellowship for Early Stage Researchers Training (contract MEST-CT-2004-514509) at the Wolfson Archaeological Science Laboratories at University College London (UCL) Institute of Archaeology, which was funded by the European Union.

I am very grateful to all of these individuals and institutions.

Contents

Abbreviations

A	Rock art sites
Ch	Chenque (stone structure, funerary cairn)
HG	Hunting ground
HS	Hunting stand
LAL	Limited activities loci
MAL	Multiple activities loci
NE	Negative evidence
OK	Occasional knapping
PP	Hunt primary processing site
Q	Quarries for extracting lithic raw materials
TC	Transient camp
V	Site of extraction of vegetal resources
W	Lithic workshops

Acronyms Used for Varieties of Lithic Raw Materials Variety

ABG	Andesite, basalt, and gabbro/diabase
Abl	Black andesite
Ag	Green andesite
Ay	Yellow andesite
Bbl	Black basalt
Bbr	Brown basalt
Bg	Green basalt
Cbbrg	Chalcedony, brown, black, red, green
Cbl	Black chalcedony
Cbr	Brown chalcedony
Cr	Red chalcedony
Cw	White chalcedony

Cy	Yellow chalcedony
Cyw	Chalcedony, white, and yellow
Gg	Green gabbro
Obl	Obsidian, black
P	Petrified wood
PWbl	Black petrified wood
PWbr	Brown petrified wood
PWw	White petrified wood
PWy	Yellow petrified wood
Rg	Green rhyolite
Rr	Red rhyolite

Chapter 1
Introduction

Abstract This chapter presents the study's problem statement, questions, and objectives. This study focuses on settlement patterns among hunter-gatherers, the detection of spatial mobility patterns, and the use of space in the Central Plateau of Santa Cruz province, Argentina. The objectives include acquiring new data and implementing and exploring a methodology that incorporates techniques successfully applied in archaeological projects worldwide, offering positive perspectives. This study pioneers the application of GIS (Geographic Information Systems) and statistics to analyze the archaeological landscape in the study region.

Keywords Geographic Information Systems (GIS) · Human spatiality · Hunter-gatherer archaeology · Landscape · Mobility

1.1 Study Problem and Objectives

In 1520, Antonio Pigafetta, the Italian chronicler of the first trip around the world, wrote about his encounter with Tehuelche indigenous people in Puerto San Julián on the central coast of Argentina's Santa Cruz province in the sixteenth century.

> ...we stayed there for two whole months without ever seeing any inhabitant. One day -when we least expected it- we saw a giant who was next to the sea, almost naked and was dancing, jumping, and singing, and at the same time, he was throwing sand and dust on his head...our captain (Magellan) called that kind of giants Patagonian... (Pigafetta, [1525] 2001:41)

Pigafetta's writings made the region world-famous and gave it its name: Patagonia. However, far from the coast, in the Santa Cruz central plateau or Deseado Massif, there are no historical documents known today about the first contact between the two cultures. Therefore, archaeological investigations in the region are fundamental to understanding ancient human occupations and their subsequent history.

The archaeological wealth of the area was first revealed in *The Stone Age in Patagonia*, a synthesis of regional archaeology written by Félix Outes. This work was based on the study of Florentino Ameghino's private archaeological collection, assembled by his brother, the explorer Carlos Ameghino. It also drew from the collections of the

L. A. Magnin, *The Landscape and Mobility of Hunter-Gatherer Societies*, The Latin American Studies Book Series, https://doi.org/10.1007/978-3-032-08127-8_1

national museum of Buenos Aires, which were formed by donations from various individuals and materials gathered by the naturalist Karl Hermann Konrad Burmeister during his travels between 1882 and 1887 (Outes, 1905). However, the archaeological study of the central plateau in the Santa Cruz province began with Francisco de Aparicio's field trip in 1933 (De Aparicio, 1933–1935), which occurred more than four centuries after the first contact between indigenous peoples and Europeans. De Aparicio toured the area, documenting his findings with photographs, maps, and writings. He became one of the pioneers of archaeological studies in the central plateau. The current image of the area's original settlers, organized into hunter-gatherer societies with high mobility strategies who maintained social ties with other groups and had extensive knowledge of the landscape, is the product of nearly 60 years of professional research.

The first field studies, which included excavations, were carried out in the caves of the Los Toldos archaeological site. These studies were directed by Osvaldo Menghin in the 1950s (Menghin, 1952) and continued with the work of Cardich et al. (1973, 1982); Cardich (1977, 1978) Durán (1986–1987); Aguerre (1987) and Miotti (1989). Recently, new questions and researchers have expanded stratigraphic studies in the area (Blanco, 2015; Carden, 2009; Cueto, 2015; Frank, 2011; Garcia Añino, 2018; González Dubox, 2024; Hermo, 2008; Lynch, 2016; Marchionni, 2013; Miotti, 1992, 2003, 2006; Miotti and Salemme, 2004; Mosquera, 2016; Paunero, 2003a, 2003b; Paunero et al., 2007; Skarbun, 2011; Valiza Davis, 2020). Some of the fundamental research topics within current archaeological studies for deepening knowledge of these societies are the age of the first settlement, subsistence, use of the landscape, and mobility strategies. These themes have been studied based on evidence from stratigraphic sites, and to a lesser extent, surface sites. This research raises the issue of the need to expand knowledge of the activities carried out by past hunter-gatherer societies throughout the entire landscape of the study area. One way to address this issue is to maximize the information obtained from surface archaeological distributions. Here, sites, assemblages, and isolated finds are used to generate models that describe the location of different types of evidence in relation to important explanatory factors.

This book is based on a doctoral thesis developed as part of broader archaeological research in the Santa Cruz central plateau (Magnin, 2010). The book contributes to regional archaeology in two main ways. First, it studies the use of space by ancient societies beyond the information provided by archaeological sites. It does so by including the range of activities performed in the landscape (Borrero et al., 1992; Foley, 1981). Second, it establishes a methodological basis that explains and quantifies the variables and conditions involved in explaining spatial distributions. Since the generated model is formalized (Köhler, 1988; Lock & Harris, 2000), it serves as a foundation for comparison with other archaeological contexts. Some of the guiding research questions are: can patterns be identified in the distribution of archaeological remains in la primavera canyon and its surroundings?; what variables may be relevant to proposing a possible explanation for that distribution?; what formation processes have occurred?; what interpretations may be considered in systemic terms?; what use was given to this sector by ancient populations?; what places in the landscape were most probably chosen for camps?; how are other kinds of evidence structured around them?; do these distributions indicate year-round or seasonal occupation?; what chronology are the surface archaeological deposits assigned to?

The first general objective of this work is to further study landscape use strategies and the mobility of the ancient inhabitants of the study area. To that end, the specific objectives are to: (1) register surface archaeological remains and characterize environmental resources in the area; (2) control for sampling biases and differential archaeological visibility; (3) critically evaluate the suitability of available environmental data as proxies for past landscapes in the analysis; (4) include variables modeling the physical and social environments to study archaeological distributions; (5) describe patterns and, based on the results, investigate particular aspects of the locations in more depth; (6) interpret the results based on regional archaeological knowledge, archaeological studies of hunter-gatherers from other areas, and ethnographic, ethnohistorical, and ethnoarchaeological information.

A second general objective is to develop an operational work methodology that can be applied with appropriate modifications in other archaeological sites in the basin and in areas with similar archaeological issues. To achieve this, the methodology and procedures must be explained in a way that allows them to be replicated in other contexts for comparison of the resulting patterns. Methodological clarity also allows for the possibility of re-studying the same data with new techniques or different adjustments to its parameters.

Chapter 2
Study Area

Abstract This chapter describes the most significant environmental and paleoenvironmental features of the Santa Cruz Province Central Plateau study area. The current landscape is analyzed by investigating the reliability of the current distribution of resources as analogs of past landscapes. Taphonomic and geoarchaeological considerations are incorporated into this analysis, demonstrating their significant potential to advance this research in two primary ways: evaluating which current landscape resources were potentially significant to past societies and providing insights into archaeological formation processes.

Keywords Archaeological visibility · Caves and rockshelters · Climate changes · Deseado Massif · Landscape units · Patatgonia steppe · Semi-desert climate · Subsistence resources · Taphonomic factors

The basis for studying human location decisions at the La Primavera locality is understanding the variability of past regional resources and evaluating archaeological formation processes in different landscape areas. This chapter explores the available georeferenced information and evaluates its potential as an indicator of past resources and on site formation history. Therefore, it is important to gather the available maps and field data and determine their potential for this study. Some of the maps used have a scale of 1:500,000 (geological provinces) and 1:250,000 (lithological units, broad geomorphological units, and floristic units maps, by the Argentine Geological Mining Service and the National Geographic Institute of Argentina SIG IGN 250). These maps provided general information about the area, which was used to construct a more detailed geomorphological map. This process included a field survey and interpretation of satellite imagery, to produce cartography at a scale of 1:50,000 (Gómez & Magnin, 2008. See the description of the methodology in Chap. 5 for more information). In the latter case, the map units inform us about the prevalent formation processes affecting the archaeological evidence, based on the available geoarchaeological and taphonomic information (Marchionni et al., 2020).

Three independent sources of information are considered regarding the resources used by past populations: archaeological studies of the area, ethnohistoric records

L. A. Magnin, *The Landscape and Mobility of Hunter-Gatherer Societies*, The Latin American Studies Book Series, https://doi.org/10.1007/978-3-032-08127-8_2

referring to the use of this environment by mobile hunter-gatherers, and the current use of natural resources by the area's inhabitants.

Previous archaeological studies have proposed that the area was favorable for human occupation (Miotti, 1992, 1995). At the end of the Pleistocene especially, resources were abundant, including water, vegetation, fauna, and lithic raw materials. These investigations, primarily conducted through excavations, provide information about the resources and how they were used.

The ethnographic and archaeological compilation gathered by Martinić (1995), as well as the chronicles of the travelers J. Claraz and G. Musters and the descriptions of M. Gusinde, who were in direct contact with Patagonian hunter-gatherers, provide important information from a more recent period. Jorge Claraz (1832–1930), a Swiss naturalist, entered the area between the Negro and Chubut rivers in 1865–1866 accompanied by a group of indigenous Pampas and Tehuelche. He wrote a detailed travel diary containing valuable geological, biological, and anthropological information. George Chaworth Musters (1841–1879) wrote the second chronicle. He traveled to the Argentine provinces of Santa Cruz, Chubut, and Río Negro for nine months in 1869. Musters was born in Naples to English parents. A captain in the British Navy, he undertook this trip in search of adventure after reading about Charles Darwin's voyages aboard the Beagle (1831–1836). Eager to explore continental Patagonia, Musters traveled from Punta Arenas to Carmen de Patagones through the interior. He followed an old indigenous path that ran parallel to the Andean foothills and coincided with part of the current National Route 40. Finally, Martin Gusinde (1882–1989), an Austrian priest and ethnologist, made four trips to Tierra del Fuego between 1918 and 1924. His work describes his experiences with the indigenous Selk'nam people, who had a mobile, terrestrial economy based on hunting and gathering.

Other sources of ethnographic information include the four interviews with Silvana Chapalala (Pati) (1901–2002) conducted by researcher Ana Aguerre between 1995 and 1997. Pati was a Tehuelche descendant on her father's side and Spanish on her mother's side. She lived in the Río Pinturas area in the traditional manner until she was sixteen. Her family belonged to the Northern Tehuelches, who moved to more remote areas farther south after the Argentinean state's military offensive against free indigenous peoples, known as the Desert Campaign (1883) (Aguerre, 2000).

The natural resources used by the local inhabitants of the Central Plateau were identified through informal interviews with current residents of the area. For example, Silvia "Chela" Pandolfi de Iribarne, a local resident deceased in 2023 who inherited her knowledge from her Tehuelche grandmother, discusses the uses of local flora.

2.1 Environmental Characteristics

The study area is defined by two archaeological scales (García Sanjuán, 2005). On a macro scale, the area extends approximately 140 km from west to east and 80 km from north to south. It is located between 47°30′53″ and 48°6′43″ S latitude and 67°36′59″ and 69°10′2″ W longitude in the province of Santa Cruz (extra-Andean southern Patagonia, Argentina). The area covers 11,200 km^2 and includes the Zanjon Rojo, Zanjon Blanco, and Zanjon del Zorro endorheic temporary stream basins. Within this macro area, the sector that was systematically prospected and sampled for this study corresponds to a mesoscale and extends 100 km^2, covering the archaeological locality (Figs. 2.1, 2.2, and 2.3).

2.1.1 *Regional Geology*

The study area is located within the Deseado Massif, also known as the Nesocratón of the Deseado, a geological province extending from the Deseado River in the north to the Chico River in the south and from the Atlantic coast in the east to the Mayo River Dorsal in the northwest (De Giusto et al., 1980:1418; Leanza, 1958; Miotti, 1989, Ramos, 1999) (Fig. 2.4). This province exhibits positive and rigid behavior, as well as high relative tectonic stability. This is in contrast to the marked subsidence observed in neighboring regions, such as the San Jorge Gulf Basin in the extra-Andean Chubut region and the Magellanic Basin (De Giusto et al., 1980). However, recent studies suggest that, following the cratonization of the pre-Permian basement, the region experienced land subsidence. Two episodes of large-scale rifting began at the end of the Paleozoic era and intensified during the Jurassic period. For this

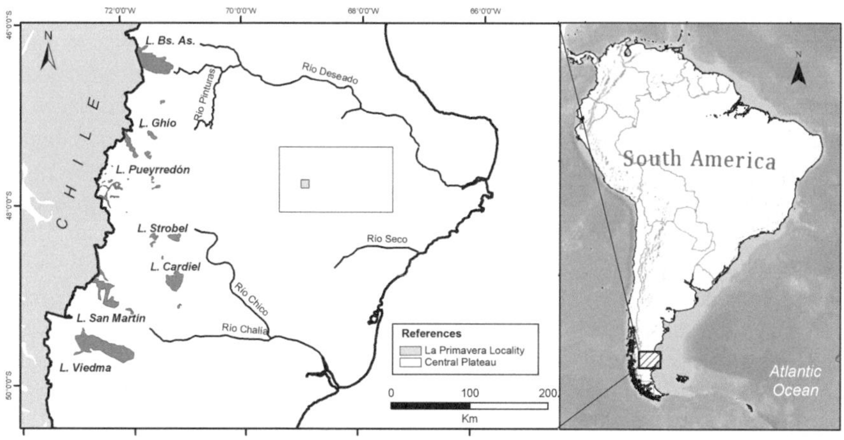

Fig. 2.1 Location of the macro study area in Santa Cruz province, Argentinean Patagonia

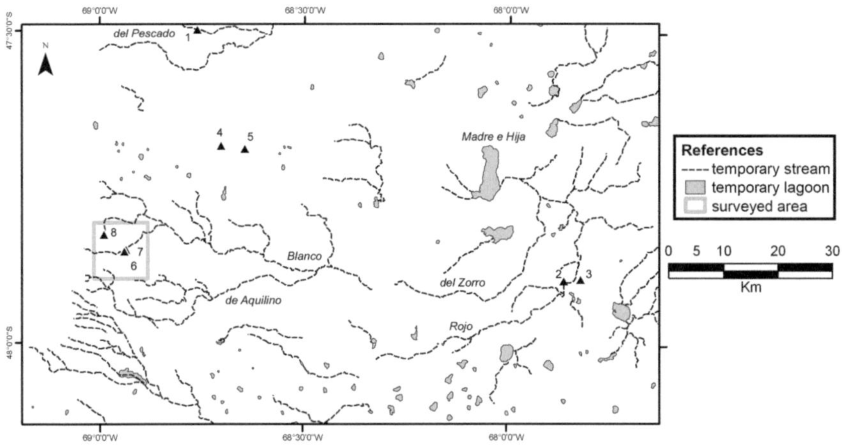

Fig. 2.2 Location of the surveyed sector (La Primavera locality) within the macro study area. The triangles represent some of the archaeological sites mentioned in this work: 1) Los Toldos, Cave 3; 2) Piedra Museo, AEP-1; 3) Chenque El Sargento; 4) Cueva Moreno; 5) Cueva La Hacienda; 6) Cueva Maripe; 7) La Quinta; and 8) Cueva Mora

reason, recent works use the term Deseado Region instead of Massif and Nesocratón (Sruoga et al., 2008). Its evolution and structure are the result of diastrophic cycles of varying intensity that occurred throughout the Phanerozoic era and are closely linked to the Andean orogeny and the development of the Mid-Atlantic Ridge (the dismemberment of Gondwana) during the Mesozoic and Cenozoic eras (Panza, 2001). The base of the geological sequence is a rigid basement that does not allow significant folding in response to tectonic deformation forces (De Giusto et al., 1980). This base supports a sedimentary sequence that began depositing in the Permian and Triassic periods, corresponding to an environment of alluvial plains and extensive, shallow lakes. Tectonic movements interrupted this sedimentary process, determining the formation of an erosive unconformity between the deposits of both periods. Additionally, intrusive events occurred, and the basement was subdivided into blocks. These subsoil blocks are visible on the surface, particularly at the 69° meridian west, where a significant fracture divides the western blocks, causing them to rise higher than the eastern ones. Throughout the Jurassic period, there were significant volcanic activities interspersed with periods of erosion and sedimentation from surrounding areas. Due to this Jurassic volcanism it is accepted that there were important rifting processes highlighted by Sruoga et al. (2008). New tectonic phases occurred between the Jurassic and Cretaceous deposits and throughout the latter. Afterwards, a period of tectonic tranquility ensued, during which only epeirogenic movements occurred. These movements allowed for marine ingressions from the Atlantic to alternate with continental deposits during times when the area was emerging (De Giusto et al., 1980). From the Upper Cretaceous through the Paleogene and Neogene periods, volcanic effusions occurred in various regions interspersed with the aforementioned deposits. The oldest levels of river terraces on the Deseado River date to this final

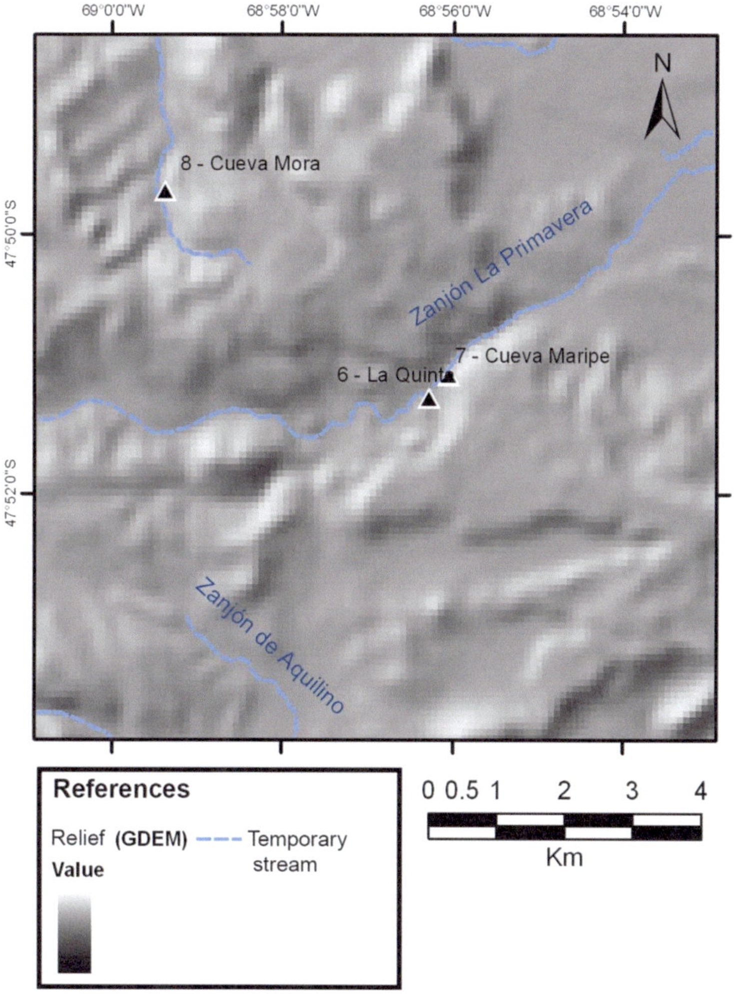

Fig. 2.3 Area surveyed in this study. The triangles indicate the locations of sites with radiocarbon dates, including the La Quinta surface site. The Digital Elevation Model used is ASTER GDEM V.1

period. Finally, during the Pleistocene, several epeirogenic ascension movements occurred, producing periods of erosion and subsequent aggradation. This resulted in the accumulation of river terrace deposits, as well as deposits covering various pediment surfaces, ancient beaches, and lake coastlines. During the Holocene, processes of river and wind erosion, as well as mass removal, affected the region. These processes originated deposits of plains and alluvial cones in current rivers, fine sediments in endorheic depressions, and products of mass removal on the slopes

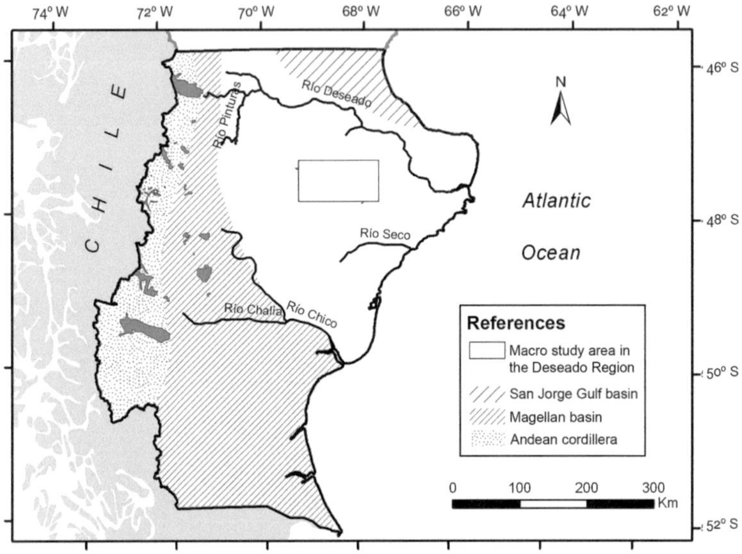

Fig. 2.4 The macro study area is outlined on the map of geological provinces (Ramos, 1999)

of the main basaltic plateaus (Panza, 2001). These geological processes shaped the landscape over thousands of years before the first human populations arrived.

2.1.1.1 Geological Resources

The region boasts a variety of rocky formations, flood basalts, and sedimentary deposits that could be valuable resources for human populations. This variety of lithologies, mapped at a scale of 1:250,000 on the geological sheet (Monumento Natural Bosques Petrificados, Panza, 2001), are presented in Fig. 2.5 and Table 2.1.

Caves and rock shelters The local geology is important for identifying the most sheltered areas of the landscape that attracted human settlement. The Chon Aike formation, which is part of the Bahía Laura Group of Jurassic-age volcanic rocks, is mainly composed of rhyolitic ignimbrites arranged in compact, thick mantles that form large walls and very abrupt ridges, as well as lapillites, lapillitic tuffs, and compact volcanic agglomerates. These rounded outcrops lack stratification planes and may exhibit columnar disjunction in their upper portions (Panza, 2001). Large cavities and caves are formed in these outcrops through weathering and wind erosion. Several of these caves have been inhabited since prehistoric times and were often used as ranch posts in historical times, such as those at Cañadón La Primavera, Zanjón de Aquilino, Los Ventisqueros, Cañadón de la Víbora, and Los Toldos. Other geological formations in the region contain caves and rock shelters that have been identified as archaeological sites. For instance, the Tito del Valle site is in a volcanic conglomerate cave (Bajo Pobre formation), and the Cueva Grande, Alero El Puesto 1 (AEP-1), and

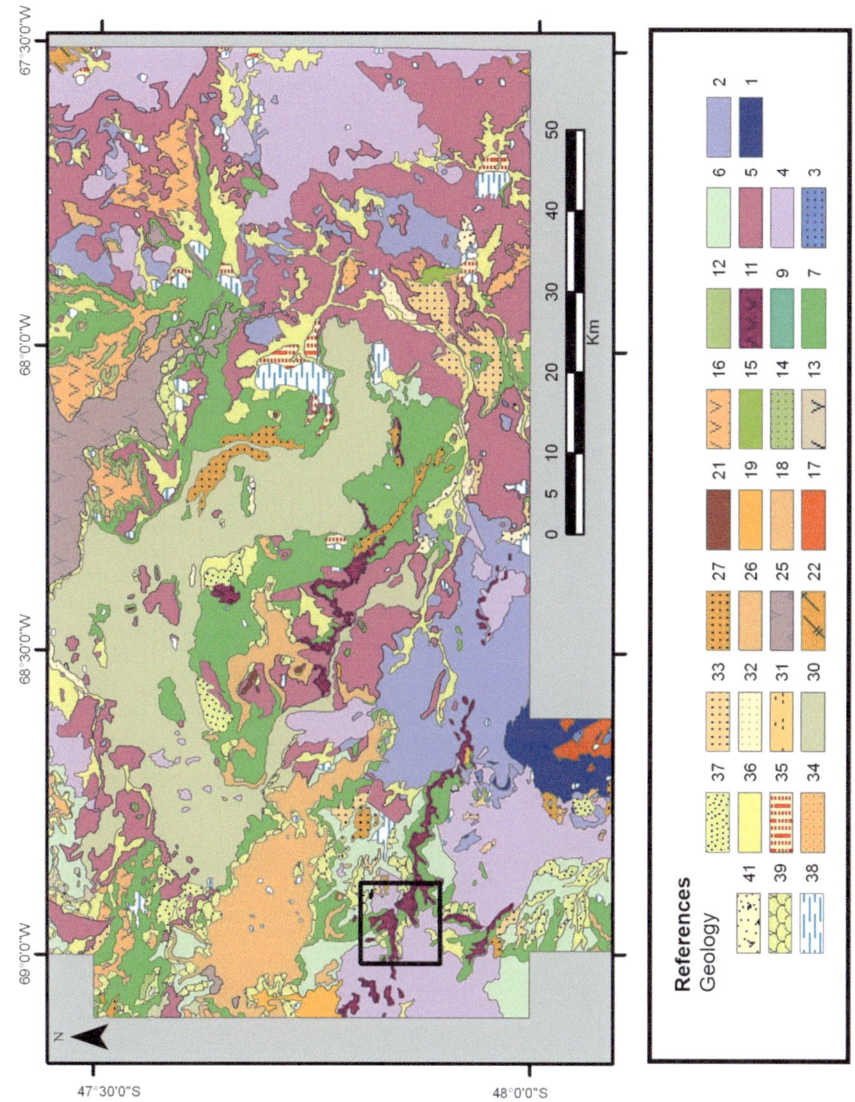

Fig. 2.5 Geology of the Macro-Study Area. The numbers refer to the formations and deposits listed in Table 2.1. This map was digitized from geological sheets 4769-III, 4769-IV and 4969-II

Alero El Galpón (AEG) sites are in El Museo formation, a coquinoid sandstone of marine origin. This information is summarized in Table 2.2 and illustrated by the photographs in Fig. 2.6.

Stones for building structures Another resource provided by the local geology is rock blocks used for building hunting blinds, structures to support residential

Table 2.1 The stratigraphic table corresponds to Geological Sheet *4769-IV Monumento Natural Bosques Petrificados* (Panza, 2001)

		Age	Formation	Composition
Mesozoic	Jurassic	Liassic	1-Roca Blanca	Tuffs, sandstones, tuffites and pelites
		Dogger	2-Bajo Pobre	Basalts, andesites, volcanic agglomerates, few sedimentites and tuffs
			3-Cerro León	Andesitic porphyries, diabases
		Dogger-Malm	4-Chön Aike	Ignimbrites, agglomerates and rhyolitic tuffs; scarce rhyolitic tuffites and porphyries (4). Epithermal quartz veins (4a)
			5-La Matilde	Tobas, chonitas, lapillites and tufites; thin intercalations of rhyolitic ignimbrites
	Upper Jurassic-Lower Cretaceous		6-Bajo Grande	Tuffs, cinerites, pelites, coarse sandstones to conglomerates
	Cretaceous	Lower Cretaceous (Barremiano-Aptiano)	7-Baqueró	Tuffs, cinerites, pelites, coarse sandstones to conglomerates
		Lower–Upper Cretaceous	8-Castillo	Tuffs, fine tuffs, tuffites

(continued)

Table 2.1 (continued)

		Age	Formation	Composition
			9-Bajo Barreal	Chonites, claystones and tuffaceous siltstones; scarce sandstones and tuffaceous conglomerates
			10-Laguna Palacios	Tuffs, cinerites; subordinate tuffites
		Lower Cretaceous	11-Basalt Las Mercedes	Olivinic basalts
Cenozoic	Paleogene	Paleocene	12-Salamanca	Fine to medium sandstones, claystones
			13-Basalt La Levadura	Olivinic basalts
			14-Río Chico	Tuffs, chonites, fine to medium sandstones, claystones
		Eocene (Middle–Upper)	15-Puesto del Museo	Coquinoid sandstones
			16-Basalt Cerro del Doce	Olivinic basalts (16), chimneys and dams (16a)
		Oligocene	17-(Group) Sarmiento	Tuffs, cinerites, tuffaceous claystones and bentonites
			18-Basalt Alma Gaucha	Olivine basalts and basanites
	Paleogene-Neogene	Upper Oligocene–Lower Miocene	19-Monte León	Claystones, tuffs, coquinas and sandstones
	Neogene	Middle Miocene	20-Basalt El Pedrero	Olivinic basalts

(continued)

Table 2.1 (continued)

		Age	Formation	Composition
			21-Madre e Hija	Phoidites (analcimites, nephelinites, phonolites)
		Upper Miocene	Deseado River terrace deposits (Levels I, II and III)	Gravels, sandy gravels and sands
		Upper Miocene-Lower Pliocene	25-Basalt Cerro Mojón	Olivinic basalts
		Lower Pliocene	26-La Avenida	Medium to coarse gravel with a sandy matrix
			27-Deposits covering pediment levels (Level I)	Gravel with a sandy matrix
	Superior-Neógene/ Cuaternary	Upper Pliocene-Lower Pleistocene	30-Basalt La Angelita	Olivinic basalts
	Cuaternary	Pleistocene	31, 32-Deseado River terrace deposits (Levels VI and VII)	Sandy gravel, subordinate fine to coarse sand, little silt and clay
			33, 34-Deposits covering pediment levels (Levels II and III)	Gravels with a sandy and silty matrix, sands
			35-Ancient beaches and lake coastlines deposits	Fine sands, silts and clays; fine gravel ridges with a medium sandy matrix
		Holocene	36-Flood landfill deposits	Fine to coarse sands, gravels, silts and clays
			37-Alluvial cones deposits	Sands and gravels; subordinate silts and clays
			38-Fine sediments of depressions and shallow lakes	Silts and clays

(continued)

Table 2.1 (continued)

		Age	Formation	Composition
			39-Landslide materials	Blocks
			41-Undifferentiated alluvial and colluvial deposits	Fine to medium sands, subordinate silts and clays; scattered shots

Table 2.2 Geology of archaeological sites in caves and rock shelters registered in the macro study area

Geological formation	Archaeological locality	Archaeological site	GPS_ID
Bajo Pobre (2)	Bajo Grande	Tito del Valle	210
Chon Aike (4)	La Primavera	La Primavera 1 and 2	389
Chon Aike (4)	La Primavera	La Primavera 5	576
Chon Aike (4)	La Primavera	Cueva Maripe	563
Chon Aike (4)	La Primavera	Alero Lucifer	1242
Chon Aike (4)	La Primavera	Cueva Lagarto	2344
Chon Aike (4)	La Primavera	Cueva Mora	2108
Chon Aike (4)	Las Mercedes	Cueva de la U (LM3)	1304
Chon Aike (4)	Las Mercedes	Cueva del Círculo 1	1319
Chon Aike (4)	Las Mercedes	Cueva del Círculo 2	1320
Chon Aike (4)	Las Mercedes	Cueva del Círculo 3	1321
Baqueró (7)	Las Mercedes	Cueva El Edificio	2533
Baqueró (7)	Las Mercedes	Cueva Mano Negra	2534
Baqueró (7)	Las Mercedes	Cueva La Escuelita	2535
Baqueró (7)	Aguada del Cuero	Cueva Moreno	156
Baqueró (7)	Aguada del Cuero	Cueva la Hacienda	76
Chon Aike (4)	Los Toldos	LT-C3	233
Chon Aike (4)	Los Toldos	LT-C13	212
Chon Aike (4)	Los Toldos	LT-C14	213
La Matilde (5)	Los Toldos	LT-C15	214
El Museo (15)	Piedra Museo	Cueva Grande	972
El Museo (15)	Piedra Museo	AEP-1	738

awnings, cairns, as well as for burials, which are locally called *chenques*. According to the area's known archaeological records, basalt blocks are the most commonly used rocks. These blocks are distributed irregularly on the surface of the plateaus. They are formed through the physical weathering or mechanical disintegration of basalt due to freezing and thawing (Panza, 2001). Some of these basalts belong to the Bajo Pobre

1

2

Fig. 2.6 Caves and rock shelters with archaeological evidence in the study area. 1) Tito del Valle Site. Cave formed in the tuffs and conglomerates of Bajo Pobre formation. 2) Detail of the conglomerates that form the roof of Cueva Tito del Valle; 3) Cueva Maripe, 4) Cueva LP1 and LP2, 5) Lucifer Rock shelter, 6) Cueva del Lagarto, 7) Cueva del Círculo 2, 8) Cueva del Círculo 3, and 9) Cueva de la U, all caves and rock shelters in Chon Aike formation; 10) Cueva El Edificio, 11) Cueva de la Mano Negra, 12) Cueva La Escuelita, and 13) Cueva Moreno, formed in Baqueró formation; 14) Caves at Cañadón de Los Toldos, in Chon Aike formation; 15) Cueva Grande, and 16) AEP-1 rock shelter, formed in El Museo formation

3

4

Fig. 2.6 (continued)

5

6

Fig. 2.6 (continued)

7

8

Fig. 2.6 (continued)

9

10

Fig. 2.6 (continued)

11

12

Fig. 2.6 (continued)

13

14

Fig. 2.6 (continued)

15

16

Fig. 2.6 (continued)

formation, the oldest of the lava flows, followed by the Las Mercedes, Cerro del Doce, Alma Gaucha, Madre e Hija, Cerro Mojón, and La Angelita formations. Numerous archaeological structures, mainly chenques and hunting blinds, as well as historical rural structures, such as cairns, fence bases, and adobe post bases, were built using these basalts (Fig. 2.7). The only rock structure in a stratigraphic context known in the study area and a big block of basalt extracted from its place of origin and relocated to a cave deserve special mention. These basalt and ignimbrite rocks structure were discovered inside Cueva Maripe. A large basalt block (measuring 70 cm long, 70 cm wide, and 50 cm high) was artificially introduced into the cave from the bottom of the canyon. Based on the stratigraphy of the site, it was determined that it was probably placed there during historical times when the cave was used as a rural post and sheep corral. However, a linear structure in the southern chamber of Cueva Maripe is likely older. This structure, built with ignimbrite blocks, is spatially associated with stratigraphic levels containing hunter-gatherer archaeological evidence.

Other types of rock used for building structures include tuffs and ignimbrites. These rocks are found in the mass rock removal sectors of the Chon Aike, La Matilde, and Baqueró formations, to name a few. In the La Lotita locality, tuffs and ignimbrites were used to build circular structures that are thought to have served as bases for awnings, as well as a wall that closes the entrance to the Casa de Piedra site, which may have had historical significance. A *real* or wall-like structure that was used as a shelter was recorded in Aguada del Cuero; a very similar structure was recorded in the Lucifer rock shelter in the La Primavera locality. In the Bajo Grande locality, the same type of rock was used to build a chenque near the Tito del Valle site. Historical constructions, such as the houses and sheds of Patagonian ranches, are usually built with tuffs and ignimbrites; however, in this case, they are edged to obtain rock bricks.

In the Piedra Museo locality, blocks of coquinoid sandstone were used to build corrals near Cueva Grande that may be associated with sheep herding.

Other materials used to construct posts for sheep caretakers include adobe or mud bricks (Miotti, 1989). Examples include the walls closing the Tito del Valle cave and the AEG site in the Piedra Museo locality (Miotti et al., 2022), as well as many rural posts in the aforementioned localities and in the La Primavera (registered in transect section I1-I2). These materials originate from fine silt and clay sediments found at the bottoms of shallow bodies of water and temporary tributaries.

Construction of walls in caves and rock shelters has also been documented in nearby areas such as the Pinturas River in the northwest of Santa Cruz province. Gradín and Aguerre (1992) proposed that the last indigenous families in the area settled on the right bank of the Pinturas River when the area was colonized by farmers in 1910. When these families were pressured by rural settlers, they occupied natural shelters such as the Charcamata rockshelter, Cueva de la Negra, Cuevas de Quesada, and Cueva Grande de Altamirano on the left bank. The male members of these families worked shearing sheep in the spring and summer, extending the occupation of these caves and shelters until 1945.

In summary, the region's architectural records indicate a preference for using immediately available rocks as construction materials, including basalts, ignimbrites, tuffs, and various sedimentary rocks. Chapter 6 analyzes the identification of pre-rural stone structures associated with a mobile hunter-gatherer lifestyle and the functionality of those structures.

Raw Material for Lithic Knapping This section discusses the distribution of raw materials used to manufacture lithic artifacts. Three sources of information were used: (1) the description and distribution of the lithological units and deposits present in

1

2

Fig. 2.7 Stone and adobe structures registered at archaeological sites of hunter-gatherers. 1) A semicircular hunting blind and 2) a chenque built with blocks of vesicular basalt in the La Primavera locality; 3) a basalt block that was probably entered into the Cueva Maripe in historical times, grids A12 and B12, south chamber of the cave; and 4) a wall of ignimbrite blocks in stratigraphic association with archaeological levels, south chamber of the cave; 5) a structure next to Cueva Moreno made with rock blocks from the Baqueró Formation; 6) a structure in Alero Lucifer that also presents stoked firewood. The blocks are tuffs detached from the Chon Aike formation; 7) a wall of stones and adobes that closes the entrance to the Tito del Valle Cave; 8) corrals made of coquinoid sandstone blocks in Piedra Museo

3

4

Fig. 2.7 (continued)

5

6

Fig. 2.7 (continued)

7

8

Fig. 2.7 (continued)

Geological Sheet 1:250,000 of the Monumento Natural Bosques Petrificados (Panza, 2001); (2) previous archaeological quarry studies in the macro study area; and (3) field studies in the La Primavera locality. Information regarding the area, quality, availability, and presentation of lithic raw materials is summarized in Table 2.3.

Primary sources

The study area is notable for its abundance of rocks suitable for knapping (Hermo, 2009). These rocks are good raw materials due to their isotropism, marked cryptocrystallinity, elasticity, homogeneity, hardness, brittleness, conchoidal fracture, and availability as clasts of an appropriate size for knapping (Cattaneo et al., 2006). These volcanic rocks have a high siliceous content and belong to the Bajo Pobre, Chon Aike, La Matilde, Bajo Grande, and Baqueró geological formations from the Jurassic to the Cretaceous periods (Panza, 2001). In the field, they are usually present as exposed veins showing evidence of extraction activities (primary sources, as defined by Nami, 1985). Some of these outcrops extend for several hundred meters, such as the 260-m-long Gris quarry (CDG) and the 300-m-long Rojo quarry (CDR) (Hermo, 2009). In other cases, the quarries are subfloor, as in the Rocky quarry and the F4 quarry, or punctual, as in the Verde quarry (CDV), which extends 5 m^2.

The quarries located in the study area thus far are described briefly below. Their geological formations were identified in GIS by overlaying their GPS locations on the 1:250,000 geological map (Panza, 2001), which was digitized from a paper map and georeferenced. In some cases, the identified geological formation differs from that inferred by field observations. There are two main reasons for this discrepancy. First, the geological sheet presents a regional scale of capture with a low degree of detail. Second, it is sometimes difficult to distinguish some formations from others in fieldwork due to their similar lithology and appearance. The unequivocal classification of the quarries according to their formation is not fundamental to this work. Current interpretations of regional geology consider the rocks of the Middle to Upper Jurassic formations (Chon Aike, La Matilde, and some authors include Bajo Pobre, Los Pirineos, and Bajo Grande) to be too intertwined and similar in petrology to clearly separate them. This is why geologists group them in the Jurassic Volcanic Province, the Chon Aike Group, or the Chon Aike Province (Moreira, 2009). In several cases, the formations are clearly differentiated by age, despite having similar lithology. For instance, the Bajo Grande Formation (Jurassic) and the Baqueró Formation (Cretaceous) have similar lithology, mainly pyroclastites and sedimentites, which makes distinguishing them in the field difficult. Additionally, much of the siliceous and sedimentary rock in the region has undergone silicification, giving it characteristics suitable for knapping. These characteristics may vary at different points within each outcrop (Cattáneo et al., 2006; Hermo, 2008). In the following paragraphs, we present the geological formations and the quarries associated with them to characterize the region of the Santa Cruz Central Plateau in terms of its resources.

Bajo Pobre Formation

Bajo Pobre is an important source of rocks for knapping. It mainly contains opals and chalcedonies that form in the tonsils of basalts and volcanic agglomerates, also

Table 2.3 Characteristics of lithic raw material sources in the Macro-study area

References	Locality	Source type		Quarry site name	Geological formation	Lithology	Extent	Quality			Presentation				Distribution		
		1°	2°					Poor	Regular	Good	Outcrop	Vein	Block	pebble	Isolated	Disperse	Concentrated
1	Piedra Museo		x		Bajo Pobre			x	x		x	x	x	x			x
1	Piedra Museo	x		PM-17/1	La Matilde	Opal, chalcedony, silicified tuff		x	x	x	x		x	x	x	x	
1	Piedra Museo		x		La Matilde drift	Opal, chalcedony, silicified tuff		x	x	x				x	x	x	x
1	Piedra Museo		x		Pediment level II	–			x	x			x	x		x	x
1	Piedra Museo		x		Pediment level III	–			x	x				x		x	x
1	Piedra Museo		x		Beaches and coastal strands	–		x	x	x			x	x			x
1	Piedra Museo		x		Alluvial plains	–		x	x	x	x		x	x	x	x	x
1	Piedra Museo		x		Alluvial cones	–		x	x		sd	sd	sd	sd	sd	sd	sd
1	Piedra Museo		x		Basaltic landslides	–					x		x	x			x

(continued)

Table 2.3 (continued)

References	Locality	Source type		Quarry site name	Geological formation	Lithology	Extent	Quality			Presentation				Distribution		
		1°	2°					Poor	Regular	Good	Outcrop	Vein	Block	pebble	Isolated	Disperse	Concentrated
4	Piedra Museo	x		Lag-D	La Matilde	Silicified tuff	The concentration tion of materials on the edge of some measuring more than 100 m			x			x				
	Piedra Museo	x		PM-CES	Bajo Pobre	Silicified tuff, ignimbrite	The outcrop is 250 × 70 m aprox						x				
1	Piedra Museo		x	17/1	La Matilde “acarreo”	Opal, chalcedony, silicified tuff		x	x	x				x	x	x	x
3	La Lotita	x		CDR	Chon Aike iii	Rhyolitic ignimbrite	Opal, chalcedony, silicified tuff 300 m long × 4 m wide and up to 10 m thick		x	x	x		x		x		
2	PNBPJ	x	x	Est-F	La Matilde	Petrified wood, Ignimbrite	570 × 485 m	x	x	x			x	x	x	x	x
	La Trabajosa	x		CLT	Chon Aike					x	x						
	Bajo Grande	x		CTDV	Bajo Pobre	Varied	The outcrop is 120 × 50 m approx						x	x			

(continued)

Table 2.3 (continued)

References	Locality	Source type		Quarry site name	Geological formation	Lithology	Extent	Quality			Presentation				Distribution		
		1°	2°					Poor	Regular	Good	Outcrop	Vein	Block	pebble	Isolated	Disperse	Concentrated
	La Primavera	x		D9	Chon Aike	Rhyolite	The outcrop is 200 × 50 m approx						x				
	La Primavera	x		CDG	Outcrop of gray silicified rock in the form of a vein introduced between the Fm Baqueró and Chon Aike	Siliceous rocks	The outcrop is 260 × 20 m approx	x	x				x				
	La Primavera	x		CF4	Baqueró	Siliceous rocks	20 m aprox						x			x	
	Las Mercedes	x		CP	Chon Aike	Siliceous rocks	50 × 25 m approx	x	x								x
	La Primavera		x	K4-K3	Pediment level I	Varied								x		x	
	La Primavera		x	G0-G1	Alluvial plains	Varied										x	
	La Primavera		x	K2-K1	Alluvial cones	Varied								x		x	

(continued)

Table 2.3 (continued)

References	Locality	Source type		Quarry site name	Geological formation	Lithology	Extent	Quality			Presentation				Distribution		
		1°	2°					Poor	Regular	Good	Outcrop	Vein	Block	pebble	Isolated	Disperse	Concentrated
	La Primavera		x	K3-K2	Pediment level III	Varied										X	
	La Primavera		x	CB	Rockslides from Baqueró Fm	Siliceous rocks										X	
	La Primavera		x	E9-E10	Alluvial plains	Varied										x	
3	La Primavera	x		CR	Chon Aike iv	Rhyolitic ignimbrite			x	x		x			x		
3	La Primavera	x		CDV	Baqueró v	Silicified breccia	5				x					x	
3	La Primavera		x	LP-P1 (K1)	Pediment level II	Silicified rhyolite		x	x	x				x		x	x

References used: Cattáneo et al. (2006); Hermo, (2009); Hermo and Miotti (2003); Magnin (2008). Quarry Names: PM-17/1: Cantera 17 de Enero (Piedra Museo locality); 17/1: Mass removal of Cantera 17 de Enero (Piedra Museo locality); LP-P1 (K1): Pedimento 1 (La Primavera locality) Est-F: Estación F (Bosques Petrificados de Jaramillo locality); CDR: Cantera del Rojo (Ea. La Lotita); CR: Cantera Rocky (La Primavera locality); CDV: Cantera del Verde (Ea. La Dora); CLT: Cantera La Trabajosa (Ea. La Trabajosa); Lag-D: Laguna D (Piedra Museo locality); PM-CES: Cantera El Sargento (Piedra Museo locality); CTDV: Cantera Tito del Valle (Ea. Bajo Grande); D9: Cantera D9 (La Primavera locality); CDG: Cantera del Gris (La Primavera locality); CF4: Cantera F4 (La Primavera locality); CP: Cantera Platense (Ea. Las Mercedes); K4-K3: Transect section (Ea. Las Mercedes); G0-G1: Transect section (La Primavera locality); K2-K1: Transect section (Ea. Las Mercedes); K3-K2: Transect section (Ea. Las Mercedes); CB: Cantera los Bloques (La Primavera locality); E9-E10: Transect section (Ea. La Primavera); *Ea.: Estancia, sheep ranch

known as porphyries. The El Sargento quarry, located within this formation, extends approximately 250 m by 70 m at its maximum (Fig. 2.8).

Fig. 2.8 Primary source of raw material on the Fm. Bajo Pobre, 1) View of the rock outcrop, 2) detail of the raw material from the El Sargento quarry and workshop at the Piedra Museo locality

Chon Aike Formation

This unit is predominantly composed of rhyolitic ignimbrites, thick volcanic agglomerates, and few tuffs, lapillites, and tuffites. Lava facies are in the minority, and hypabyssal rocks (rhyolitic porphyries) are also present (Panza, 2001). In the La Primavera locality, close to the Cueva Maripe site, the Rocky quarry exploited a dark brown, subfloor rhyolitic ignimbrite. The site covers 3500 m^2 (Hermo, 2009). The Rojo quarry in the La Lotita locality is an outcrop of this formation extruding from the Las Mercedes basalt (Hermo, 2008). The Platense quarry, located in the Las Mercedes locality, measures approximately 30 m by 10 m and appears as a primary outcrop in the form of a vein. In sector D9, an ignimbrite crest measuring 50 m by 200 m was registered, around which an extensive, almost continuous layer of red siliceous rock artifacts is distributed. Among the numerous knapping debris, some pieces are very large and have a cortex similar to that of the outcrop. Based on the presence of formatted artifacts and debitage, as well as the large number of non-formatted potential base forms for knapping, it is concluded that D9 functioned as a quarry and lithic workshop. In addition to these larger veins, other recorded veins are less than two meters in length. These veins consist of siliceous material that filled fissures and cavities within the volcanic rocks of the Chon Aike formation. Additionally, the boulders that constitute the clasts of the volcanic conglomerates are potential sources of useful rock (Fig. 2.9).

Obsidian, another important knapping raw material, may be sourced from this geological unit. Obsidian appears to have held special significance in every technological context in which it was used. Its excellent quality for knapping has been highlighted in the bibliography from a techno-economic view (Civalero, 1999; Nami, 1992), as well as for symbolic motives associated with its visual characteristics, such as brightness and color homogeneity (Hermo, 2008). From a social perspective, the limited regional availability of obsidian and the ability to study chronological differences in its use allow us to analyze the potential mobility of human groups to obtain this material or the potential social interaction and exchange of goods in a large region (Civalero, 1999; Stern, 1999). This raw material was used in various forms throughout central and southern Patagonia, as evidenced by obsidian artifacts found in archaeological sites across Santa Cruz province (Molinari & Espinosa, 1999).

Our study area is no exception. Obsidian is present at many archaeological sites, and no local sources have been detected (see Fig. 2.10). Geochemical studies of five black obsidian samples from the Piedra Museo site (stratigraphic levels 1 and 2) date back more than 7600 years B.P. These studies yielded the first sourcing information in the area. Processed using trace element analysis and neutron activation techniques, the samples showed correspondence to types Pampa del Asador I and II, as defined by Stern (1999, 2004). This suggests that the analyzed obsidians originated 150–220 km northwest of the province. Black obsidian is available in an outcrop in Cerro Pampa and in pebbles up to 10 cm in diameter in extensive fluvioglacial deposits extending eastward in Pampa del Asador, Pampa de La Chispa, and the alluvial fan at the foot of Cerro Bayo (Belardi et al., 2006; Espinosa & Goñi, 1999; Franco et al., 2023;

1

Fig. 2.9 The primary sources of raw material are found in the Chon Aike formation. 1) General view of the available raw material in the Rojo quarry; 2) detail of the raw material; 3) General view of the Platense quarry outcrop, and 4) detail of the volcanic breccia rock with flake negatives; 5) General view of the silicified ignimbrite seam, and 6) detail of nodular flakes with cortex from quarry D9; 7) General view and 8) detail of a flint vein in La Primavera locality; 9) Siliceous nodules in situ in Cueva Mora in the tuff where they were generated; and 10) siliceous nodules detached from the rock outcrop

Stern 1999, 2004). The total area where obsidian is available is 188 km^2 (Fig. 2.10; Table 2.4).

As Stern's (1999) study and subsequent ones show, it is widely accepted that the Pampa del Asador area and its surroundings are the origin of black obsidian used as a raw material in a large region. This area extends from Magallanes in the south to the Aysén region in the north in Chile and to the province of Santa Cruz in Argentina. It also extends to Peninsula Valdés in the province of Chubut. In the '90s, some authors proposed an alternative hypothesis (Molinari & Espinosa, 1999) that other unknown sources of obsidian may exist. These authors noted the presence of obsidian artifacts in 34.5% of a sample of 340 surface sites throughout Santa Cruz province. Recognizing the widespread distribution of this raw material fostered the search for other obsidian sources. For instance, in Ea. 17 de Marzo, south of the Deseado Massif, a concentration of small black obsidian pebbles (less than 5 cm in diameter) of the PDA type was interpreted as a secondary source (Franco et al., 2017). Additionally, an obsidian outcrop has been identified and proposed as a potential source in the Tres Cerros, on the north Atlantic coast (Franco et al., 2023). However, discussions indicate that the nodules in 17 de Marzo are significantly small and that the obsidian in Tres Cerros is of lower quality for knapping. To date, no evidence of extraction exists (Ambrustolo, 2020).

2

Fig. 2.9 (continued)

3

4

Fig. 2.9 (continued)

5

6

Fig. 2.9 (continued)

7

8

Fig. 2.9 (continued)

9

10

Fig. 2.9 (continued)

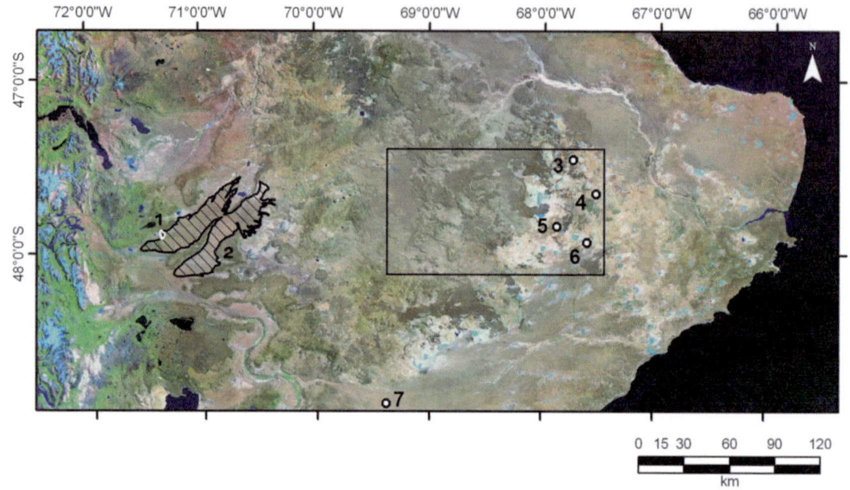

Fig. 2.10 The location of the Piedra Museo site and obsidian sources is in the northern part of the Santa Cruz province. The base image is a NASA Landsat GeoCover mosaic. References: 1) Cerro Pampa; 2) Pampa del Asador, Pampa de la Chispa, and Cerro Bayo; 3) Estancia La Escondida; 4) Estancia La Magdalena; 5) Piedra Museo; 6) Tres Cerros; and 7) 17 de Marzo

Table 2.4 Distances between the geographic center of the study area and localities with obsidian outcrops in northern Santa Cruz

From	To	Distance in km
Center of the study area	Cerro Pampa	221.97
	Cerro Bayo alluvial fan	154.05

Other varieties of obsidian used in southern Patagonia come from Sierra Baguales, south of Argentino Lake (gray and green banded), and Seno Otway in the Magellan Region of Chile (green obsidian).

According to the geological chart (Panza, 2001), black and gray-black obsidian outcrops have been documented in the east. La Escondida and La Magdalena, east of the Central Plateau. However, these areas have not yet been surveyed for archaeological research purposes (Fig. 2.10). During our surveys at the La Primavera locality, we located volcanic glass in situ (Fig. 2.11). However, the sample size is too small to be considered a supply source. In conclusion, the obsidian artifacts and nodules found in the study area are considered non-local.

La Matilde Formation

The La Matilde formation is composed mainly of tuffs, tuffites, and highly silicified ignimbrites, and it contains petrified Araucariaceae trunks (opals and chalcedonies) (Panza, 2001). Certain sectors of these trunks were quarried for high-quality rocks suitable for knapping. These raw materials were valued because they produced thin

Fig. 2.11 Volcanic glass in La Primavera locality, 1) photograph of the find place on the eastern slope of the hill on the Chon Aike formation, 2) the sample has a glassy texture and a uniform, crystalline-free mass with a glassy shine

artifacts with long edges (Cattaneo, 2000; Hermo, 2005). Examples include the Estación-F quarry-workshop in the BPJNP locality and 17 de Enero qarry in the Piedra Museo locality. The tuffs of the La Matilde formation have a high degree of silicification and sometimes exhibit conchoidal fractures (Panza, 2001); therefore, they are also suitable for knapping. However, the distribution of rocks within this unit is heterogeneous since it varies from isolated to dispersed to concentrated (Cattáneo et al., 2006) (Fig. 2.12).

Bajo Grande Formation

The CDV quarry (volcanic breccia) in the La Primavera locality is associated with a formation composed of tuffs, tuffites, sandstones, and conglomerates (Fig. 2.13).

Baqueró Formation

This formation contains tuffs, cinerites, chonites, coarse sandstones, and conglomerates, all of which are suitable for knapping. The conglomerates of Baqueró formation conform to paleochannels within the tuffaceous sequence of Blanco Creek, west of the estancias San Miguel, La Porfiada, and La Argentina. These conglomerates are highly consolidated due to the presence of opal (siliceous cement). In the areas surrounding the estancias La Dorita, Bajo Grande, Las Mercedes, and Aguada del Cuero, as well as in the BPJNP, this formation only has imprints of petrified trunks, which are poorly preserved. At the Cueva Moreno site in Aguada del Cuero, these logs are part of the cave ceiling (Miotti et al., 2005). However, they are not suitable for knapping because their opal clasts have internal fracture planes that cause them to fragment in multiple directions. Therefore, past populations did not use the petrified wood from the Cueva Moreno site as raw material for manufacturing useful artifacts. Within the La Primavera locality, the Gris quarry also contains flint of regular and poor quality. Nevertheless, some cores adequate for knapping were found along the 100-m-by-20-m outcropping seam. The F4 quarry is also in a sector of the Baqueró Formation. It is accessible from the surface, where erosion of the Las Mercedes basalt reveals the underlying rock (Fig. 2.14).

Secondary sources

Some geological deposits are secondary sources for extracting raw materials (Nami, 1992). These deposits cover pediment levels, coastal ranges, plains, and alluvial cones, as well as landslide material.

Deposits that cover levels of pediments

These deposits consist of nodules of various lithologies from different primary formations that have been weathered and transported by fluvial and gravitational forces. They are angular to sub-rounded boulders up to 15 cm in diameter that are embedded in a sandy matrix. The predominant rocks are pyroclastics and silicified vulcanites (Panza, 2001). These flat outcrops of varying size have a distinctive finger-like design. In the countryside, they appear as pampas covered by shrubby and herbaceous vegetation. They formed in three cycles between the Pliocene and Pleistocene, creating

1

2

Fig. 2.12 Silicified wood in BPJNP locality. 1) Estación-F quarry-workshop, in an outcrop of La Matilde formation with fossil flora; 2) Detail of the knapping blanks left on one of the petrified trunks. Photograph taken by L. Miotti

levels I, II, and III of the pediment. The oldest level is in a higher topographic position than the youngest level.

At the Piedra Museo locality, blocks and pebbles are concentrated in space (more than 10 rocks per m^2) and distributed homogeneously within the deposition units, but not all of them are suitable for knapping (Cattáneo, 2000; Cattáneo et al., 2006). In La Primavera, some evidences of qarring were found on these deposits to the north of the locality (Hermo, 2009). The LP-P1 quarry, for example, is located on Pediment Level I deposits that concentrate clasts from different geological units (Hermo, 2009). This deposit was sampled during fieldwork in 2008 along transects K4-K3 (see Fig. 2.15 for the layout of the transects and Fig. 5.2 in Chap. 5). Another deposit presenting

1

Fig. 2.13 Volcanic breccia in the La Primavera locality; 1) Archaeological artifacts and raw material samples collected in the Verde quarry by L. Miotti and D. Hermo in 2003; 2) view of the quarry. The red dots indicate the location of the outcropping rocks. Photograph taken by D. Hermo

2

Fig. 2.13 (continued)

similar evidences was recorded along the K3-K2 transects in 2008. These deposits cover pediment levels III (Panza, 2001).

Deposits of coastal ranges, plains, and alluvial cones

These Quaternary deposits consist of fine to coarse sands, gravels, silts, and clays (Panza, 2001). Deposits from ancient beaches and coastal ranges formed on the eastern edge of shallow lakes such as Grande, El Baldío, and La 10. These deposits are very fine and correspond to times when the beaches covered a larger surface area than they do today. Their formation occurred approximately in the Upper Pleistocene-Holocene. Two or three coastal range systems with an approximate north–south orientation are present in these deposits. These systems consist mostly of psephitic accumulations (fine pebbles) with an abundant sandy matrix. Floodplain deposits are found in the channels of the main ephemeral waterways. Finally, in sectors where the slope drops, materials are deposited without selection by the channels, forming alluvial cones (Panza, 2001) (Fig. 2.16).

In the Piedra Museo locality, accumulations of high-quality psephitic clasts have been recorded in the coastal ranges of a paleolake to the south of the eponymous site. This type of deposit was essential to the lithic supply in the area (Cattáneo, 2002). In the alluvial plains of the same area, raw materials are distributed more variably, either in isolation or in concentrations. Furthermore, their quality varies from poor to excellent (Cattáneo et al., 2006).

However, no coastal ridge deposits were found in the La Primavera locality, although boulders suitable for knapping were recorded in the alluvial cone deposits.

1

2

Fig. 2.14 Primary sources of raw material on the Baqueró formation; 1) Outcropping vein in the Gris quarry, located south of Cañadón La Primavera; 2) detail of the edged rock; 3) General view; and 4) detail of the nodules that emerge in the F4 quarry, on the slope of the Las Mercedes basalt. A lava window in the basalt allows exposure of the underlying Baqueró formation rocks

3

4

Fig. 2.14 (continued)

Fig. 2.15 The deposits that cover the pediment levels contain nodules that are useful for knapping, 1) general view of the deposit in the K4-K3 transect section, 2) detailed view of an artifact found on the surface of the deposit in the K3-K2 transect section. Both are located in the lower section at the mouth of La Primavera canyon

1

Fig. 2.16 Alluvial cone deposits in La Primavera, 1) located in the transect section J2-J1, and 2) detail photo of knapped flint nodules and cores; 3) Alluvial cone deposits located in the transect sections K2-K1, and 4) a siliceous core. Unformatted knapping debris and artifacts occur continuously but dispersed throughout these deposits

The rounded clasts covering pediments, alluvial plains, and alluvial cones were useful as raw material for knapping and as hard hammers for knapping and grinding slabs used in processing leather and plants.

Landslide material

Landslide deposits are accumulations of loose materials on the slopes of basaltic plateaus formed by mechanical destruction. In the field, they cover part of the slopes and are crescent-shaped. Under river erosion, they soften and form rounded hills. They are composed of angular to sub-angular fragments ranging in size from 10 cm to over 1 m (Panza, 2001). At the Piedra Museo locality, Cattaneo et al. (2006) state that landslide deposits are a source of raw materials of poor to regular quality. An example of these deposits is the Los Bloques quarry (CLB) in La Primavera (see Fig. 7.7, Chap. 7 for its location). In this case, blocks of Fm. Baqueró have fallen from the plateau slope and provide materials for the alluvial cone deposit. At this site, nodules of various lithologies and degrees of rounding were used as a source of raw materials (Fig. 2.17).

Analysis of thin sections

The raw materials available in the La Primavera locality were the subject of specific studies (Hermo, 2008; Hermo et al., 2015). Petrographic microscope analyses

2

3

Fig. 2.16 (continued)

Fig. 2.16 (continued)

Fig. 2.17 Landslide materials at Primavera locality; 1) General view of Los Bloques quarry, and 2) detail of the siliceous raw materials available

2

Fig. 2.17 (continued)

revealed the variety of rocks in La Primavera (from the Rocky and Verde quarries and Pediment 1) and in the surrounding area (the Rojo quarry, located 11 km south of Cueva Maripe). Tests on 14 samples from these sites revealed eight rock groups (Table 2.5), four of which coincided with the raw materials of the knapped artifacts from Cueva Maripe's stratigraphic levels. Each quarry presents a particular rock group, except for Pediment 1. As previously mentioned, this deposit contains several types of rocks with different compositions and origins. Four of the rock samples from Pediment 1 matched the artifacts from Cueva Maripe.

Raw materials used to manufacture artifacts through chipping, abrasion, and polishing The hardest and most resistant rocks are suitable for manufacturing artifacts by chipping, abrasion, and polishing. In the La Primavera locality, the raw materials used for manufacturing artifacts with these techniques are plutonic or hypabyssal rocks. Three basaltic porphyry boleadora fragments were found in the stratigraphy at the Cueva Maripe site (Hermo, 2008), while two andesite and three gabbro/diabase boleadora fragments were found in surface surveys. Besides the secondary deposits, the possible sources of these rocks in the study area are the Cerro León formation, which includes olivine diabase and basaltic porphyry; the Cerro del 12 basalt, which includes diabase and basaltic porphyry; and the Alma Gaucha basalt, which includes diabase and basaltic porphyry (Panza, 2001). Finally, andesites are common volcanic rocks in the Roca Blanca, Bajo Pobre, and Chon Aike formations (Panza, 2001).

Table 2.5 Petrographic determinations for lithic artifacts from Cueva Maripe and rock samples from lithic quarries in the La Primavera locality (table modified from Hermo, 2008)

	Chalcedony group1	Opals group 1	Silicified ignimbrites group 1	Silicified ignimbrites group 2	Siliceous rocks group 3	Distance to Cave Maripe (km in straight line)
LP-P1 (4 samples)	x	x	x	–	–	2.5
CDR (3 samples)	–	–	x	–	–	11.0
CR (1 sample)	–	–	–	x	–	0.1
CDV (1 sample)	–	–	–	–	x	6.1
Cueva Maripe (4 samples)	–	x	x	x	x	–

Other useful minerals Pyrite and chalcopyrite can be used to start a fire, as documented by Gusinde (1982–1989) among the Selk'nam of Tierra del Fuego. According to the geological chart, these minerals may be found in the Bajo Pobre formation (Panza, 2001), though there are no records of them yet.

Other minerals, such as hematite, ochre pigments, and gypsum, are components used in painted rock art. This was first verified in archaeological contexts in the Pinturas River area, located 145 km northwest of the study area. The mineralogical composition analysis of paintings on the supporting rock and artifacts recovered in the archaeological layers at the Cueva de las Manos and Cerro Casa de Piedra 5 localities determined that paints of different colors (ranging from reddish to ochre) contain hematite, maghemite, natrojarosite, and illite. The black pigment is likely charcoal or manganese oxide (Gradín et al., 1987; Iñíguez & Gradín, 1977). These mineral pigments were mixed with pure or crystalline gypsum, which functions as a fixative. For application, a perishable vehicle that did not preserve would have been used.

In La Primavera, compositional X-ray diffraction studies were performed on paintings covering the walls and ceiling of Cueva Maripe ($n = 5$, colors: red, light red, and yellow) and on pigment samples taken from the same site's stratigraphy ($n = 5$, colors: red, pale red, and pale yellow) (Fig. 2.18). These studies were also performed on samples of ocher outcrops registered in La Primavera and in the nearby localities of Piedra Museo and Bosque Petrificado de Jaramillo National Park (BPJNP) ($n = 7$, colors: red, yellow, and brown) (Table 2.6) (Carden et al., 2014). During fieldwork, natural sources of ochres useful for making paints were found at contact points between basaltic flows and underlying rocks, such as tuffs. These ochres have the appearance and consistency of bricks and vary in color from yellow to orange and red (Fig. 2.19). The results showed that all the samples contain some of the following coloring minerals: hematite (red), lepidocrocite (yellow), illite, and gypsum (white).

Fig. 2.18 Ocher chalks found in stratigraphy in Cueva Maripe

One of the red paint samples from the wall of the cave at the North Chamber (quadrat 6B) (Chap. 8, Fig. 8.3) coincided with a surface sample from an outcrop on the plateau slope, 3.3 km from the site. This compositional similarity led Carden et al. (2014) to hypothesize that the source of the raw materials used for cave painting was located near the site. This implies that the pigments were collected within short-distance mobility circuits, corresponding to daily return shipments (Binford, 1982; Politis, 2006). Regarding gypsum, the wall sample contains abundant gypsum, whereas the surface sample does not. This suggests that gypsum may have been added during paint preparation, as proposed for the Río Pinturas and Perito Moreno National Park area (Aschero, 1983; Barbosa & Gradin 1986–87; Iñíguez & Gradin, 1977), though natural deposition of gypsum on the cave wall cannot be ruled out (Carden et al., 2014).

This preliminary information shows that local pigments were used in rock art painting, although other possible uses include body paint or its application on leather and wooden artifacts should also be considered. The lack of coincidences between the samples analyzed in the study of Carden et al. (2014) raises the need to continue searching for other potential sources of raw materials, both in the vicinity of the cave and at greater distances. In relation to that, the geological sheet mentions the presence of *iron oxide* and *ferruginous material* in several units, as the Bajo Pobre, Baqueró, Laguna Palacios, and Salamanca formations. Gypsum is mentioned in the geological sheet as part of the Salamanca and Sarmiento formations, where it occurs in crystallizations with a fibrous habit, forming twinned or rosette crystals.

Finally, the clay useful for pottery is found in Quaternary deposits near depressions and creeks. At the Cueva Maripe site in La Primavera, only two ceramic sherds were found in the stratigraphy. Although no dates are available, the sherds appear in the

Table 2.6 Ochers found in the macro-study area

Reference	Locality (Sitie)	GPS_ID	Fm
Miotti (2006)	Piedra Museo (El Sargento)	1172 1176 1181	Bajo Pobre
	BPJNP (El Abra)	903	Bajo Pobre
	BPJNP (Madre e Hija)	804	Monte León/Madre e Hija
	La Primavera	1204	Baqueró/Chon aike
	La Primavera	1207	Baqueró/Chon Aike
Carden (2009)	La Primavera	1227	Contact between Basalt Las Mercedes and Baqueró
	La Primavera	1295	Basalt Las Mercedes
	La Primavera	2048	Chon Aike
	La Primavera	2067	Chon Aike
	La Primavera	2234	Baqueró
	La Primavera	2288	Basalt Las Mercedes

most superficial part of the sequence in grid B12 of the south chamber (see the cave plan in Chap. 8, Fig. 8.3). Ceramics are found very sparsely in the province

1

Fig. 2.19 Outcrop of ocher 1) in the Bajo Pobre formation in the BPJNP; 2) the slope of Co. Madre e Hija, BPJNP; 3) in an outcrop next to the El Sargento chenque, in the Piedra Museo locality; and 4) in the La Primavera basaltic plateau

Fig. 2.19 (continued)

4

Fig. 2.19 (continued)

of Santa Cruz. Most records are in the northwest, in the basins of Salitroso and Posadas lakes, Cerro Pampa, Perito Moreno National Park, and the basins of Cardiel and Strobel lakes (Cassiodoro, 2011). Sparse records also exist along the Atlantic coast north of the province (Roumec et al., 2020) and in the BPJNP (APN, 2018). These pottery shards constitute the southernmost records of this technology. They are mainly associated with residential camps but also with hunting blinds and burial sites. Their dates range from 2010 ± 50 BP to 170 ± 40 (Cassiodoro, 2008).

2.1.1.2 The Structure and Availability of Lithic Resources

Useful rocks and minerals for hunter-gatherer populations were available in the landscape. Due to the wide extent and ubiquity of the geological units containing them, their distribution was considered homogeneous in the area (Cattaneo et al., 2006). Therefore, in general, we know that sourcing lithic raw materials was not a critical issue and was probably linked to other activities. However, obtaining further information on raw material sourcing strategies is difficult due to the regional geology's nature (Hermo, 2008, 2009). Specifically, the ability to identify a source depends on diagnostic features in the rocks and their discrete, known distribution in the landscape (Bayón et al., 1999; Flegenheimer & Bayón, 2008). This geology contrasts with that of other areas, which have more heterogeneous resource distribution. For example, consider the Pampean region (Bayón et al., 2006; Berón, 2006; Vecchi,

2011, among others). On a regional scale, obsidian is the only raw material whose source can be determined to be between 60 and 100 km from the archaeological finding context. In this area, there are plenty of good to excellent raw materials for knapping. The medium-to-long-distance precedence of obsidian requires an explanation that goes beyond a merely functional justification (Bayón & Flegenheimer, 2004). In this regard, social and symbolic justifications have been considered, as well as mechanisms for their acquisition that include long-distance travel and exchange with other groups (Hermo, 2008; Hermo & Miotti, 2011).

On the other hand, the structure and natural availability of rocks and minerals yield information when analyzed jointly with archaeological systematic evidence at a more detailed scale (Magnin, 2015). Within the La Primavera locality, useful rock availability is not homogeneous. The outcrops and secondary deposits within the geological units show heterogeneity in raw material qualities (Hermo, 2008). For this reason, Chap. 7 analyzes the distribution of these resources in greater detail through a field survey of the micro-study area. We recorded the distribution and quality of raw materials suitable for knapping and used this data to discuss patterns of raw material use and circulation in the area.

2.1.2 Physiography and Geomorphology

The Deseado Massif has a gently irregular morphology with little relative relief development. Panza et al. (1994) argue that fluvial action predominates as a landscape modifier from a geomorphological point of view. However, it is a semi-desert region, so wind, mass removal, and volcanic activity significantly influence the landscape in some areas. While the landscape is undulating on a local scale, the region's relief is characterized by extensive plateau plains with a gentle slope toward the east. The highest elevations consist of Jurassic volcanic rocks covered by basaltic mantles that shield the lower, more easily eroded sedimentary layers from erosion. In some cases, the basalts are arranged in elongated flows that indicate the channeling of lava along pre-existing river courses, and relief inversion occurs. Due to the low relief, the extensive Deseado Massif, and the aforementioned lava flows filling ancient riverbeds, drainage is only partially integrated into multiple endorheic systems. These ephemeral streams carry water only during the rainy season, which occurs in winter and early spring. In the study area, the main creeks that drain the western sector are Zanjón de Aquilino and Zanjón Blanco. The Zanjón La Primavera, one of the Zanjón Blanco's tributaries, crosses the study area from southwest to northeast, draining terrain with altitudes between 1000 and 200 m above sea level (Fig. 2.3). The Zanjón del Zorro and the Zanjón Blanco both drain east into Laguna Grande, which sits 50 m above sea level (see Fig. 2.2). This shallow lake and La Porfiada are the most significant depressions. Both shallow lakes, as well as many of the ravines and creeks that form the drainage network, have structural origins (Panza, 2001). These features particularly follow fracture lines in the underlying rock that form an orthogonal grid with predominant NW–SE and SW-NE directions (De Giusto et al.,

1980). As previously mentioned, the ditch basins are generally disconnected from each other and have poorly developed runoff. In contrast, the Zanjón del Pescado, located north of the study area, was once a tributary of the Deseado River, but today it drains into an endorheic shallow lake (Fig. 2.2).

The fluvial action gradually dissected the plain surfaces of the region, transforming them into isolated remnants. Numerous dead-end depressions of hydro-aqueous origin open on them. Their margins present abrupt walls surrounded by rough reliefs produced by landslides and rockfalls. Other sectors are characterized by rugged, irregular morphology with landscapes of hills and volcanic cones alternating with rocky outcrops, hills (mostly Jurassic), and small plateaus. Among the most abrupt and visible elevations in this area is the Cerro Madre e Hija or Horqueta. The landscape is completed by a large number of endorheic depressions with muddy plains and temporary shallow lakes.

2.1.3 Climate and Paleo-Climate

The study area has a cold, arid, temperate plateau climate. Its latitudinal location means the region's surface receives the sun's rays at a high angle, placing it in a cold temperate climate zone. Average annual temperatures range from 8 to 10 °C. Minimum temperatures in July range from 0 °C in hilly environments and volcanic plateaus (at heights between 500 and 1000 m above sea level) to 3 °C in lower-altitude plateau sectors near the coast. The region receives a significant amount of light energy with large seasonal differences, evident in the notable variation in day and night length between summer and winter. This is accentuated by the region's great continentality and intense westerly winds, which prevent the sea from moderating temperatures. The winds originate in the permanent South Pacific Anticyclone and reach the region after releasing their moisture on the western edge of the Andes mountains. The winds reach their greatest intensity during spring and summer, drying out the surface and affecting soil development, vegetation distribution, and landscape modeling. Precipitation is generally less than 200 mm per year, mostly occurring in winter. As a result of the invasion of polar air every year, snowfall and frost are recorded. On average, there are 15.1 days with snowfall and 95 days with frost per year (Oliva et al., 2004). Humidity is higher in the Andean range to the west of the province where annual rainfall varies between 300 and 900 mm. Most of the mountain range has a temperate-cold climate, but the northern portion has a temperate climate with average annual temperatures between 12 and 20 °C. Despite its proximity to the sea, the eastern part of the study area remains arid due to the dominant action of westerly winds, with an average annual rainfall of less than 300 mm. The climate is temperate along the coast of the San Jorge Gulf (Soto & Vázquez, 2000).

In recent years, numerous paleoclimatic studies from different disciplines have been carried out on the climate of the past in the Patagonian steppes east of the Andes. Examples include the palynological studies of Mancini (1998) and De Porras et al. (2009), as well as the studies of lacustrine sediments by Stine and Stine

(1990) in Lake Cardiel. These studies provide information about the conditions that occurred during the periods of human occupation recorded in the area since the Final Pleistocene/Holocene transition. General trends regarding significant environmental changes have been reconstructed based on contributions from numerous authors. Following Salemme and Miotti's (2008) synthesis work, the main climatic changes that occurred in Patagonia can be analyzed within the following chronological intervals: the Pleistocene-Holocene Transition (PL/H), which extends from ca. 13,000–9500 years before present (BP) and lasting 3500 years; the Early Holocene (EH), lasting approximately 2000 years between 9500 and 7500 BP; the Middle Holocene (MH), lasting 4000 years between 7500 and 3500 BP; and the Late Holocene (LH) from 3500 BP to the present.

Table 2.7 summarizes the main climate changes that will be discussed throughout this work. At the beginning of human occupation of the region, conditions of climatic instability dominated, including colder temperatures and variations in humidity. A significant regional-scale fact is that the Deseado River was permanent. It originated from the continental ice located in the current Lake Buenos Aires. Consequently, its entire basin featured extensive grasslands and plentiful water. However, at the beginning of the Holocene, the river's sources were interrupted by the geomorphic changes caused by deglaciation, and its waters began to drain into the Pacific slope. This situation coincided with other significant changes, including the extinction of megafauna and rising sea levels. The aridification and warming that triggered these changes occurred throughout the EH and continued until the beginning of the MH. The trend of constant aridification has continued until the present. Around 7000 years ago, there was a change in this area toward a humid and warm climate. In the region, this period is called the Optimum Climaticum (Salemme & Miotti, 2008). Climatic cooling and aridization began in the LH and was interrupted around 1000 C14 BP by the Medieval Climate Anomaly. This anomaly was caused by changes in atmospheric circulation and resulted in high temperatures and dry weather. The climate is drier than during the middle and early Holocene.

A final cold and humid period began 500 years ago, between 1550 and 1850 AD, during the Little Ice Age. This period is considered an event of increased humidity that led to the formation of soils in Cabo Vírgenes and Tierra del Fuego, and had a supra-regional scope (pedogenetic event of the Late Holocene) (Barberena & Borrero, 2010, see Favier Dubois, 2004, for a previous interpretation). Finally, the current arid and dry climate began approximately 100 years ago. The region's environmental and economic conditions have resulted in sparse human populations on the Central Plateau and a few residents in ranches. The population density in the Deseado department is 1.1 inhabitants/km^2 (INDEC, 2008). Recent archaeological research has sought to study the impact of these climate changes on human populations using multiple lines of evidence (Guraieb et al., 2015; Pérez et al., 2016). The results show that the population size increased steadily from the initial settlement in the Patagonia region until the Late Holocene, and involved changes in mobility strategies and settlement.

Table 2.7 Paleoclimate, glacial advances and changes in sea level

(Pl/H) *Ca.* 13.000–9500 C14 AP	(EH) *Ca.* 9500–7500 C14 BP	(MH) *Ca.* 7500–3500 C14 BP	(LH), *Ca.* 3500–1200 C14 BP
13.000 BP Glacial advance Cold and Dry Climate instability 11,000 BP Cold and wet 10,500 BP Llanquihue 3. Cold and Dry 9500 BP Cold and arid. The coast is 200 km east of the current line The coast is 200 km east of the current line	The Deseado River drains into the Pacific slope Around 9000 BP megafauna extinction 8000 AP Increase in temperature and aridity The mountain range is free of ice The Strait of Magellan forms	Temperature and aridity continue to increase Northern Patagonia and Cuyo: the increase in snowfall makes them uninhabitable 7500 BP Flandrian marine transgression begins Since 7000 BP *Climatic Optimum* in the Deseado Massif: Warm and wet 4000 BP the sea reaches its current level	3500 BP First cooling pulse 3000 BP Cold and arid climate 2000 BP Second cooling pulse 1000 BP Medieval climate anomaly: hot and dry 500 BP Little ice age: cold and wet 100 BP Increase in temperatures and dry

Acronyms of the time periods (used from now on): Pl/H: Pleistocene/Holocene transition; EH: Early Holocene; HM: Middle Holocene; LH: Late Holocene (Salemme & Miotti, 2008 and works cited there)

2.1.4 Volcanism and Seismicity Since the Final Pleistocene

Jurassic volcanism in this area was of fundamental importance and had a great impact on the landscape, flora, and fauna that was unmatched in later geological times. Throughout the Quaternary, volcanism and tectonic processes had a lesser impact, with a very low seismic coefficient. However, their effect on human populations cannot be dismissed. Of the volcanoes in the Andes Range, the Hudson volcano has most directly affected the study area. It is located in the southern Andes of Chile at 45°54′ S and 71°58′ W. During the Holocene, there were at least 12 eruptions; the main ones occurred around 6700 and 3600 years BP (Table 2.8). The ejected pyroclastic materials were at least twice the size of those emitted in 1991 and could be detected at great distances from the source of emission (Naranjo & Stern, 1998). In the first case, ash carried by winds fell more than 100 km from the volcano and formed deposits less than 10 cm thick. In the second case, the ash formed deposits less than 5 cm thick. In relation to the 1991 volcanic event, ash layers between 6 and 10 cm thick were recorded in the study area (Vázquez, 2012). Two explosions occurred within eight days, resulting in consequences for the closest human populations. The people suffered from irritation of the eyes, respiratory tract, and digestive tract, and they had to be evacuated. The explosions also greatly impacted the animals. There was high mortality among livestock (400,000 sheep) and wild animals because the ash covered pastures and dried up waters, shallow lakes, and streams, preventing

Table 2.8 Volcanism and tectonic events (Corbella & Lara, 2008; Naranjo & Stern, 1998; Salemme & Miotti, 2008)

(Pl/H) *ca.* 13,000–9500 BP	(EH) *ca.* 9500–7500 BP	(MH) *ca.* 7500–3500 BP	(LH) *ca.* 3500–1200 BP
Eruptions of the Reclus volcano in 12,000 and 10,500 BP affect the Magellan Basin	9000 BP Massive landslides Eruption of Mt. Burney around 8000 BP affects the Magellan Basin	Volcanism events (Hudson) affect the northwest of Santa Cruz (Lake Buenos Aires) and its ash reached the Deseado Massif. The main ones occurred at approximately ca. 6700 BP and 3600 BP	3000 BP Aguilera Volcano affects the Magallanes Basin 2000 BP Collapses are recorded in caves 1991 AD last major event (Hudson)

them from feeding. The 1991 explosion caused a drop in temperatures due to an ash cloud that blocked sunlight for two weeks. Three years after the event, the Santa Cruz central plateau showed a 50% increase in aridity compared to pre-eruption values because it is an unstable ecosystem with little rain and few pastures. The human population decreased significantly as the economy was seriously affected. However, 17 years after the eruption, local fauna populations had recovered. Although previous explosive events were twice as violent (Naranjo & Stern, 1998), these historic eruptions inform on the magnitude of the impact of these events on human populations, and how they respond to these catastrophic events, as well as the archaeological visibility of these natural events.

2.1.5 Water as the Critical Resource

Due to the semi-desert climate, water is a critical resource. It is only available in temporary streams and bodies of water for short periods. The most significant temporary shallow lakes are Laguna Grande (approximately 10 by 4 km in size), La 10 (5 by 3 km), and La Porfiada (5 by 2.3 km). They are all located east of the study area (Fig. 2.20). During periods of greatest humidity, these three shallow lakes were probably important sources of water for humans and animals. Based on the regional climate variations summarized in Table 2.7, humid periods are proposed for 11,000–10,500 BP, 7000–3000 BP, and 500–100 BP. Laguna Grande and La 10 have extensive lacustrine littoral barriers, which indicate the presence of a large, permanent body of water in the Pleistocene/Holocene (Fig. 2.21). Other smaller shallow lakes in the eastern and central sectors also have littoral barriers, which indicates that they were once larger bodies of water, probably permanent ones. Examples include El Baldío (3.5 by 2 km) in Estancia Las Martinetas and the unnamed shallow lakes of Ea La Paloma (2.2 by 1.8 km), Ea Las Barrancas (2 by 0.7 km), and Ea San Miguel (1.7 by

1.3 km). The last one is close to the Piedra Museo site. Archaeological and paleoenvironmental studies in that site indicate that the area was inhabited during the Pleistocene/Holocene transition and possibly until the Middle Holocene. Ancient hunter-gatherers performed specific activities, such as hunting and dismembering animals, near the ancient paleolake (Miotti, 1992; Miotti & Hermo, 2000; Zárate et al., 2000).

Apart from these larger shallow lakes, others in the area are permanent water sources. Most of these smaller lakes are located in the northern half of the region and are probably fed by groundwater (Mazzoni, 2001). There are no specific studies evaluating their hydrological dynamics. According to Clapperton (1993), many of the depressions may have formed during glacial periods when strong winds blew down from the ice sheet that covered the mountain range. The low temperatures could have eliminated vegetation and increased the mechanical weathering of rocks (Mazzoni et al., 1999). Based on this information, we can assume that the depressions were already formed when the first human groups arrived (Figs. 2.20 and 2.21).

Today, the landscape looks very different. In the prospected sector (see the white box in Fig. 2.20), there are numerous temporary shallow lakes that are smaller than the ones mentioned above. These typically form on volcanic plateaus, filling the bottom of dead-end shallows (Fidalgo, 1973). The amount of water in these shallow lakes depends on rainfall, resulting in significant fluctuations in their levels. In the area, the only permanent water sources are springs, which provide excellent-quality water. These springs are critical water sources for animals and local people in this landscape, as they were in the past. Most of the springs are located on the edges of the plateaus (De Giusto et al., 1980), where a series of topographic, hydrological, and geomorphological conditions occur. For a detailed study of these conditions in the BPJNP, see Melendi (2023) and Mazzoni and Rabassa (2018). The plateaus act as rainwater reservoirs, allowing water to infiltrate and flow underground until it reaches the edges of the plateaus, where it emerges permanently and forms wet meadow ecosystems called *mallines*. These are high-density grasslands formed by reeds and grasses (Mazzoni & Vázquez, 1999). Due to the arid landscape, they are highly visible from great distances despite their limited extent. The dynamics of the scoria-spring-malline systems have been studied in other volcanic regions (Mazzoni, 2007; Mazzoni & Rabassa, 2007) and in Santa Cruz (Mazzoni & Rabassa, 2018; Melendi, 2023). A statistical analysis of nearly 200 volcanic plateaus in Mazzoni's (2007) study of the Piedra del Águila area in the province of Neuquén showed an association between the size of basaltic plateaus and the volume of water provided by springs (Mazzoni, 2007). The size of the basaltic plateau defines the water-capturing surface. According to the study's results, plateaus larger than 50 km^2 have abundant mallines, which indicates the presence of springs. Plateaus with smaller areas have few or no springs. In the case of the La Primavera locality, the basaltic plateaus do not exceed 21.1 km^2, and it is estimated that their area has not substantially changed in the last 10,000 years (Mazzoni, 2007; Mazzoni & Rabassa, 2007, 2018). Based on the above, it is estimated that water springs have been available since the inversion of the relief occurred after the great basaltic effusions of the Miocene and Pliocene, leaving these lava flows in an elevated position. This process likely took several million

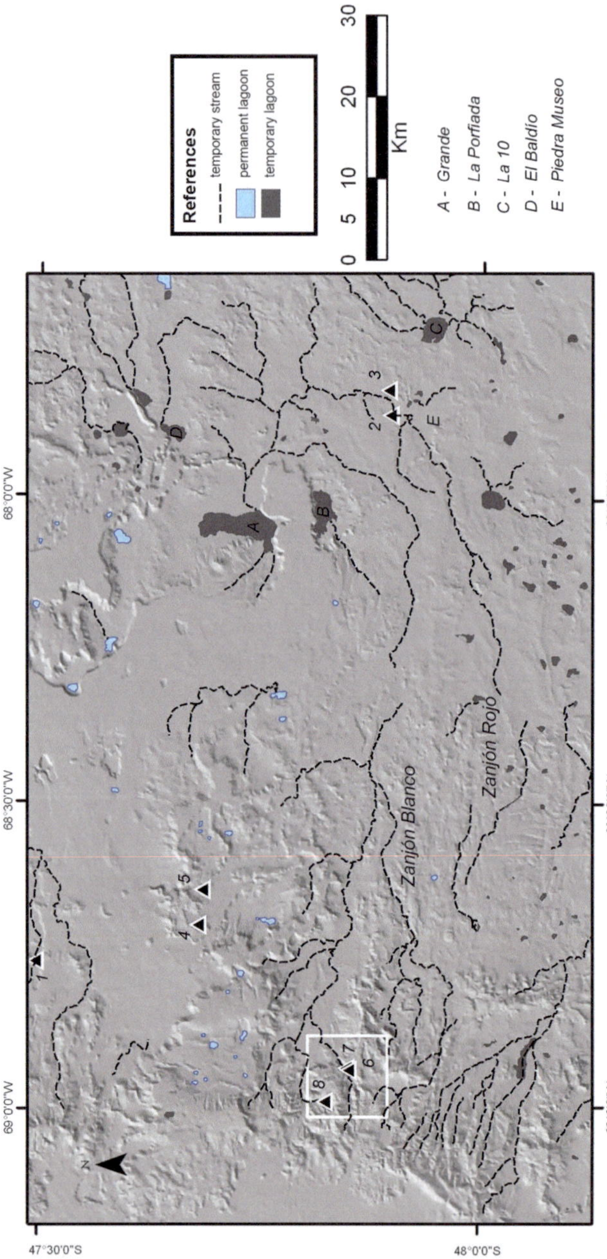

Fig. 2.20 Relief and hydrography map of the macro study area at present. The sampled sector is represented by a white box. The triangles symbolize some of the archaeological sites mentioned in this Chap.: 1) Cave 3 of Los Toldos, 2) Piedra Museo, 3) Chenque El Sargento, 4) Cueva Moreno, 5) Cueva de La Hacienda, 6) Cueva Maripe, 7) La Quinta, and 8) Cueva Mora. The vector shapefiles of temporary tributaries and shallow lakes used to generate the map belong to IGN SIG:250. The base relief was generated from the USGS DEM from which a hillshade processing was derived. All maps presented in this work are POSGAR 2007 Argentina 2 cartographic projection EPSG:5344

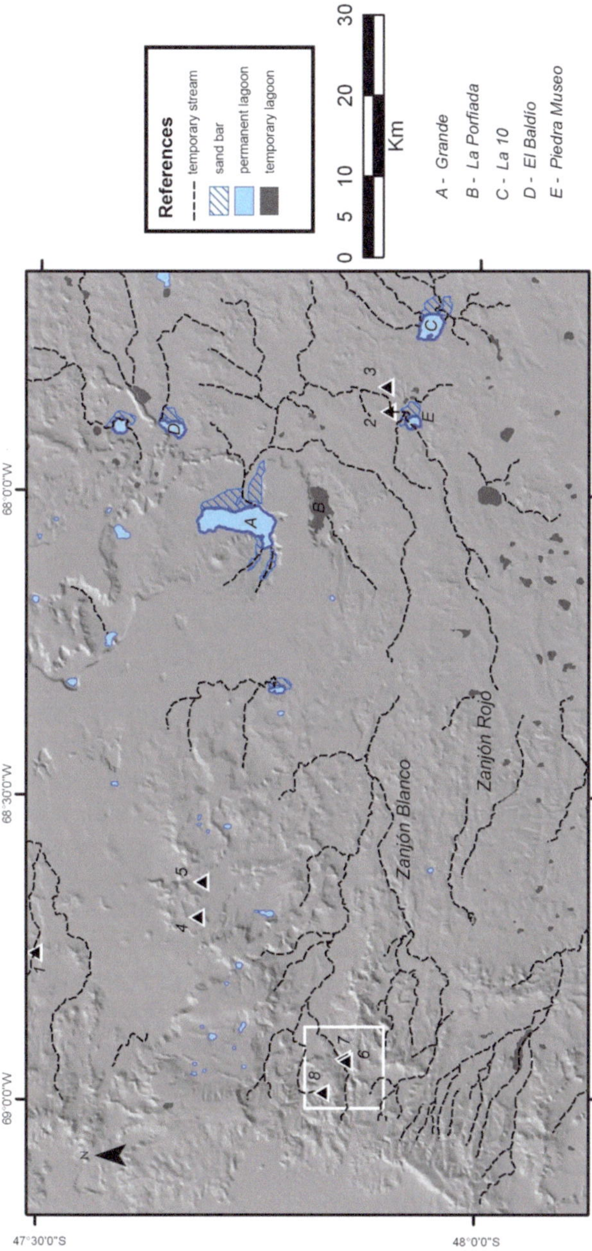

Fig. 2.21 Relief and hydrography map of the macro-study area during the Pleistocene. The extensive lacustrine litoral barriers formed by the large, shallow lakes indicate a greater presence of surface water in the past. On this map, they are represented as permanent shallow lakes. For illustrative purposes only, the current locations and sizes of temporary shallow lakes and Pleistocene lacustrine ridges are marked as permanent lakes

years, allowing us to propose a relationship between the age of the lava flow and the presence of springs and mallines along its borders. The oldest escorials likely have more developed mallines and spring systems than the younger ones. For example, the Las Mercedes basalt, which dates from the Upper Cretaceous (approximately 145 million years ago), is expected to have highly developed mallines compared to the La Angelita basalt, which formed in the Upper Pliocene/Pleistocene (2.0 ± 0.05 million years ago) (Panza, 2001) (Fig. 2.11). Studies are needed to contrast the hypothesis of differential availability of mallines and springs at the macro-area scale, as well as to adjust their chronologies. However, given the current state of knowledge, we can assume that, when the first settlers arrived in this area, the basaltic plateaus were already in their current state (Mazzoni, personal communication). Currently, the main springs experience seasonal fluctuations in water volume, depending on the extent of the escorial and precipitation (Mazzoni, 2007).

2.1.5.1 The Main Water Sources for Past Populations

In conclusion, the springs have been the main water sources in the present as well as in certain arid periods of the past when conditions were similar to those currently prevailing. They had permanent water flows, though the volume varied depending on rainfall. The springs were active even in winter since they generally do not freeze. When they do freeze under very cold conditions, animals can still access water. Current settlers of the area told us that guanacos break the ice with their paws to drink (Ramón Cano, personal communication). Other sources include temporary shallow lakes that retain water for varying periods, depending on annual rainfall. Finally, transient and ephemeral tributaries provide water immediately after rains, but are less stable sources available only during the rainy season. During periods of higher effective humidity, the landscape had more available surface water, as proposed by geomorphological, pollen, and stratigraphic studies of other Patagonian sectors (Mazzoni, 2001; Stine & Stine, 1990). In the Pleistocene era, particularly between 11,000 and 10,500 years before present (BP), there could have been significant permanent water bodies near the mouth of the drainage system. This is evidenced by lacustrine littoral barrier deposits in the eastern shallow lakes, which were generated by the action of waves during storm events. These deposits show that the water bodies were more extensive than they are today (Panza, 2001; see Figs. 2.20 and 2.21). However, further research is needed to understand changes in water availability throughout the year, both currently and in the past. The surveyed study area, on the other hand, is located in the source area of the main creeks, where there are no significant shallow lakes. The regional information and field observations presented above allow us to suggest that the springs are the only consistently available water source in the study area and are therefore the most important. The small shallow lakes on the plateaus and temporary creeks provide water after rainfall and are only important during wetter periods. However, the water quality is always better in the springs because it is neither salinized nor containing clay.

2.1.6 Landscape Units: Sediment Burial Potential

As is widely recognized in contemporary research, archaeological evidence is preserved differently under varying conditions. Therefore, before interpreting regional archaeological distributions in terms of past human behavior, the predominant surface processes in different environments must be evaluated through geoarchaeological work (Butzer, 1982; Ravesloot & Waters, 2004; Waters & Kuehn, 1996).

Reviews of archaeological studies of large regional areas show that a common procedure is to delimit different environmental, landscape, or geomorphic units. This is useful for studying the preservation and differential visibility of archaeological materials (in regional Patagonian archaeology see Belardi, 2005; Belardi & Carballo Marina, 2003; Belardi et al., 2006; Bellelli et al., 2000; Carballo Marina, 2006; Carballo Marina et al., 2000; Favier Dubois, 2001; Figuerero Torres et al., 2004; Scheinshon, 2001, and for this study area: Gómez & Magnin, 2008 and Magnin, 2013; Marchionni et al., 2020). The procedure involves defining units that are characterized by their degree of surface stability (Wells, 2003) and archaeological visibility (Belardi, 2005). This contributes to interpreting archaeological distributions in the landscape.

In order to characterize the current landscape and the main processes that shape the land surface in La Primavera, this study defines a map of landscape units based on fieldwork and published information. The existing cartography is compiled in a GIS environment as base data. The land use base coverage is a vector shapefile created by the National Geographic Institute (IGN) at a scale of 1:250,000 (SIG IGN 250). Each unit class on this map was defined based on differences in texture and color in Landsat TM satellite images (Fig. 2.22). The IGN's graphic information was supplemented with geomorphological data (Gómez & Magnin, 2008; Magnin, 2013; Panza, 2001). Each map unit defines an environment in which predominant geomorphic processes occur. This has particular implications for recording archaeological evidence on the surface (Butzer, 1982). These expectations can be summarized by assigning a sediment burial potential value ranging from high to low, null, or differential (when different burial conditions occur within a landscape unit) (Table 2.9). Figure 2.22 shows that the La Primavera Locality extends across two landscape units: a basaltic plateau and rocky outcrops. Sediment burial potential in these units is low to nonexistent, and archaeological visibility is high. The best chance for the burial of archaeological materials occurs in caves formed in the rocky outcrops. Since these units are defined at a regional analysis scale, they are mapped with low detail. As the processes affecting archaeological visibility, preservation, and integrity are local, small-scale geomorphological maps must be generated (Behrensmeyer, 1978; Belardi & Carballo Marina, 2003; Ebert & Kohler, 1988). Therefore, the landscape units presented here are a preliminary analysis of trends in sediment burial potential and archaeological visibility. In Chap. 6, more detailed cartography is generated with geomorphological units defined at a higher level of detail.

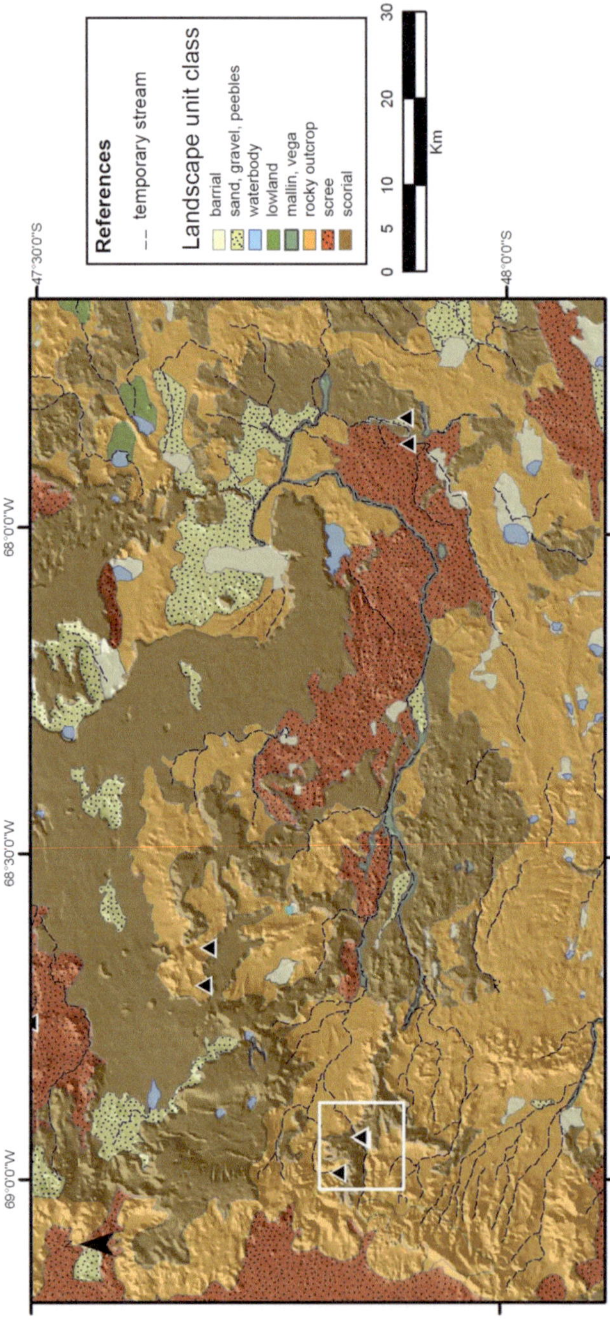

Fig. 2.22 This map shows the main landscape units present in the macro study area. The IGN 250 GIS land use vector shapefile was used for this purpose. Vector shapefiles of land uses at a scale of 1:250,000 are presented with 40% transparency. A raster file derived from the USGS 90 m DEM is used as a base map

Table 2.9 Regional taphonomy: sediment burial potential in each landscape unit

	Landscape units	Land surface use units (IGN cartography) corresponding to each landscape unit	Description of the land surface use unit (IGN cartography)	Sediment burial potential*
1	Plateaus, Lava flows	Scorial	Land covered in scoria (volcanic lava)	Low/null
2	Hills	Rocky outcrop	Superficial manifestation of strata, vein or mass of any type of rock that appears on the surface of the ground. They cover more or less large areas	Null (except caves)
3 4	Pediment/Mass removal	Scree	Land covered with loose stone (rocks of different sizes and composition), its origin is very varied	Low/null
5	Endorheic depression	Barrial	Low, flat terrain, generally wet, covered with mud. In dry seasons the mud transforms into an almost impalpable dust	Differential
6	Alluvial deposits	Sand deposit with gravel and pebbles	Sandy area that presents materials from decomposition, quarry accumulation, etc.	Differential
7	Bañado	Lowland	Flooded land that contains water throughout the year. This area is generally muddy or swampy, covered with shallow water, coming from the overflow of a lake, river, stream, or from the invasion of sea waters	High
8	Shallow lake	Waterbody	Surface covered with water of natural or artificial origin that occupies a large area of land	High
9	Mallín	Mallín, vega	Fertile, humid and low soil, with natural herbaceous vegetation	High
10	Paleochannel	Paleochannel or abandoned channel	Old area of semi or completely deactivated watercourse, which may or may not have water in certain climatic conditions, and may become a flooded and swampy area	High

*This was defined based on Panza's (2001) geomorphological description, previous works in the area (Gómez & Magnin 2008; Magnin, 2013), and other works in other similar areas (Belardi, 2005; Borrero et al., 1991)

As the table shows, the lowest landscape sectors, which are natural sedimentary traps, have a high potential for burying artifacts once they are deposited on the ground surface, either intentionally or accidentally. These lower sectors are located to the east and north of the macro study area (Fig. 2.22). In contrast, the potential for burial in the hills is zero; artifacts deposited there tend to be exposed to weathering or claimed by people over time. Depending on their position and orientation with respect to the wind, some caves and rock shelters formed in these units may constitute natural traps for sediments and eventually archaeological remains.

In the case of plateaus, lava flows, pediments, and mass removal sectors, the potential is low to nonexistent, given the substrate's characteristics with little to no soil development (Belardi & Carballo, 2003). In endorheic depressions and alluvial plains, the processes that take place are variable (see Chap. 6).

2.1.7 Soils and Phytogeography

Given its climate, soils, and vegetation, the study area falls within the Central Plateau ecological unit (Oliva et al., 2004). The soils are generally Aridisols with a sandy or clay loam texture and low organic matter content. Many of these soils are degraded by overgrazing. In these cases, subsurface clay horizons are exposed and covered with small pebbles due to freezing and thawing, forming erosion pavements (Oliva et al., 2004). According to Cabrera (1971), the Central Plateau of Santa Cruz corresponds to the phytogeographic territory known as the Patagonian Province, which is part of the Andean-Patagonian Domain. In the study area, the dominant vegetation is shrub and subshrub steppe with areas of herbaceous steppe where extreme adaptations to wind and drought are prevalent (Table 2.10; Fig. 2.23) (Baetti, 2004). Steppes with very low vegetation cover, consisting of dwarf and creeping shrubs, are common. The largest shrubs are distributed along runoff lines and underground drainage networks. Areas of grasses and coironales, as well as concentrations of bushes at the bottom of ravines and ephemeral vegetation covering erosion pediments, are also frequent (Oliva et al., 2004). As previously mentioned, wet meadows, called *mallines*, form in association with springs or where water accumulates (Cabrera, 1971). Among the creeping shrubs, the colapiche (*Nassauvia gromerulosa*) is dominant in the subshrub steppes that cover most of the area. It grows in impoverished soils and eroding pavements, competing with grasses, completely displacing them. Some of the important dwarf shrubs are the neneo (*Mulinum microphyllum*) and ephedra (*Ephedra frustillata*). Among the common subshrubs in degraded areas with clay soils and erosion pavements, are the manca perro (*Nassauvia ulicina*) and the cat's claw (*Chuquiraga aurea*). Within taller shrubs, the black bush (*Junellia tridens*) is common in the steppe; it is also found on terraces, plateaus, and at the bottom of canyons or river valleys near mallines. The canyons also have bushes of yellow bush (*Anartrophyllum rigidum*), molle (*Schinus polygamus*), and calafate (*Berberis heterophylla*). In the slightly degraded sectors of the subshrub steppes, it is common to find species such as the bitter coirón (*Stipa speciosa*) and the feathered coiron (*Stipa neai*). On the other hand, in sectors of sand accumulation, which indicate soil degradation, other bitter coirones appear: *Stipa humilis, Stipa chrisophylla*, and *Stipa ibari*. On high sedimentary and basaltic plateaus, subsists the white coirón (*Festuca pallescens*), which is one of the best forages, along with the poa coirón (*Poa dusenii*) (Kofalt & Mascó, 2004; Oliva et al., 2004). The La Primavera locality extends into semi-desert and sub-bush steppe, as well as a smaller area of bush steppe (Fig. 2.23).

The vegetation units in the map of Fig. 2.23 were reclassified based on average vegetation cover percentages (Table 2.10) to create the vegetation cover map in Fig. 2.24. This map will be used later to estimate regional archaeological visibility.

2.1.7.1 Pollen Studies

The distribution of past vegetation in the area and its changes over time are closely associated with climate fluctuations. Pollen studies in the Piedra Museo locality (Borromei, 2003) indicate conditions of low water availability around 13,000 years BP, followed by a climate change around 11,000 BP. During this time, there was an increase in effective humidity due to increased precipitation and decreased temperatures. This change coincided with the replacement of steppe vegetation with *Asteraceae* and *Ephedra* species, which represent very arid conditions, by a grass steppe (*Poaceae sp.*) that indicates higher humidity (Páez et al., 1999). The second climate change, which occurred around 9500 BP, is interpreted as an increase in temperature and a decrease in precipitation, reaching the modern range. During this time, shrub steppe began to dominate, replacing grass steppe in Piedra Museo until around 7500 BP. In Los Toldos, this process occurred between 10,000 and 8750 BP, and in La Martita Cave, it occurred between 8000 and 6000 BP (Mancini, 1988).

The latest studies by De Porras et al. (2008) in Los Toldos, which cover the entire history of occupation of the site, indicate that the current vegetation was established in that locality around ca. 3750 AP. Then the shrub-steppe *Asteraceae subf. Asteroideae* with *Schinus, E. frustillata*, and a high proportion of grasses, was replaced by the shrub steppe of *Colliguaja integerrima* and *Asteraceae subf. Asteroideae*. Likewise, studies in the La Martita and Alero Cárdenas caves present a change toward the current floristic conformation during the Late Holocene (Mancini 1998).

A pollen and sedimentological study by Mosquera and Mancini (2021) in a wet meadow at La Primavera and the diatom study at Cueva Maripe (Fernandez, 2013) confirm these tendencies from more humid to dry conditions from the Mid Holocene to Late Holocene period. They also analyze the coincidence of radiocarbon hiatuses in the area and their coincidence with thc increase of aridity human occupation in the region.

Table 2.10 The main species that form the subunits (types) within the physiognomic domains present in the study area are listed below

Units	Adesmia campestris (mamuel choike)	Senecio filaginoides (mata mora)	Festuca pallescens (coirón blanco)	Festuca argentina (huecú)	Poa ligularis (coirón poa)	Poa dusenii (coirón poa)	Schinus marchandii (molle)	Berberis heterophylla (calafate)	Mulinum spinosum (neneo)	Nassauvia gromerulosa (colapiche)
EA 1 & EA 8	x	x	X		x					
EA 10		x					x	x	x	
ES 1						x				x
ES 2			X	x	x					x
ES 3			X	x	x	x				x
S 2										

Units	Nassauvia ulicina (manca perro)	Chuquiraga aurea (uña de gato)	Stipa humilis	Stipa speciosa (coirón amargo)	Stipa chubutensis	Junelia tridens (mata negra)	Nardophyllum obtusifolium	Annual or herbaceous species	Plant cover (%)
EA 1 & EA 8									40–70
EA 10									30–40
ES 1	x	x	x	x					30–50
ES 2	x				x	x	x		30–40
ES 3	x	x				x	x		30–60
S 2	x	x		x				x	10–30

References EA1 and EA8 = grassy shrub steppe; EA10 = shrub steppe 10; ES1, ES2, and ES3 = grassy subshrub steppe; EG2 = grassy steppe 2; and S2 = subshrub semidesert (Baetti, 2004)

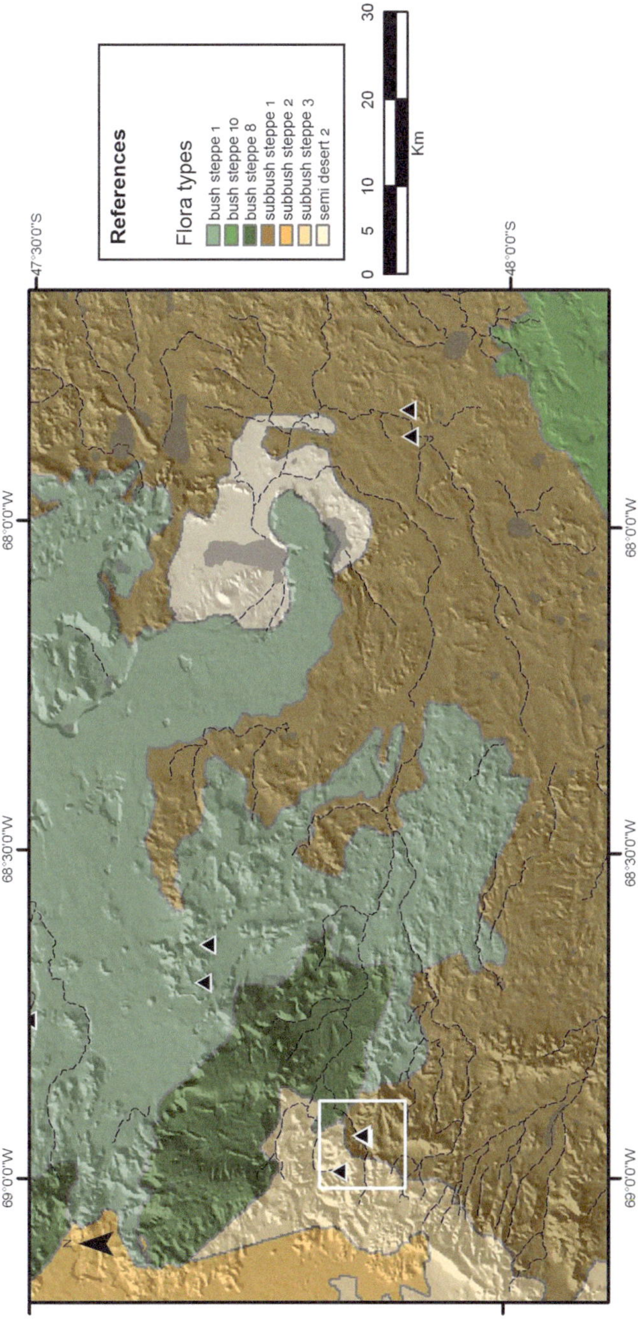

Fig. 2.23 This map shows the vegetation units in the macro-study area. These sub-units are located within the large physiognomic domains defined for Santa Cruz (Baetti, 2004)

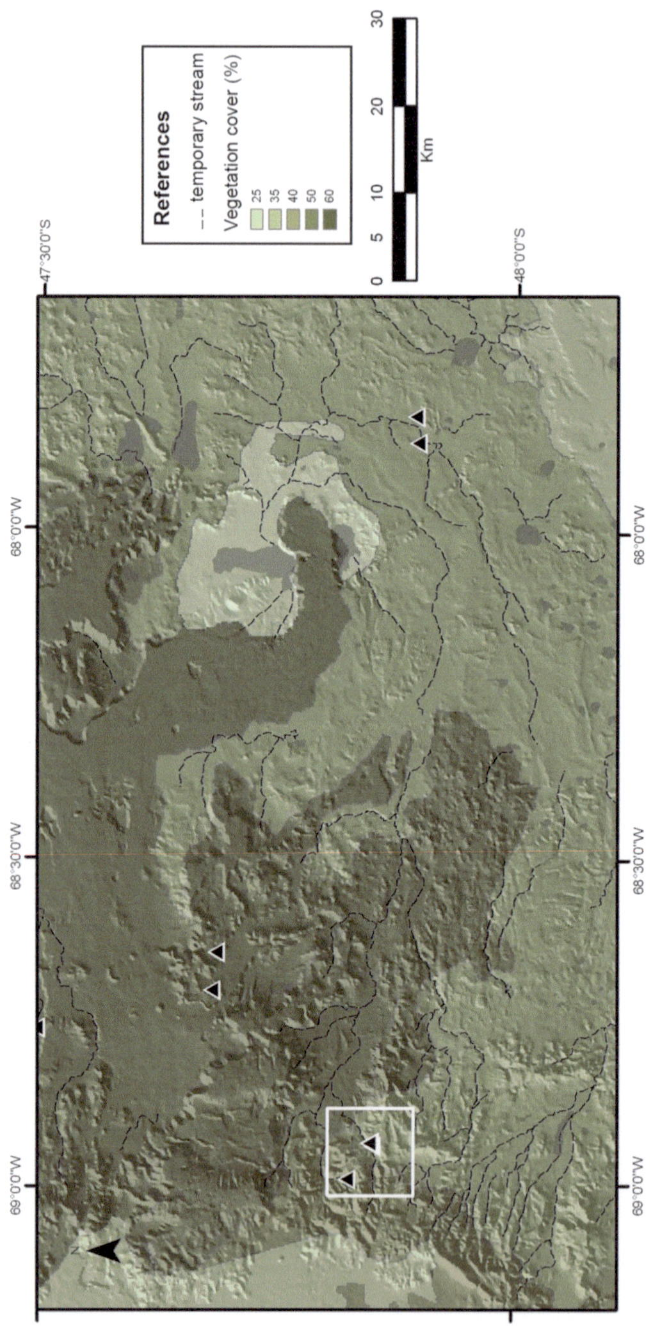

Fig. 2.24 Map of percent plant cover. It is derived from the map of vegetation units in the macro-study area (Fig. 2.23) and was reclassified according to the average vegetation cover values within each sub-unit (Table 2.10)

2.1.7.2 Potential Uses of Flora

The native plant species that characterize the Central Plateau present a wide amount of potential uses that were informed by the local residents (Chela Pandolfi de Iribarne, personal communication) and published works (Correa, 1971; Crivelli Montero et al., 1992; Gradín & Aguerre, 1992; Haloua, 1997; Kofalt and Mascó 2004; Pérez de Micou et al., 1992). Among them some food plants useful for making infusions such as pampas tea (*Satureja darwinii*); the tunas or bloodsucker (*Maihuenia patagonica*); plants with edible berries such as the calafate (*Berberis buxifolia*), with which sweets, syrups and wine can be prepared. Its wood also has berberine, a substance with medicinal properties. Other medicinal plants are paramela (*Adesmia boronioides*), mamuel choike (*Adesmia volckmsnni Ph*illipi), alfilerillo (*Erodium cicutarium*), ajeijo (*Polygonum brasilens*), blonde mallow or toad grass (*Marrubium vulgare*), goat's beard (*E. frustillata*), thyme (*Pleurophora patagonica*) and leño de piedra (*Azorella monantha*). Others can be toxic to humans or animals such as the pichaga (*Trifolium* sp.) and the coirón huecú (*Festuca argentina*).

Among the plants with pigments used to dye wool are calafate, green bush (*Lepidophyllum cupressiforme*), mora bush (*Senecio filaginoides*) and duraznillo (*C. integerrima*). Mamuel choike, calafate, leño de piedra, and molle (*Schinus marchandii*) are currently used as firewood. Although some of these plants have specific uses, others, such as calafate (*Berberis ilicifolia*), have the potential to have been used for many tasks. As documented by the Austrian ethnologist Martin Gusinde (1982–1989) in his travels to Tierra del Fuego, the branches of this plant were used by the Selk'nam for the manufacture of shafts for arrows (see also Ratto, 1994), rods to dry leather and to make fire tongs, given that *its wood is not prone to catching fire and allows burning embers to be handled without danger* (Gusinde, 1982–1989: 187). The same author mentions that the bows were made from trunks of Nothofagus Antarctica (ñire) or *Nothofagus betuloides* (guindo or *Magallanes coihue*). In the case of the northern sector of the Isla Grande (Gusinde, 1982–1989:213), and the Magallanes area (Martinić, 1995); Where these trees were scarce, bow stems were acquired by barter with their southern neighbors. Given that these species are not found in the study area either (in the present or in past times), the stems useful for making elements such as bows, spears, and even for assembling the structure of awnings and windbreaks must have been made from other species or have been obtained by barter or through trips to mountain range sectors where these trees forests exist since 4000 years BP (Mancini, 2002; Mancini et al., 2005). Among the locally available species, the one that reaches the largest size, the molle, reaches three meters in height; others such as the duraznillo (*C. integerrima*) and the yaoyín (*Lycium chilense*) reach two meters and do not have thorns, which makes them more accessible. However, the characteristics of the woods useful for manufacturing these elements vary in relation to the function they must fulfill, and experimental work is necessary to have greater certainty in this regard. The use of molle (*S. marchandii*) was recorded archaeologically by Pérez de Micou et al., (1992) at the Campo Moncada site. In this work the authors mention that *molle resin was used to adhere stone to wood, in arrows, leather scrapers and as chewing gum* (Pérez de Micou et al., 1992:110). Haloua (1997) also mentions its use as a mastic, since it was used by the Tehuelches to attach arrows to the shafts. Another possible use of molle resin is oral hygiene (Musters [1871], 1997). On the other hand, Crivelli Montero et al. (1996) in their studies in the Epullán Grande cave (Neuquén) documented the use of molle as firewood and considered that this was the best firewood existing in the steppe (see also Pérez de Micou & Ratto, 2004). In the Santa Cruz Central Plateau, its use as firewood was documented at the Piedra Museo AEP-1 and Cerro Tres Tetas sites since ca. 11,000 years BP (Steele & Politis, 2009).

The archaeological use of *Fuegian coirón* (*Festuca gracillima*) has also been recognized in the preparation of occupation surfaces. They were found arranged in *beds* in different strata of the northern chamber of Cueva Maripe and in the *Patagonian* occupation floors of Cave 3 of Los Toldos (Miotti, 1989). The same was recorded in archaeological sites outside the Central Plateau, such as in the Pinturas River area (Gradín & Aguerre, 1992). On the other hand, there are ethnographic chronicles (Claraz, 1988) that document the preparation of occupation surfaces by placing plants.

The use of reeds to make baskets was documented among the Selk'nam. Their use was common among inhabitants near the coast, who used them to collect marine products, although they were also used to make bracelets for personal adornment (Gusinde, 1982–1989).

Finally, the carob or algarrobillo (*Prosopis denudans*) is mentioned by Musters ([1871] 1997:316):

> …I saw the carob for the first time. There were two kinds with a black fruit, which…is poisonous, and the other with yellow carob beans, which even when they were a little dry…we picked and ate all…We sat around the fire and had a mate and a little Indian cake, a yellow paste made from crushed carob and mixed with water…

2.1.8 Zoogeography

From the zoogeographic point of view, the area corresponds to the Patagonian Province, located in the Neotropical Region and the Patagonian Subregion sensu Cabrera and Willinck (1980). All the mammals that inhabit this sector are adapted to aridity and even rarely drink water. Among the larger ones (body length more than 45 cm) are the herbivores: guanaco (*Lama guanicoe*), mara (*Dolichotis patagonum*), orange chinchillon (*Lagidium wolffsonhi*) and the mountain chinchillon (*Lagidium viscacia*). Among the carnivores, the puma (*Puma concolor*), the red fox (*Pseudolopex culpaeus*), the grassland cat (*Lynchailurus pajeros*), the wild cat (*Puma geoffroyi*); and the gray fox (*Pseudalopex griseus*). Among the smaller mammals are the Patagonian weasel (*Lestodelphis halli*); the Patagonian skunk (*Conepatus humboldti*), the hairy skunk (*Chaetophractus villosus*), and the piche (*Zaedyus pichiy*), all of which are omnivorous; and the tuco-tuco (*Ctenomys magellanicus*) and the small cuis (*Microcavia australis*) among the herbivorous species (Magnin and Moreira 2024). Among the birds are the terrestrial rhea (*Pterocnemia pennata*), the black eagle (*Geranoaetus melanoleucus*), the southern bandurria (*Theristicus melanosis*), the jotes (*Cathartes aura* and *Coragyps atratus*), the ñacurutú (*Bubo magellanicus*) and the caracara (*Caracara plancus*). Other minor birds in this environment are the martineta or copetona (*Eudromia elegans*), the chimango (*Milvago chimango*), the barn owl (*Tyto alba*), the Patagonian snipe (*Attagis malouinus*), and the small partridge or pale inambú (*Nothura darwinii*). Among the birds of aquatic environments are the black-necked swan (*Cygnus melancoryphus*), the flamingo (*Phoenicopterus chilensis)*, the heron (*Egretta alba*), the coscoroba (*Coscoroba coscoroba*), the biguá (*Phalacrocorax olivaceus*), the common, royal and red goose (*Chloephaga picta, Ch. poliocephala, Ch. rubidiceps*), the white egret (*Egretta thula*), and the raven (*Plegadis chichi*). Among the smaller ones are the maicero and overo, barcino and grebe ducks (*Anas georgica, A. sibilatrix, A. flavirostris, Oxyura vittata*), the macaes (*Podiceps rolland, Podiceps occipitales, Podilymbus podiceps*) and the common rockfish (*Pardirallus sanguinolentus*). The birds that live in both types of environments are the Ashy's Hawk (*Circus cinereus*), the patagonian quiula (*Tinamotis ingoufi*), the little coot (*Fulica leucoptera*), the red-leaved coot (*F. Armillata*) and the common woodcock (*Gallinago paraguaiae*). Finally, numerous passerines live in different environments (Magnin & Moreira, 2024).

2.1.8.1 Distribution of Animal Species in the Macro-Study Area

Information about the habits of native species currently inhabiting the region was tabulated and organized to create a thematic map showing the distribution of animal species within different landscape units (Magnin & Moreira, 2024). The map in Fig. 2.25 was generated using the proportions of the most common species within each of the landscape units defined for the macro area (16 mammals and 33 birds in total). The 60 units on the map resulted from combining the landscape units (11 from map 2.22) and the vegetation units (eight from map 2.23). The map units derived from these combinations were characterized in the analyzed bibliography (Cabrera & Willink, 1980; Harris, 2008; Redford & Eisenberg, 1992) to define the sectors of the landscape that best describe the habitat of each species. The final number of species in each unit is expressed through graduated-color symbols divided into five categories based on natural breaks and represented as percentages of the total. Consequently, the map depicts the landscape units with a higher percentage of mammal and bird species in more intense colors. These sectors coincide with endorheic depressions, which are important attractors of fauna. The lowlands have a great diversity and abundance of birds, especially in the spring and summer, when species from northwest Argentina come to nest in Patagonia and converge with birds from the Northern Hemisphere that are in their non-reproductive stage. Added to these species are resident birds that only make small-scale movements (Mazzoni et al., 1999). Within the total area, the La Primavera locality presents high and intermediate values of species richness (Fig. 2.25).

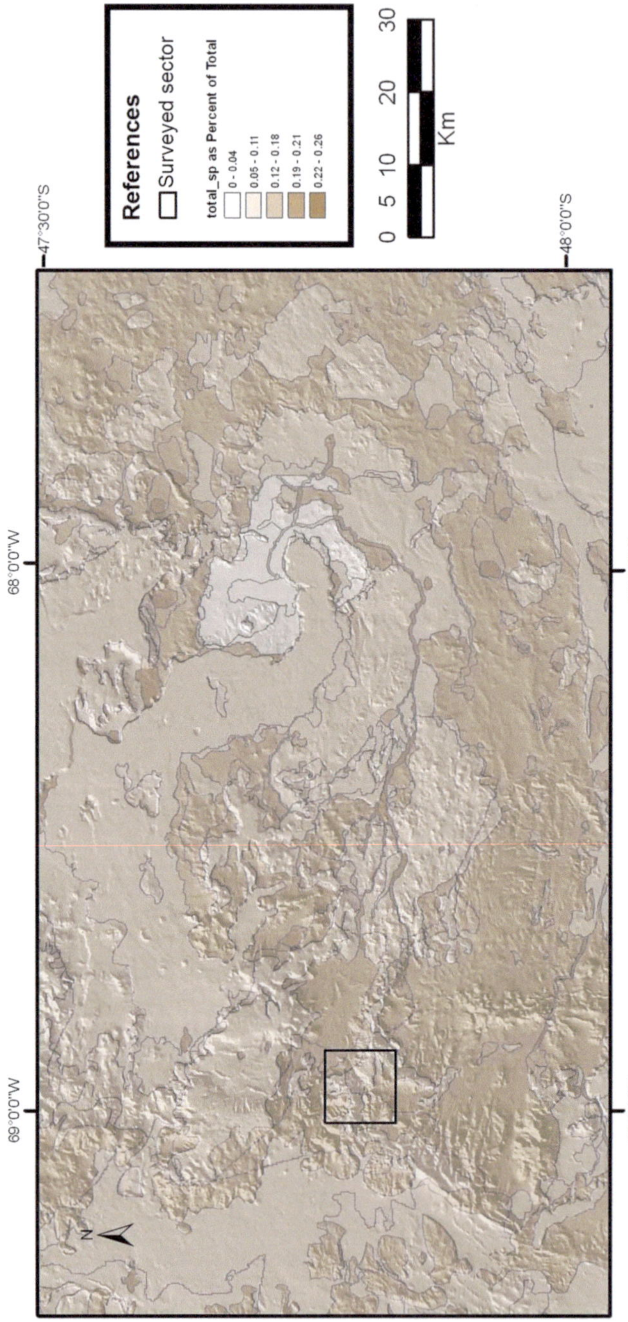

Fig. 2.25 Distribution map of mammal and bird species in each landscape unit

2.1.8.2 Extinct Fauna or Displaced to Other Regions

In addition to the current fauna, it is important to note that when the first humans arrived, a number of animals lived in the region that are now extinct or displaced to other northern regions. The time period during which these animals lived in the region varies depending on each species. Those found in archaeological sites include the American horse (*Hipidion saldiasi*) (in Los Toldos and El Ceibo sites, Cardich 1987; AEP-1 site, Miotti et al., 2003); two camelids: *Lama vicugna gracilis* (in Los Toldos and El Ceibo sites, Cardich, 1987; in AEP-1 site, Miotti et al., 2003; in Casa del Minero 1 site, Paunero et al., 2007), and *Hemiauchenia paradoxa* (in Casa del Minero 1 site, Paunero et al., 2007); a cervid, the huemul (*Hippocamelus sp.*) that could have inhabited the region even until the arrival of Europeans (Miotti, 1989); the common rhea (*Rhea Americana*), which is now found at the Pampas region and other parts of South America (in Casa del Minero 1 site, Paunero et al., 2007); and also a feline that disappeared from the region (*Pantera onca*), depicted in the El Ceibo cave paintings (Cardich, 1987).

2.1.8.3 Domestic and Introduced Fauna

The first record of a domestic animal species is found at the Cueva 3 archaeological site in Los Toldos. The final *Casapedrense* settlers probably used dogs (*Canis familiaris*) as an auxiliary in hunting (Cardich, 1987). In post-Hispanic times, the European hare (*Lepus europeaus*), sheep (*Ovis aries*), also goats (*Capra hircus*), cattle (*Bos taurus*), and horses (*Equus caballus*) were introduced (Navas, 1987). The first cattle drives (cows) in the province were carried out from Chubut to Puerto Deseado and Puerto Santa Cruz in 1884 (Godoy, 1997). The1937 agricultural census indicates that the Deseado area reached a historical maximum for the entire province, with over 1,900,000 heads of sheep (Oliva et al., 1995).

2.1.8.4 Zooarchaeological Studies and Ethnographic Records. Uses of Fauna

A general model of subsistence strategies and their changes regarding the faunal resources that supported hunter-gatherer populations in the past is presented in Table 2.11. Archaeological studies (Borrero, 1985; De Nigris & Mengoni Goñalons, 2004; Miotti, 1989; Miotti & Salemme, 1999, 2004a; Miotti et al., 1999a; Salemme & Miotti, 1987) have shown that guanacos (*L. guanicoe*), complemented by rheids, have been the primary component of the diet since humans first settled in the area (Borrero, 1985; Cardich & Miotti, 1983; Miotti, 1989; Miotti & Salemme, 1999, 2004a; Miotti et al., 1999a; Salemme & Miotti, 1987). Prior to the extinction of Pleistocene megafauna (ca. 9000 BP), the diet was more varied and included Pleistocene horses and extinct camelids. Then, during the Middle Holocene, the archaeological record shows evidence of specialization in guanaco hunting. In the Late Holocene, the

Table 2.11 Model of subsistence strategies throughout the settlement process of Patagonia

Exploration and colonization	Territorial consolidation	Effective Occupation	Settler colonialism
(Pl/H y EH) *ca. 11,000–7500 BP*	*(MH)* *ca. 7500–3500 BP*	*(LH),* *ca. 3500–1200 BP*	*(recent LH)* *ca. 1.200-* *(Equestrian period)* *ca. 250 BP* *sixteenth century*
Generalized strategy: guanacos, extinct Lama genus species and extinct horses, supplemented by rheids and other mega mammals	Specialization in guanaco hunting complemented with choique (*Pterocnemia pennata*)	Generalized: guanacos complemented by rheids, birds, small mammals and marine resources	*Idem* + horse (for food, transportation and for hunting) food storage and preservation

Sources consulted: Borrero (1985), Miotti (1989, 1999), Miotti and Salemme (1999, 2004a), Alberdi et al. (2001), Carden (2009), Goñi et al. (2004), Hermo (2008), Salemme and Miotti (2008)

diet expanded again to include rheids, birds, small mammals, and marine resources (*Mytilus sp.* and *Fisurella sp.*). During the historical period, the camelids and reiforms that currently inhabit the area were the main source of sustenance, and among them, the choique is mentioned as the preferred one (Musters [1871], 1997). According to the chronicles, the Tehuelches collectively hunted guanacos during equestrian times (Claraz, 1988; Musters [1871], 1997), although there are also records of solitary hunts, in which the participation of canids was crucial (Aguerre, 2000). In all cases, the animals' meat, fat, viscera, blood, and marrow were used for food. The leather, bones, and tendons of these animals provided fundamental raw materials and the necessary artifacts for working with them. Feathers were also highly valued for trade (Miotti, 1989). At archaeological sites such as Los Toldos (Cave 3, associated with Toldense occupations), remains of pumas (Felis sp.) and foxes (Canis sp.) were found, probably valued for their skins. Additionally, the presence of birds of prey and martinets in archaeological sites, such as Los Toldos Cave 3, suggests their use, though it cannot be confirmed whether it was for food or for the use of their feathers (Miotti, 1989). The chronicles also indicate that rhea feathers were also highly valued as a means of exchange (Musters [1871], 1997).

2.1.8.5 Fauna and Anthropic Factors as Taphonomic Agents

Finally, studying the area's fauna is important from a taphonomic point of view (Wood & Johnson, 1978). The piche (Zaedyus sp.), peludo (*Chaetophractus sp.*), cuis (*Microcavia sp.*), and tuco-tuco (*Ctenomys sp.*) are among the fauna inhabiting this area that have digging habits (Durán, 1991) (Table 5 in Magnin & Moreira, 2024) (Fig. 2.26). The peludos, given their body size and the extensiveness and depth of their burrows, can generate the greatest modifications. At surface sites, this has been observed in the exposure of archaeological materials from subsurface layers at entrances. For example, this occurred at the workshop site D9. In general, these animals inhabit areas with soft, non-floodable soils. Conversely, other animals, such as guanacos, create wallowing areas where they repeatedly take dust baths (Harris, 2008) (Fig. 2.27). As was recorded at Site Laguna C, this behavior disturbs the evidence when it coincides with archaeological sites, clearing the central sector of the wallow of materials and depositing them at its limits. However, the main modification to the archaeological evidence (and to the landscape in general) was caused by the introduction of sheep. As numerous studies have shown, the way these animals feed (they uproot grasses instead of cutting them, as other local herbivorous species do) and walk (they walk on hooves, unlike guanacos, which walk on thickened pads located on the soles of their feet) has favored the loss of vegetation and consequent soil erosion (Oliva et al., 1995). It has also been mentioned that rubbing these animals' bodies when taking refuge in caves with rock art causes significant damage (Carden, 2009). The introduction of horses also had an important impact on archaeological sites (APN, 2018). As detailed in Chap. 6, field surveys were conducted to systematically record the impact of these and other animals on surface archaeological evidence.

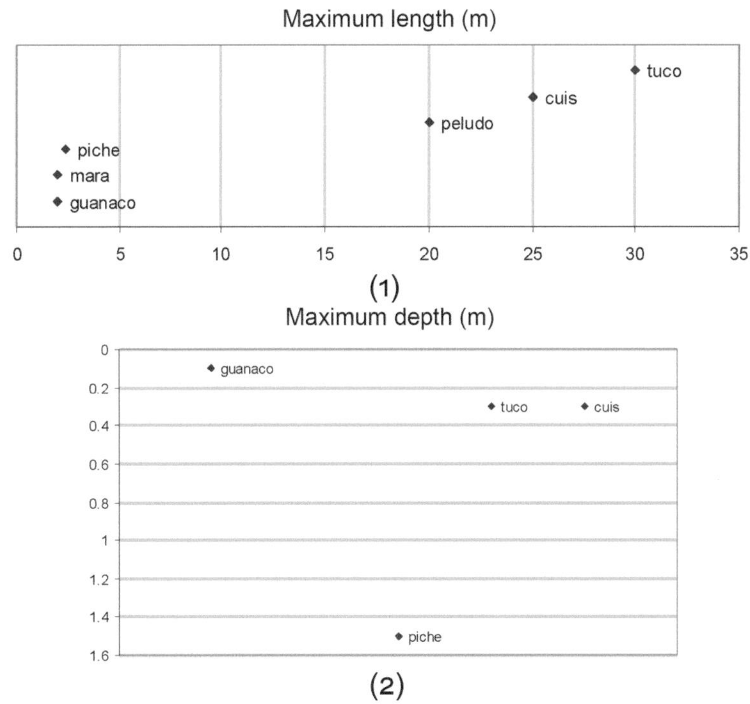

Fig. 2.26 Potential bioturbation in the study area: 1) The length and 2) the depth of burrows made by burrowing animals and guanaco wallows

1

Fig. 2.27 Bioturbation observed in surface sites: 1) Burrow of *Zaedyus sp.* affecting the archaeological site D9, 2) Guanaco wallow at the Laguna Cósmica site, La Primavera Locality

2

Fig. 2.27 (continued)

In addition to the introduction of livestock, anthropic factors include the action of current archaeological object collectors on surface assemblages (Miotti & Podgorny, 1995). The sectors most accessible from the local entrance road to the estancias La Primavera and Las Mercedes probably have the greatest impact. However, we believe this factor has decreased in importance over the last twenty years, as the area is now nearly unpopulated. The construction of Provincial Route No. 12 may have affected surface sites, such as the Los Blocks quarry located nearby. The greatest impact recorded during the 2008 fieldwork was caused by the excavation of a water extraction pit and the opening of roads next to an adobe stand in La Primavera. According to the camp tender of the neighboring ranch, the water was used for the construction of the road (Ramón Cano, personal communication). Additionally, the modification of archaeological stone structures was documented. For instance, evidence was found of the dismantling and rearrangement of structures (possibly burial sites) to support a metal pipe cross on a plateau (see Chap. 7). In the interior areas of the plateaus, far from the roads, evidence of modern modifications was only found where wire fences were laid. All cases of possible disturbance were recorded during fieldwork to better interpret the evidence.

2.2 Archaeological Visibility

The information gathered was used to estimate archaeological visibility in the macro-study area. Based on Belardi's (2005) scheme, we estimated the archaeological visibility in different landscape sectors through the relationship between vegetation cover and sediment burial potential. We combined the data from the percentage vegetation cover map (Fig. 2.24) and the map of landscape units (Fig. 2.22) into a map of archaeological visibility (Fig. 2.28). The coincidence of burial potential with null-to-low values and an average vegetation cover of 35% allows us to conclude that visibility conditions in the La Primavera locality are very good to excellent (Belardi, 2005). It is also necessary to consider that these surfaces are subject to erosion processes that favor the formation of palimpsests. Therefore, the possible mixture of evidence deposited at different times cannot be discarded. Therefore, the systematic recording of environmental context conditions is essential for subsequent interpretation.

2.3 Resources for Human Occupation

Despite the changing and sometimes harsh climatic conditions, this area has a high availability of potentially important resources for hunter-gatherer societies. Among these resources are water springs that allow wet meadows, or mallines, to grow. Water is a critical resource in this sector of extra-Andean Patagonia, and these springs were and still are extremely important for human populations (Miotti, 1989; Miotti & Salemme, 2003, 2004; Melendi, 2023). Animal life and flora undoubtedly provided

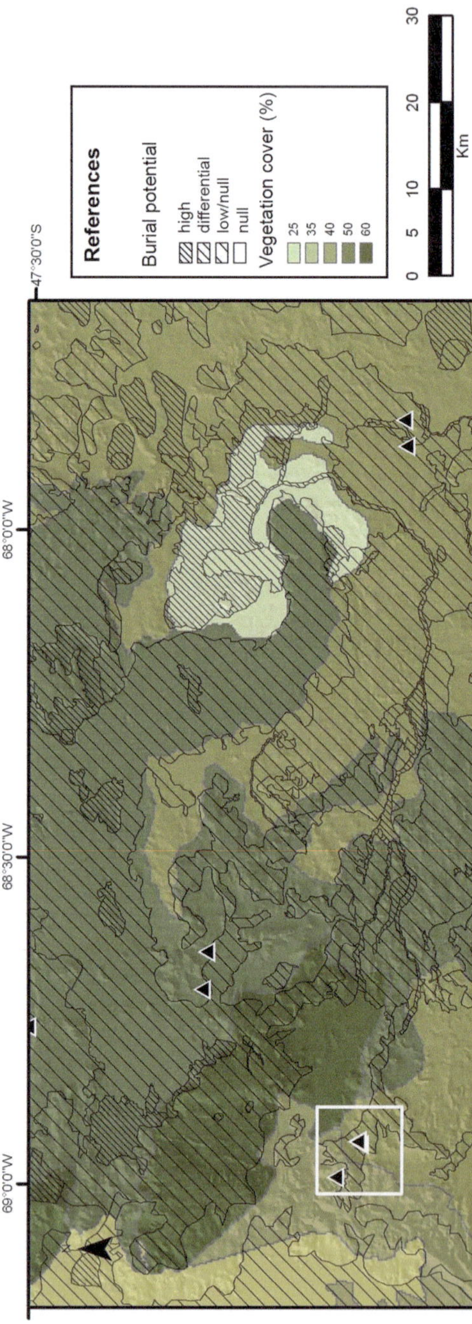

Fig. 2.28 Archaeological visibility map of the macro-study area. It is derived from the overlay of previous landscape unit maps (Fig. 2.22) and the average vegetation cover map (Fig. 2.24)

efficient support to the populations, who must have understood their various cycles and properties. Additionally, this region has an abundance of lithic raw materials for making knapped artifacts and shelters formed in rock outcrops (Cattaneo, 1999, 2002; Hermo & Miotti, 2003; Miotti, 1989, 2006; Miotti et al., 1999; Miotti & Cattaneo, 2003; Miotti & Salemme, 2003a; Panza, 2001). Based on the presented information, it can be concluded that the natural environment's characteristics were favorable for human occupation since the arrival of the first colonizers at the end of the Pleistocene. This is clearly documented by environmental data and existing archaeological evidence in the macro study area. However, substantial questions remain that go beyond this environmental analysis and will be addressed in the next chapters. When was the La Primavera locality populated? What possibilities does the environment offer for annual or seasonal occupation? Has this changed over time? What patterns can be detected in archaeological distributions? How can they be interpreted? Answering these questions required undertaking fieldwork at the La Primavera locality, changing the scale of the analysis.

Chapter 3
Archaeological Background

Abstract This chapter outlines the broader research context within which this work is situated. Background archaeological studies of hunter-gatherers in Patagonia provide knowledge based on various lines of analysis. These baseline studies provide insights into when the first colonizers arrived in the region, the technologies they used, their diet, their communication methods, their mobility strategies, and how they inhabited the landscape. This work contributes new information on landscape use and mobility in the La Primavera locality. One of the most significant contributions to the methodological aspects of this research is the Spatial Archaeology approach and its quantitative perspective, which is mediated by the application of GIS (Geographic Information Systems) to identify spatial patterns.

Keywords Deseado Massif archaeology · Geographic Information Systems (GIS) applications · Hunter-gatherer mobility models · Patagonia settlement

3.1 Research Context

This work is based on Magnin's (2010) doctoral thesis, which built upon previous research in this area by Miotti (1989), Cattaneo (2002), Hermo (2008), Carden (2009), Skarbun (2011), and many other researchers. Magnin's thesis also laid the groundwork for subsequent contributions by Marchionni (2013), Ciampagna (2014), Blanco (2015), Mosquera (2016), Lynch (2016), García Añino (2018), and Flores Coni (2019). This study builds on these solid archaeological studies, which examined the settlement process of large areas of the Central Plateau by mobile hunter-gatherer groups. It contributes to the field by analyzing the spatial distribution of surface archaeological evidence. In the study area, the interaction between human groups and the landscape was examined at various spatial scales and through multiple lines of analysis. Starting with detailed scales and moving to broader processes, the analysis of activity areas within sites was based on bone refitting and taphonomic studies (Giardina et al., 2000; Hermo & Miotti, 2003; Marchionni, 2013; Miotti, 1996; Miotti et al., 1999a), as well as the distribution of rock art (Carden, 2009). The study

L. A. Magnin, *The Landscape and Mobility of Hunter-Gatherer Societies*, The Latin American Studies Book Series, https://doi.org/10.1007/978-3-032-08127-8_3

also examined the exploitation territories of sites (sensu Roper, 1979) and the topographic location of sites in relation to critical and basic resources, including studies of lithic raw material sourcing (Cattáneo, 2002; Hermo, 2005; Hermo & Miotti, 2003; Magnin, 2013; Miotti, 1991, 1996b; Miotti et al., 1999a, 1999b). This expanded the analysis to include the surroundings of sites.

Other studies focused on site location, visual dominance evaluation, and identifying special topographies for particular practices (Magnin, 2013, 2009; Miotti, 1995, 2006; Miotti et al., 1999b). In a regional scale of analysis, previous studies analyzed the distribution of rock art and its potential as a symbolic marker of permeable territories (Carden, 2009; Miotti et al., 2007). At an extra-regional scale, other studies have focused on early archaeological contexts, incorporating biogeographic, paleodietary, and ethnoarchaeological perspectives (Miotti, 1989, 1992, 1995, 1996a, 1996b, 2001, 2003; Miotti & Salemme, 2003, 2004a, 2004b; Salemme & Miotti, 2008).

3.2 Regional Hunter-Gatherers Archaeological Studies

3.2.1 Chronology of Human Occupations in the Area

The region's archaeological research history has been built on the available database of radiocarbon dates. In the Santa Cruz Central Plateau, the earliest evidence dates back to the Pleistocene-Holocene transition (12,890 $\pm$ 90 14C years BP) and was found at the AEP-1 site in the Piedra Museo locality in the lower sector of the Elornia Creek basin. For the Early, Middle, and Late Holocene, dating comes from sites throughout the macro-study area (Table 3.1 and Fig. 3.1). The latest dates obtained for stratified contexts are from the Cueva Maripe site (1078 $\pm$ 40 14C years BP), and the latest dates obtained for open-air contexts are from the El Sargento burial site in the Piedra Museo locality (727 $\pm$ 48 14C years BP) and the La Quinta site (939 $\pm$ 59 14C years BP), in the La Primavera locality. In total, the documentation of human activity in the macroarea of the Central Plateau spans a maximum period of approximately 12,100 years. Based on existing data, the temporal range of occupation at the La Primavera site corresponds to 8579 years (Chap. 7). Considering this, the archaeological distributions recorded on the terrain's surface are part of a remnant settlement pattern (Dewar & McBride, 1992), resulting from occupation of the area from the Early to Late Holocene and subsequent formation processes. As surface evidence, the pattern is expected to reflect the Late Holocene more intensely. However, due to natural processes (Chap. 2), it cannot be ruled out that some of the evidence dates back to the Middle or Early Holocene.

The information obtained from the distributions analyzed in this study will be interpreted based on settlement, subsistence, and mobility models. These models were generated by different Patagonia research teams using chronological information and other lines of evidence.

Table 3.1 Radiocarbon dates for the study area (Santa Cruz Central Plateau)

N	Locality, site	Lab code	Material	14C Date ± uncertainty	Cal BP from_95.4	Cal BP to_95.4	References
1	Cerro Tres Tetas, Cave 1	LP525	Charcoal	11,560 ± 140	13,745	13,160	Paunero (2003a)
2	Cerro Tres Tetas, Cave 1	AA22233	Charcoal	11,100 ± 150	13,238	12,742	Paunero (2003a)
3	Cerro Tres Tetas, Cave 1	AA39368	Charcoal	11,015 ± 66	13,073	12,769	Paunero (2003a)
4	Piedra Museo, AEP-1	AA27950	Charcoal	11,000 ± 65	13,071	12,760	Miotti et al. (2003)
5	La María, Cueva Casa del Minero I	AA37207	Charcoal	10,999 ± 55	13,069	12,761	Paunero (2003b)
6	La María, Cueva Casa del Minero I	AA37208	Charcoal	10,967 ± 55	13,053	12,745	Paunero (2003b)
7	Piedra Museo, AEP-1	OxA8528	Bone	10,925 ± 65	12,971	12,733	Miotti et al. (2003)
8	Cerro Tres Tetas, Cave 1	OxA9244	Charcoal	10,915 ± 65	12,970	12,731	Paunero (2003a)
9	Cerro Tres Tetas, Cave 1	AA39366	Charcoal	10,853 ± 70	12,915	12,693	Paunero (2003a)
10	Cerro Tres Tetas, Cave 1	LP781	Charcoal	10,850 ± 150	13,095	12,484	Paunero (2003a)
11	La Gruta, La Gruta 1	AA84224	Charcoal	10,845 + 61	12,897	12,694	Franco et al. (2010b)
12	La Gruta, La Gruta 1	AA84223	Charcoal	10,840 + 62	12,898	12,691	Franco et al. (2010b)
13	La Gruta, La Gruta 1	UGAMS 7538	Charcoal	10,790 + 30	12,749	12,701	Franco et al. (2013)
14	La Gruta, La Gruta 3	UGAMS 12,428	Charcoal	10,720 + 30	12,736	12,630	Brook et al. (2015)
15	La Gruta, La Gruta 1	AA76792	Charcoal	10,656 + 54	12,730	12,489	Franco et al. (2010a)
16	La Gruta, La Gruta 1	UGAMS 17,353	Charcoal	10,520 ± 30	12,616	12,199	Franco et al. (2020)
17	La María, Cueva Túnel	AA824926	Bone	10,510 ± 100	12,692	12,000	De Porras (2010)
18	La Gruta, La Gruta 1	AA84225	Charcoal	10,477 + 56	12,614	12,063	Franco et al. (2010b)
19	Piedra Museo, AEP-1	GRA9837	Charcoal	10,470 ± 60	12,612	12,021	Miotti et al. (2003)

(continued)

Table 3.1 (continued)

N	Locality, site	Lab code	Material	14C Date ± uncertainty	Cal BP from_95.4	Cal BP to_95.4	References
20	Piedra Museo, AEP-1	OxA9249	Charcoal	10,470 ± 65	12,613	12,018	Miotti et al. (2003)
21	La María, Cueva Túnel	LP1965	Bone	10,420 ± 180	12,733	11,430	Paunero (2009)
22	La María, Cueva Túnel	AA71147	Charcoal	10,408 ± 59	12,575	11,960	Paunero (2009)
23	Piedra Museo, AEP-1	AA8428	Bone	10,400 ± 80	12,610	11,845	Miotti et al. (2003)
24	La María, Cueva Túnel	AA71148	Bone	10,400 ± 100	12,592	11,889	Paunero (2009)
25	Piedra Museo, AEP-1	OxA8527	Bone	10,390 ± 70	12,579	11,890	Miotti et al. (2003)
26	Cerro Tres Tetas, Cave 1	LP800	Charcoal	10,260 ± 110	12,473	11,345	Paunero (2003a)
27	La María, Casa del Minero 1	AA45705	n/d	10,250 ± 110	12,470	11,340	Paunero (2003)
28	Piedra Museo, AEP-1	LP859	Bone	9710 ± 105	11,248	10,708	Miotti et al. (2003)
29	La Gruta, La Gruta 3	UGAMS 13,611	Bone	9560 + 30	11,075	10,664	Brook et al. (2015)
30	La Gruta, La Gruta 1	UGAMS 38,383	Charcoal	9530 + 30	11,069	10,587	Franco et al. (2020)
31	La Primavera, Cueva Maripe	AA65175	Charcoal	9518 ± 64	11,086	10,566	Miotti et al. (2007)
32	La Gruta, La Gruta 3	UGAMS15766	Bone	9470 + 30	10,990	10,565	Brook et al. (2015)
33	Río Pinturas, Cueva Grande del Arroyo Feo	CSIC514	Charcoal	9410 ± 70	11,060	10,300	Alonso et al. (1984–1985)
34	Río Pinturas, Cueva Grande del Arroyo Feo	CSIC396	Charcoal	9330 ± 80	10,685	10,251	Alonso et al. (1984–1985)
35	Río Pinturas, Cueva de las Manos	CSIC138	Charcoal	9320 ± 90	10,692	10,245	Gradín et al. (1976)

(continued)

Table 3.1 (continued)

N	Locality, site	Lab code	Material	14C Date ± uncertainty	Cal BP from_95.4	Cal BP to_95.4	References
36	Río Pinturas, Cueva de las Manos	CSIC385	Charcoal	9300 ± 90	10,680	10,240	Aguerre (1977)
37	Piedra Museo, AEP-1	LP949	Bone	9230 ± 105	10,658	10,184	Miotti et al. (2003)
38	La Primavera, Cueva Maripe	AA95129	Charcoal	9177 ± 56	10,492	10,204	Miotti et al. (2014)
39	La María, Cueva La Mesada	Beta135963	Charcoal	9090 ± 40	10,336	9968	Paunero (2000)
40	La Gruta, La Gruta 1	UGAMS 18,361	Charcoal	9020 + 30	10,237	9920	Franco et al. (2020)
41	La Primavera, Cueva Maripe	AA65179	Charcoal	8992 ± 65	10,242	9804	Miotti et al. (2007)
42	El Verano, Cueva El Verano1	I.13,797–1	n/d	8960 ± 140	10,366	9550	Durán (1986–1987)
43	La Gruta, La Gruta 1	UGAMS 38,382	Charcoal	8930 + 30	10,189	9820	Franco et al. (2020)
44	La Primavera, Cueva Maripe	AA99070	Bone	8827 ± 87	10,154	9552	Miotti et al. (2014)
45	La Primavera, Cueva Maripe	AA65178	Charcoal	8762 ± 50	9898	9545	Miotti et al. (2007)
46	Río Pinturas, Cueva Grande del Arroyo Feo	CSIC515	Charcoal	8610 ± 70	9738	9431	Alonso et al. (1984–1985)
47	La Gruta, La Gruta 3	UGAMS115465	Charcoal	8540 ± 30	9542	9466	Brook et al. (2015)
48	Río Pinturas, Cueva Grande del Arroyo Feo	CSIC516	Charcoal	8410 ± 70	9532	9140	Alonso et al. (1984–1985)
49	Cueva Maripe	AA65174	Charcoal	8333 ± 63	9468	9033	Miotti et al. (2007)

(continued)

Table 3.1 (continued)

N	Locality, site	Lab code	Material	14C Date ± uncertainty	Cal BP from_95.4	Cal BP to_95.4	References
50	La Gruta 1	UGAMS7540	Charcoal	8090 + 30	9088	8771	Mancini et al. (2013)
51	La Martita, Cave 4	CSIC506	Charcoal	8050 ± 90	9132	8596	Aguerre (1987)
52	La Primavera, Cueva Maripe	AA95126	Bone	8012 ± 80	9025	8594	Miotti et al. (2014)
53	La María, Cueva La Ventana	Beta 135,965	Charcoal	7970 ± 40	8986	8605	Paunero (2000)
54	Río Pinturas, Alero Cárdenas	AC497	Charcoal	7750 ± 125	8983	8214	Alonso et al. (1984–1985)
55	La Primavera,Cueva Maripe	AA65177	Charcoal	7703 ± 47	8585	8377	Miotti et al. (2007)
56	Piedra Museo, AEP-1	LP450	Bone	7670 ± 110	8687	8181	Miotti (1996a)
57	La María, Cueva La Ventana	AA35237	Charcoal	7665 ± 75	8594	8216	Paunero (2000)
58	La Gruta, La Gruta 2	UGAMS9113	Bone	7560 ± 30	8405	8201	Franco et al. (2013)
59	Piedra Museo, AEP-1	LP850	Charcoal	7470 ± 90	8390	8031	Miotti et al. (1999a)
60	Río Pinturas, Alero Cárdenas	AC499	Charcoal	7300 ± 200	8429	7680	Alonso et al. (1984–1985)
61	Río Pinturas, Cueva de las Manos	NOVA117	Charcoal	7280 ± 60	8180	7945	Gradín et al. (1976)
62	La Primavera, Cueva Maripe	AA99069	Charcoal	7153 ± 50	8024	7799	Miotti et al. (2014)
63	Río Pinturas, Cueva Grande del Arroyo Feo	CSIC518	Charcoal	6000 ± 90	7155	6555	Alonso et al. (1984–1985)

(continued)

Table 3.1 (continued)

N	Locality, site	Lab code	Material	14C Date ± uncertainty	Cal BP from_95.4	Cal BP to_95.4	References
64	La María, Cueva Túnel	AA81423	Charcoal	5577 ± 44	6441	6215	Paunero (2009)
65	Río Pinturas, Cueva Grande del Arroyo Feo	CSIC519	Charcoal	5550 ± 50	6436	6200	Alonso et al. (1984–1985)
66	La Primavera, Cueva Maripe	AA99071	Charcoal	5376 ± 45	6279	5994	Miotti et al. (2014)
67	Río Pinturas, Alero Charcamata	CSIC800	Charcoal	5290 ± 60	6261	5907	Gradin and Aguerre (1994)
68	Cerro Tres Tetas, Cueva 1	LP538	Charcoal	5220 ± 70	6184	5746	Paunero (1994, 2003a)
69	Casa del Minero 1	LP 1552	Bone	5190 ± 80	6181	5663	Paunero (2003)
70	La Primavera, Cueva Maripe	AA65173	Charcoal	5084 ± 49	5913	5610	Miotti et al. (2007)
71	Río Pinturas, Alero Charcamata	CSIC801	Charcoal	5040 ± 60	5897	5602	Gradin and Aguerre (1994)
72	Río Pinturas, Cueva Grande del Arroyo Feo	CSIC397	Wood	4900 ± 50	5721	5475	Alonso et al. (1984–1985)
73	Río Pinturas, Puesto El Rodeo	AC1075	Charcoal	4900 ± 50	5659	5330	Gradin and Aguerre (1994)
74	Los Toldos, Cave 3	LP136	Charcoal	4850 ± 90	5732	5320	Cardich (1984–85)
75	Viuda Quenzana, Viuda Quenzana 8	UGAMS9111	Bone	4770 ± 25	5581	5326	Franco et al. (2013)
76	Viuda Quenzana, Viuda Quenzana 8	UGAMS9112	Bone	4740 ± 25	5575	5321	Franco et al. (2013)

(continued)

Table 3.1 (continued)

N	Locality, site	Lab code	Material	14C Date ± uncertainty	Cal BP from_95.4	Cal BP to_95.4	References
77	La Martita, Cueva 4 (inferior component)	CSIC505		4520 ± 50	5312	4886	Aguerre (1987)
78	Cueva La Mesada	Beta135964		4500 ± 40	5299	4884	Paunero (2000)
79	Río Pinturas, Cueva Grande del Arroyo Feo	CSIC521	Charcoal	4480 ± 60	5289	4873	Alonso et al. (1984–1985)
80	La Martita, Cave 4	I11904	Charcoal	4475 ± 95	5316	4846	Aguerre (1987)
81	La María, Cueva Túnel	AA81417	Charcoal	4177 ± 43	4827	4528	Frank (2011)
82	La Primavera, Cueva Maripe	AA65181	Charcoal	4113 ± 39	4814	4422	Miotti et al. (2007)
83	Río Pinturas, Cueva Grande del Arroyo Feo	CSIC520	Wood	4050 ± 50	4798	4297	Alonso et al. (1984–1985)
84	La Primavera, Cueva Maripe	AA99067	Charcoal	4002 ± 43	4567	4245	Miotti et al. (2014)
85	La Primavera, Cueva Maripe	AA99068	Charcoal	3791 ± 42	4289	3933	Miotti et al. (2014)
86	Los Toldos, Cueva 13	LP1524	Charcoal	3750 ± 80	4385	3836	De Porras et al. (2009)
87	Cueva Mora	AA85461	Bone	3678 ± 65	4219	3725	Magnin (2010)
88	La Primavera, Cueva Maripe	AA95123	Bone	3535 ± 62	3969	3586	Miotti et al. (2014)
89	La Gruta, La Gruta 1	AA4226	Charcoal	3487 ± 38	3837	3585	Franco et al. (2013)
90	Río Pinturas, Alero Cárdenas	AC598	Charcoal	3450 ± 105	3965	3403	Alonso et al. (1984–1985)

(continued)

Table 3.1 (continued)

N	Locality, site	Lab code	Material	14C Date ± uncertainty	Cal BP from_95.4	Cal BP to_95.4	References
91	Los Toldos, Cueva 13	Beta183025	Charcoal	3380 ± 40	3692	3455	De Porras et al. (2009)
92	Río Pinturas, Cueva de las Manos	NOVA116	Charcoal	3380 ± 90	3830	3383	Gradín et al. (1976)
93	Río Pinturas, Cueva Grande del Arroyo Feo	CSIC398	Charcoal	3330 ± 50	3682	3394	Alonso et al. (1984–1985)
94	Río Pinturas, Cueva Grande del Arroyo Feo	CSIC522	Charcoal	3260 ± 50	3570	3274	Alonso et al. (1984–1985)
95	La Primavera, Cueva Maripe	LP1497	Charcoal	3210 ± 60	3558	3219	Miotti et al. (2007)
96	Aguada del Cuero, Cueva Moreno	OS23754	Charcoal	3000 ± 110	3390	2850	Miotti et al. (2005)
97	La Primavera, Cueva Maripe	AA95122	Charcoal	2655 ± 39	2850	2521	Miotti et al. (2014)
98	PNBPJ, Hunhing blind Meseta La Horqueta 1	LEMA 1889.1.1	Bone	2430 ± 30	2696	2339	Magnin et al. (2023)
99	Los Toldos, Cave 13	LP1516	Charcoal	2390 ± 80	2706	2143	De Porras et al. (2009)
100	Aguada del Cuero, Cueva de La Hacienda	OS23753	Charcoal	2250 ± 70	2349	2017	Miotti et al. (2005)
101	La Martita	AC604	n/d	2190 ± 115	2420	1832	Aguerre (1987)
102	Cerro Tres Tetas, Cave 1	LP541	Charcoal	2190 ± 70	2335	1939	Paunero (1994)
103	La Primavera, Cueva Maripe	AA95124	Bone	1907 ± 48	1925	1638	Miotti et al. (2014)

(continued)

Table 3.1 (continued)

N	Locality, site	Lab code	Material	14C Date ± uncertainty	Cal BP from_95.4	Cal BP to_95.4	References
104	La Gruta, La Gruta 1	AA83474	Charcoal	1888 ± 39	1884	1700	Franco et al. (2013)
105	Río Pinturas, Cueva Grande del Arroyo Feo	CSIC523	Charcoal	1880 ± 50	1890	1612	Alonso et al. (1984–1985)
106	La Gruta, La Gruta 1	AA83476	Charcoal	1829 ± 47	1827	1587	Franco et al. (2013)
107	Cerro Tres Tetas, Cave 1	LP1117		1740 ± 60	1740	1427	Paunero (2003a)
108	La Martita, Cave 4	AC603	n/d	1620 ± 90	1698	1306	Aguerre (1987)
109	Río Pinturas, Cueva de las Manos	NOVA115	Charcoal	1610 ± 60	1583	1316	Gradín et al. (1976)
110	Río Pinturas, Cueva Grande del Arroyo Feo	CSIC399	Charcoal	1600 ± 50	1560	1318	Alonso et al. (1984–1985)
111	Río Pinturas, Alero Rosamel	CSIC798	Charcoal	1570 ± 50	1528	1315	Gradin and Aguerre (1994)
112	Río Pinturas, Alero del Búho	CSIC511	Charcoal	1520 ± 50	1513	1294	Alonso et al. (1984–1985)
113	La Gruta, La Gruta 1	AA83475	Charcoal	1452 ± 38	1373	1183	Franco et al. (2013)
114	Río Pinturas, Alero del Búho	CSIC512	Charcoal	1450 ± 50	1409	1179	Alonso et al. (1984–1985)
115	Los Toldos, Cave 1	Beta183024	Vegetal remains	1410 ± 40	1352	1177	De Porras et al. (2009)
116	La Primavera, Cueva Maripe	AA95118	Bone	1403 ± 46	1350	1177	Miotti et al. (2014)
117	Río Pinturas, Puesto El Rodeo	AC943	Vegetal remains	1380 ± 90	1411	998	Gradín and Aguerre (1994)

(continued)

Table 3.1 (continued)

N	Locality, site	Lab code	Material	14C Date ± uncertainty	Cal BP from_95.4	Cal BP to_95.4	References
118	Cerro Tres Tetas, Cave 1	LP1180	Charcoal	1340 ± 50	1300	1079	Paunero (2003a)
119	Piedra Museo, AEP-1	AA20125	Charcoal	12.890 ± 90	1302	964	Miotti et al. (1999a) (*)
120	Río Pinturas, Alero Cárdenas	AC500	Charcoal	1180 ± 85	1272	855	Alonso et al. (1984–1985)
121	Río Pinturas, Cueva Grande del Arroyo Feo	CSIC524	Charcoal	1170 ± 50	1178	930	Alonso et al. (1984–1985)
122	Río Pinturas, Alero del Búho	CSIC513	Charcoal	1170 ± 50	1178	930	Alonso et al. (1984–1985)
123	PNBPJ, De la Ostia	LEMA 1890.1.1	Bone	1096 ± 25	1048	923	Magnin et al. (2023)
124	La Primavera, Cueva Maripe	AA65176	Charcoal	1078 ± 40	1054	804	Miotti et al. (2007)
125	Río Pinturas, Alero del Búho	CSIC584	charcoal	960 ± 50	925	734	Alonso et al. (1984–1985)
126	La Primavera, La Quinta	AA85460	Bone	939 ± 59	926	685	Magnin (2010)
127	Río Pinturas, Alero del Búho	CSIC545	Charcoal	900 ± 50	905	675	Alonso et al. (1984–1985)
128	Cerro Tres Tetas, Cave 1	LP770	Charcoal	830 ± 60	897	568	Paunero (2003a)
129	Piedra Museo, Chenque El Sargento	AA65180	Bone	727 ± 48	719	555	Miotti (2008)
130	La Gruta, La Gruta 3	UGAMS13609	Charcoal	530 ± 20	540	501	Brook et al. (2015)

(continued)

Table 3.1 (continued)

N	Locality, site	Lab code	Material	14C Date ± uncertainty	Cal BP from_95.4	Cal BP to_95.4	References
131	Río Pinturas, Cueva de las Manos	CSIC137	Charcoal	430 ± 50	513	322	Gradín et al. (1976)
132	La Gruta, La Gruta 1	UGAMS7541	Charcoal	400 ± 20	494	326	Franco et al. (2013)
133	La Gruta, La Gruta 3	UGAMS13610	Charcoal	390 ± 20	492	323	Brook et al. (2015)
134	La Gruta, La Gruta 3	UGAMS12429	Bone	290 ± 20	439	155	Brook et al. (2015)
135	La Gruta, La Gruta 3	UGAMS12430	Bone	290 ± 20	439	155	Brook et al. (2015)

Calibration: OxCal 4.4, curve: SHCal20. (*) For discussion see also Miotti et al. (2003), Steele and Politis (2009), Goebel et al. (2008), Mosquera (2016, 2018)

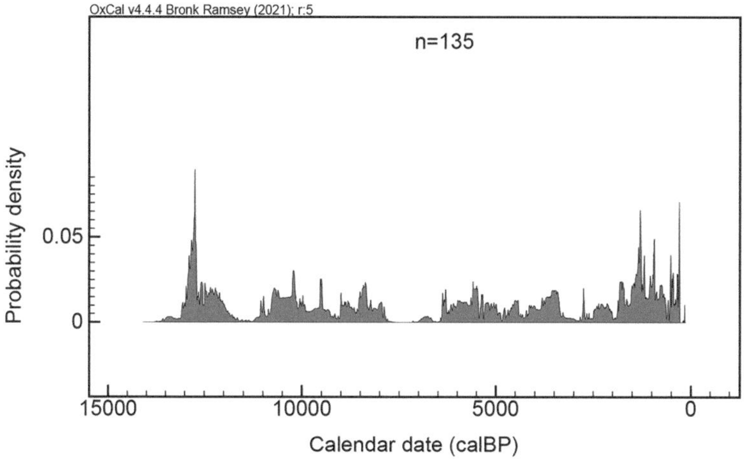

Fig. 3.1 Plot of probability sum for the Santa Cruz Central Plateau radiocarbon dates

3.2.2 *Model of the Settlement Process of Patagonia*

The process of inhabiting southern Patagonia was complex and involved changes in mobility, space use, and settlement over time. The first model to explain this phenomenon was developed by Borrero (1985, 1994–95) and was subsequently supported by empirical evidence from archaeology, biogeography, paleoenvironmental studies, ethnography, and ethnoarchaeology (Table 3.2). Archaeological studies in the area of the Santa Cruz Central Plateau have provided valuable information to approach the issue (Aschero, 1996; Borrero, 1985, 1994–95; Borrero et al., 1998; Goñi et al., 2004; Miotti, 1989, 2003a; Miotti & Salemme, 1999; Salemme & Miotti, 2008). For this reason, the inferential processes that led to each of the discussed aspects have only been expressed schematically here. To delve deeper into these previous investigations, consult the scientific work of Borrero (1985, 1994–95), Miotti (1989, 2003a), Aschero (1996), Borrero et al. (1998), Miotti and Salemme (1999), Goñi et al. (2004), Salemme and Miotti (2008), and Rodríguez and Horlent (2016), among others.

3.2.2.1 Pleistocene/Holocene Transition (Pl/H) and Early Holocene (EH)

The initial exploration and colonization of the Deseado area began with the arrival of the first human groups during the Pleistocene/Holocene transition and the Early Holocene (Pl/H, EH) (12,000–7500 BP).This period was marked by climatic instability due to the shift into the final Llanquihue 3 glacial period, followed by the Interglacial period. Paleoenvironmental conditions included low temperatures. Due to the large amount of water retained in the mountain glaciers, the sea was approximately

Table 3.2 Model of mobility, settlement, communication networks, and territoriality throughout the settlement process of the Central Plateau (Bibliographic references are in the text)

Period	Exploration and colonization	Territorial consolidation	Effective occupation/ Indigenous-hispanic contact/ Settler colonialism
Temporal phase	*(Pl/H and EH) Ca. 11.000–7500 BP*	*(MH) Ca. 7500–3500 BP*	*(LH), Ca. 3500 AP/sixteenth century/ nineteenth century*
Subsistence and mobility	High residential mobility. The resources provisioning strategy is collector from the beginning	High residential mobility and collector provisioning strategy Exchanges between the Atlantic coast, the plateau and the Andean range	Lower residential mobility and logistical strategy for using different spaces Demographic growth./ Eighteenth century: extension of action ranges by the adoption of European horses. Trade with white people./ Nineteenth century: the surviving people to the State military offensive were used as semi-slave labor
Settlement pattern in the Central Plateau	Variability and complementarity of sites	Variability and complementarity of sites. Greater occupancy redundancy	Variability and complementarity of sites. Caves collapse reduce the available residential spaces./ Nineteenth century: forced relocation and confinement in indigenous reserves. Incorporation to rural life and towns founded by the colonizers
Communication networks	Extensive communication networks, to minimize the risk in the colonization of new spaces, where demographics are low	Broader communication networks, including long-distance contacts with other territories	*Idem* (in a context of more occupied territories and with horse adoption)./ In the eighteenth century commercial parties for exchanging of goods, holding meetings or assemblies between different groups to define the politics to follow with the white lead colonizer./ Nineteenth century: cultural silencing
Territoriality	Permeable territories	Band fission: occupation of new permeable territories	Effective occupation of territories./ Nineteenth century: indigenous territories were taken by the State in the military offense called "Desert Conquest" and granted to commercial companies and sold to estancia founders

Abbreviations Pl/H: Pleistocene/Holocene transition; EH: Early Holocene; MH: Middle Holocene; LH: Late Holoceneff

200 km to the east of the current shoreline. Therefore, when the first groups arrived, extensive grass steppes covered the landscape. Together with the high availability of permanent water, these steppes supported a rich animal biomass (Salemme & Miotti, 2008). The main game animals were guanacos and other extinct species of the genus Lama, as weell as extinct horses (*Hippidion saldiasii*) complemented by rheids, mega mammals, smaller mammals and birds (Alberdi et al., 2001; Miotti & Salemme, 1999; Miotti et al., 2003). According to Binford (1980), the first exploration forays into this initially unknown environment, but where the availability of resources is homogeneous, involved a high residential mobility (*forager* or searcher) (Binford, 1980). This was complemented by a generalized economic strategy for acquiring a variety of resources. The expected archaeological record resulting from these strategies includes low variability of evidence, as well as indicators of base camps and isolated finds dispersed over large areas. The base camps are neither large nor indicative of continuous occupation. These conditions result in very low archaeological visibility and resolution for open-air sites from the early period (Aschero et al., 1992). However, the initial settlements in this particular area occurred in canyon sectors with caves and rock shelters (Miotti, 1989; Miotti & Salemme, 2004b), as suggested by PL/H climatic conditions. The choice of location for the camps would have been directed towards sheltered places in the landscape, even if they were not previously known. Over time, greater knowledge of the territory would have accentuated collector or logistical mobility (Binford, 1980), resulting in greater variability and complementarity in the archaeological evidence (Miotti, 2003). According to this idea, differentiation in the archaeological record, such as base camps, locations of limited activities, field camps, and isolated finds, is to be expected. It is not known how long it took the first explorers to become familiar with the environment and transition from a foraging strategy with mainly residential mobility to a collector strategy with greater variability in the resulting archaeological evidence, but this transition was relatively brief (Miotti, 2003). In addition to their high mobility, these human groups established communication networks with other groups, for social reasons and to obtain information about environmental resources, thus minimizing the risk and uncertainty of unknown territory (Miotti, 1994). Due to the low demographic rate, information must have circulated extensively through these networks. This context suggests that permeable territories were established in the Early Holocene (EH). These spaces facilitated interaction and the sharing of information about resources (Carden, 2009; Hartley, 1992; Ingold, 1986). The Patagonian population occurred in nodes connected by mobility corridors. In southern Patagonia, the most important corridor was the Central Massif. Along with the Pali Aike Field, these geographical areas were always available for human occupation. Última Esperanza and the north of Tierra del Fuego, however, became available only after deglaciation (Manzi & Borrero, 2007; Miotti & Salemme, 2003; Politis & Borrero, 2024). Around 9300 BP, new climate and geographic changes occurred, characterized by increased aridity and retreating glaciers. Consequently, the mountain buttresses became accessible, enabling discontinuous occupation in this region (Salemme & Miotti, 2008). At the same time, the Strait of Magellan formed in the south due to the flooding of the old

glacial valley. This event isolated part of the population belonging to the same colonization pulse that initially occupied the Santa Cruz Central Plateau. The profound climate changes that took place during this period contributed to the extinction of Pleistocene megafauna. However, the extent to which human action contributed to this event is currently under debate (Borrero, 2009; Cione et al., 2009; Prates & Perez, 2022).

3.2.2.2 Middle Holocene (MH)

Territorial consolidation occurred during the Middle Holocene (MH), which extends from 7500 to 3500 BP. The paleoclimatic record shows an improvement during the transition to the MH, followed by an increase in temperature and humidity. This led to a stage of stability, or climatic optimum, around 7000 BP in the Santa Cruz Central Plateau. However, this improvement is regionally variable. For instance, in Northern Patagonia and Cuyo, increased humidity during this post-glacial period caused increased snowfall, making the area uninhabitable. Higher temperatures caused a rise in sea level and a reduction in continentality (Miotti, 1989; Salemme & Miotti, 2008). Zooarchaeological studies indicate that the mode of subsistence became specialized in hunting guanacos (*Lama guanicoe*), supplemented by rheas (*Pterocnemia pennata*) (Miotti, 1989; Miotti & Salemme, 1999, 2004a). During this period, the groups maintained high residential and logistical mobility strategies. The Central Plateau's settlement pattern remained varied and complementary, with a diversity of sites indicating differential use of the landscape, which must have been very well known by then (Miotti & Salemme, 1999). Evidence of residential basecamps continues to be found in caves. However, an occupational hiatus was detected between 7000 and 6000 years BP, which can be correlated with volcanism and tectonic events in the region (Table 3.1 and Fig. 3.1). For instance, at AEP-1 (Piedra Museo locality), a hiatus in habitation coincided with the collapse of these shelters. In the Andean range, at CCP7 (Cerro Cueva, Casa de Piedra 7, Perito Moreno National Park), volcanic ash strata suggest catastrophic events related to volcanism. This does not imply depopulation, but rather changes in mobility that could have led to temporary vacancy of sites during this period, encouraging movement to other areas, such as the coast (Brook et al., 2015; Miotti, 1989; Salemme & Miotti, 2008). Recent studies by Mosquera (2016, 2018) determined that the middle Holocene hiatus in the Deseado Massif is between 7828 and 6434 calibrated years before present (cal BP).

During this period, the groups continued to employ strategies of high residential and logistical mobility. Studies of lithic assemblages belonging to the stratigraphic components of the Middle Holocene on the Central Plateau also indicate significant patterns in their artifactual structure (Hermo, 2008). The increased frequency of lithic scraper production was interpreted as an indicator of intensive guanaco hide processing (Aschero, 1975, 1983). These hides were likely used to make windbreaks and other leather goods (Hermo, 2008). These portable shelters facilitated the establishment of temporary camps for logistical activities, such as taking advantage of local resources, seasonal movements, residential changes, and exploring new environments.

In addition to environmental evidence, dating distributions, and variations in lithic technology, other lines of evidence could indicate long-distance movement or exchange of material objects or ideas between the Atlantic Ocean coast, the Central Plateau, and the Andean Range. For instance, bivalves, limpets, and sea snails were found in plateau regions far from the coast (Miotti, 1989; Miotti & Salemme, 1999, 2004 a), and obsidian from the western mountain range was used as a raw material for knapping artifacts (Cattaneo, 1999, 2002; Hermo, 2014; Miotti, 1989; Stern, 1999). The broader dispersion of rock art, compared to the previous period, has also been taken as an indicator of high mobility and strengthened social networks for the exchange of information (Carden et al., 2009). According to Gamble (1982), groups undergo fissions to occupy new permeable territories. These territories are not understood as two-dimensional spaces that exclude others but rather as places, where resources are concentrated, and the paths that connect them. During this period, rock art exhibits greater diversity and broader spatial distribution, which supports the intensification of social relationships (Carden et al., 2009), occurring alongside progressive demographic growth (Borrero, 1994–95). While groups made intermittent exploration movements from the Deseado Massif to the Andean Range buttresses around 8000 years ago (Early Holocene), in the Middle Holocene, they made seasonal movements to the Atlantic coast and the Andean Cordillera (Salemme & Miott, 2008). In the Central Plateau, the archaeofaunal records of Caves 3 and 13 of Los Toldos in the MH levels (Casapedrense) provide evidence of seasonal occupations. In the first cave, the most common prey hunted is *Lama guanicoe*. Sixty percent of the sample is juvenile individuals. Within this category, neonatal individuals between 18 and 36 months of age were observed based on epiphyseal fusion, tooth eruption, and dental wear. Since guanaco births occur in November, this evidence suggests that hunting and consuming guanacos occurred in different seasons of the year: autumn and spring (Cardich & Miotti, 1983; Miotti, 1989). Current research projects are studying the seasonal movements in more detail.

3.2.2.3 Late Holocene (LH)

In the Late Holocene (the last 3500 years), the Patagonian region acquired its definitive environmental configuration in terms of fauna and vegetation (Salemme & Miotti, 2008). This period also saw stable occupations in some areas. However, this does not imply climatic stability (Borrero, 1989, 1994–1995). Holocene glaciation must have directly affected the habitability of spaces such as the main lake basins and their surrounding plateaus (Borrero, 1989,). Available information from the Deseado Massif area points to a hiatus in habitation between 3005 and 2710 cal BP in the Late Holocene (Mosquera, 2016, 2018). This could indicate population movements to other regions or the formation of processes, or it could be a bias in the research design of the Deseado Massif, which focused on caves and underrepresented open-air sites. All of these hypotheses are under study.

During the period of the Medieval Climate Anomaly (MCA), which was characterized by significant aridization around 1000 BP, the main freshwater reservoirs became attractive locations for human groups. Cardiel and Salitroso Lakes, along with other depressions such as the Ibáñez River area, concentrated most of the region's population. The effective occupation of these spaces involved reduced residential mobility and logistical use of the surrounding plateaus. Residential use of lake basins is evidenced by the presence of multiple activity sites equipped for future use. The logistical use of the surrounding areas is demonstrated by hunting blinds and quarries for extracting lithic raw materials on nearby plateaus (Aschero et al., 2005; Cassiodoro, 2008; Cassiodoro et al., 2004; Goñi, 2000–2002; Goñi et al., 2004). These territorial rearrangements were related to population growth, environmental changes, and the organizational characteristics of hunter-gatherer societies (Goñi et al., 2004).

Another line of analysis reveals that rock art provides clues about social changes during this period. Currently, the rock representations of animal footprints are dispersed and frequent throughout the province of Santa Cruz. This is interpreted as an indicator of greater intergroup competition for resources amid climate stress, when tracking game animals could have been crucial for securing this resource in an environment of greater uncertainty (Carden et al., 2009). During this time, the territories were more populated, and demographic growth may have led to intergroup competition and the emergence of territorial markers. These territories are open to other human groups but are marked by rock art and symbolically appropriated as part of group identity (Curtoni, 2007). In the macro-area under study, the existence of these permeable territories is evidenced by the restricted distribution of high-frequency rock motifs, such as those of matuastos (small reptiles) and guanacos. These motifs coexist with other widely distributed and frequent motifs that refer to open social interaction networks (Carden et al., 2009). This stylistic regionalization has also been observed in other areas (Belardi & Goñi, 2006).

During the period of Spanish-Indigenous contact in the sixteenth century, the existing tendency to reduce mobility for climatic reasons was accentuated by the pressure exerted by colonizer groups external to the hunter-gatherer societies. Various authors agree that residential mobility decreased and logistical mobility increased among the patterns of space use (Goñi, 2000; Martinić, 1995; Miotti, 1989). This change is related to the adoption of the horse at the end of the seventeenth century, as documented by the Gouin de Beuchense expedition to the Deseado district (Martinić, 1995). The horse facilitated transportation and became part of the diet, as well as a new hunting technology. During this period, strategies were also developed to store food, such as preparing charqui (dried meat) and fusing and storing guanaco, rhea, and horse fat (Miotti, 1989).

In the final period, the clash between Indigenous and European cultures was the main and indisputable factor of cultural change and population displacement (Salemme & Miotti, 2008). Some authors (Goñi, 2000, 2000–2002) also highlight aridification as the environmental condition that prompted indigenous groups to retreat to areas with concentrated resources prior to the arrival of Europeans. Accessing distant resources from a nucleated population was achieved by adopting horses in a strategy called extensification (Goñi, 2000). Aboriginal populations experienced a demographic collapse due to pressure from Europeans (Goñi, 2000, 2000–2002). The effects of subjugation by settler colonialism are still experienced, resisted, and fought by resilient indigenous peoples today (Rodríguez & Horlent, 2016).

3.3 Methodological Background

The earliest antecedent of surface archaeological surveys in the area is the work of archaeologist and geographer Francisco de Aparicio. He directed the first extensive investigations in the Central Plateau from 1933 to 1935. The expedition, facilitated by Yacimientos Petrolíferos Fiscales S.A. (YPF, the National Oil Company), consisted of an exploration trip carried out with Dr. Joaquín Frenguelli and engineer José Brandmayr. The expedition began in Comodoro Rivadavia and covered more than 2000 km. It passed south of the Deseado River and reached the archaeological sites of Los Toldos, Aguada del Cuero, and Piedra Museo on the Central Plateau. Then, it traveled to the Atlantic coast and continued north to Puerto Deseado. Then, they traveled up the Deseado River to the vicinity of Jaramillo and returned to Comodoro Rivadavia. During the expedition, various archaeological evidence was documented, including painted and engraved rock art, chenques (burial sites), and shell middens on the coast. De Aparicio also excavated burials. His work includes detailed descriptions, excellent drawings, photographs, and a map that categorizes the types of archaeological sites found. This is the first archaeological map of the study area. Despite its regional scale, it records the area's archaeological richness (Fig. 3.2). This initial investigation led the Austrian prehistorian Osvaldo Menghin to the significant site of Los Toldos. Between 1951 and 1952, he excavated Caves 2 and 3, establishing a master regional stratigraphic sequence that revealed the temporal depth and types of archaeological evidence present in the area (Menghin, 1952). Subsequent research directed by Augusto Cardich in the same area provided the first radiocarbon dates, confirming the region's ancient human occupation (Cardich, 1984; Cardich et al., 1973).

Later research in the area was built on these pioneering works. It expanded the study of archaeological sites with significant sedimentary sequences that offer valuable insights. However, multidisciplinary and regional research within archaeological programs has allowed research on landscapes to develop in recent decades. In Patagonia, surface archaeological materials, not only those recorded in stratigraphic contexts, have gained importance in archaeological research with different theoretical and methodological emphases. Although caves and rock shelters, as sedimentary traps, provide important information not generally found in open-air sites (Farrand, 1985; Kornfeld et al., 2007), they are distributed sparsely across the landscape, which prevents a comprehensive view of landscape use. Conversely, data from surface and shallow sites, as well as from less dense artifact assemblages and isolated finds,

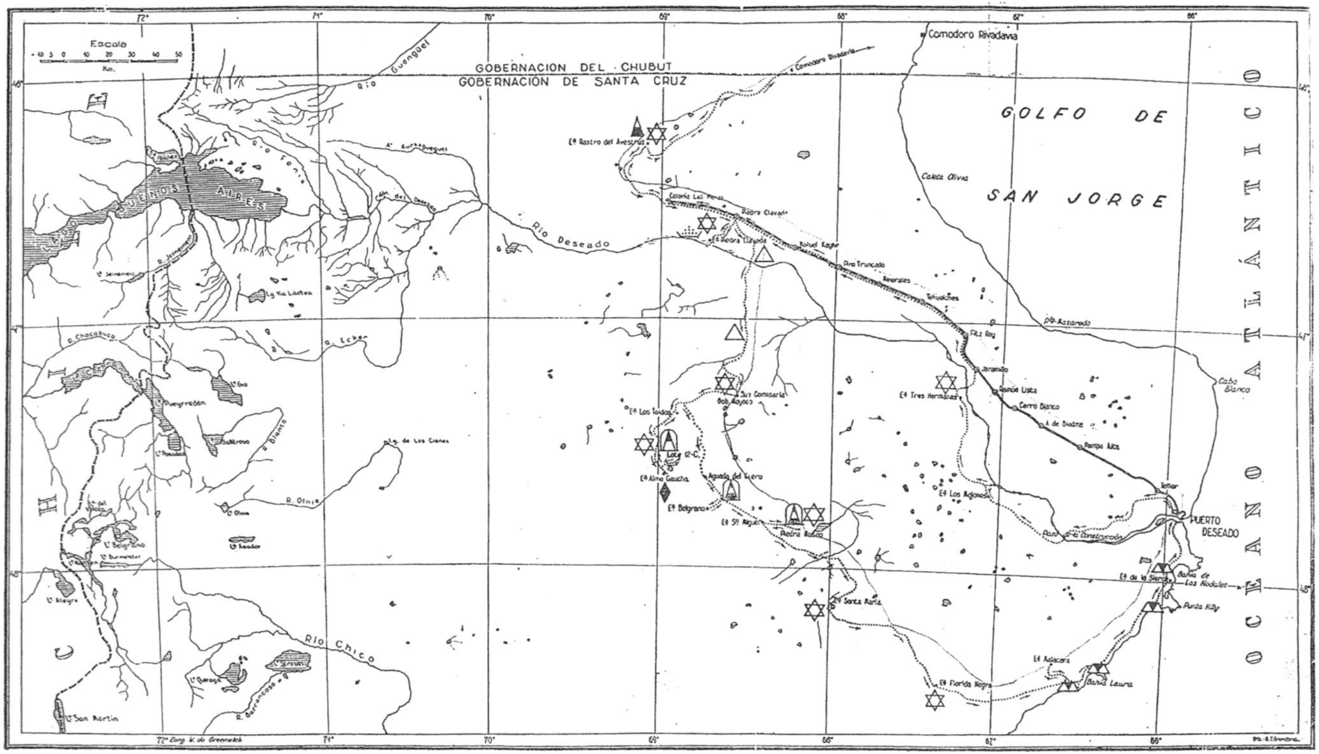

Fig. 3.2 Map taken from the published report of Francisco de Aparicio's 1933 expedition through the Deseado area (De Aparicio, 1933–1935). The symbols indicate the locations of campsites, *chenques*, workshops, quarries, isolated finds, hearths, caves with lithographs, and shell middens (Vignati, 1946)

provide complementary information that must be integrated into research to gain a more complete understanding of past hunter-gatherer landscapes (Allen, 1991; Borrero, 1995; Borrero et al., 1992; Belardi & Borrero, 1999; Foley, 1981; Gamble, 1990; Magnin, 2013). This type of evidence is usually more difficult to interpret than that from highly structured sites due to high occupational redundancy (Tainter, 1998). For this reason, surface evidence has often been overlooked in investigations. However, as explained in Chap. 2, environmental factors determine that surface evidence constitutes the majority of the archaeological record in our case study. In Patagonian archaeology, some antecedents that include surface evidence in their analyses as an important part of the research are Borrero et al. (1992), Carballo Marina and Sáenz (1992), Gradín and Aguerre (1992), Pérez de Micou et al. (1992), Belardi (1992, 2003), Belardi et al. (1992), Espinosa (2000), Favier Dubois (2001), Castro et al. (2001), Borrazzo (2004), Carballo Marina (2006), Hermo (2008), and Oría (2008). For a discussion of the spatial perspective in archaeology, see Scheinsohn (2001). Among these studies, spatial analyses have been carried out at different scales. In the Central Massif, Miotti (1988, 1989) created a map of Los Toldos and neighboring areas that shows the different types of sites, resources, and potential exploitation territories from the main residential sites.

3.4 Antecedents in GIS Applications

The methodological procedures for studying archaeological landscapes were first developed in other study areas and have since become established. Nielsen (1994) first applied GIS in Argentine archaeology in a presentation at the XI National Congress of Argentine Archaeology. He argued the advantages of its use with a case study example in Karwani, southern Bolivia. In Patagonian archaeology, the first application corresponds to Manzi (1999). In her work, she modeled the environmental and social variables that were most important in determining the locations of the Selk'nam ethnic groups that inhabited Tierra del Fuego Island at the time of European contact. The variables included in the analysis are based on ethnographic and ethnohistorical information (Manzi, 1999). Over the next decade, GIS applications in Patagonian archaeology increased. Mateucci and Sheinsohn (2004) generated a predictive model using the analysis of the current landscape (variables obtained from the digital processing of satellite images) and the locations of known archaeological sites to identify other areas of the landscape with potential for discovering new sites. The deductive procedure was used since the predictive variables were based on a theoretical model (Binford, 1990, 2001). The first specialized symposium was held at the XV National Congress of Argentine Archaeology in Río Cuarto that same year. Among the applications of GIS in the archaeology of Patagonia, Andolfo and Gómez (2004) modeled landscape conditions on the northern coast of Santa Cruz, which were important for choosing population locations and changes over time. Manzi et al. (2004) worked in the Palli Aike area and addressed database construction, a topic of special interest with GIS implementation. Cattaneo et al. (2006)

quantified the area of geological formations that are sources of lithic raw materials in the Piedra Museo locality. Scheinsohn and Mateucci (2004) created a predictive model in Cholila to identify areas with the highest probability of site presence, directing intensive research to those areas. Figuerero Torres et al. (2004) used GIS at the site scale in the Posadas Lake basin to evaluate formation processes. At the same academic event, I analyzed location patterns concerning different geoforms in the Deseado Central Plateau. This study aimed to operationalize (objectify) a settlement model (Magnin, 2004) and establish a basis for comparison (Köhler, 1988).

This study focuses on the settlement patterns in La Primavera and complements consolidated research that has provided information from other lines of analysis.

3.5 Conclusion

This chapter presented the archaeological research that informed models of the settlement, subsistence, and mobility of human groups that have occupied Patagonia since its initial colonization. These societies were highly mobile hunter-gatherers with economies based mainly on hunting guanacos. This characterization is based on the body of archaeological knowledge concerning human biogeography, environmental changes affecting population development and distribution, and general models applied to hunter-gatherers. Along with formation processes and visibility considerations, these constitute the background for interpreting the archaeology of the La Primavera locality.

Assuming surface materials validly and informatively reflect aspects of landscape use that have not been previously studied in depth, this work highlights their potential for understanding a broad spectrum of human behaviors. What observable spatial patterns exist at the locality scale in this arid region of Patagonia? The following chapters present the theoretical framework and methodology used to study landscape use by these societies. This will provide insight into location strategies for residential sites quarries and lithic resource supply; rock art sites hunting activities; and chenuqes, probable funerary tombs, in the La Primavera locality.

Chapter 4
Theory

> *...being able to measure things per se does not make the measurement meaningful. Meaning comes only from a body of theory (Aldenderfer, 1998:93)*

Abstract From a multidimensional perspective, the landscape consists of meanings and social relationships that interact with the physical space. Therefore, human spatial behavior is motivated by physical, economic, social, and symbolic factors simultaneously. In this chapter outlines the analytical process used to study the archaeological landscape at the La Primavera locality. The process has three stages: First, data from archaeological surveys are classified based on general mobility and space use models, ethnographic information about Patagonian hunter-gatherers, previous archaeological studies, and differential preservation considerations. Second, geographic information systems and statistics are used to detect patterns in the location of the studied sector (exploratory data analysis). Third, particular hypotheses are proposed to further investigate the patterns found.

Keywords Cognitive-procesual perspective · Explanatory framework · Landscape approach · Locational decisions

4.1 The Multidimensional Landscape

This study focuses on the human groups that inhabited the Santa Cruz Central Plateau and their socio-environmental relationship with the landscape. Rather than viewing the landscape as a physically based substrate on which certain activities were developed, this study conceptualizes the landscape as a complex and multidimensional entity. The theoretical starting point of this research is conceptualizing the landscape as an entity linked to different aspects of human life (Fig. 4.1).

> Landscapes, which may be defined as the assemblages of real-world features _ natural, semi-natural and wholly artificial _ give character and diversity to the earth's surface and form the physical framework within which human societies exist. They are closely linked to all aspects of human life, for not only are there economic bonds _the majority of human beings

L. A. Magnin, *The Landscape and Mobility of Hunter-Gatherer Societies*, The Latin American Studies Book Series, https://doi.org/10.1007/978-3-032-08127-8_4

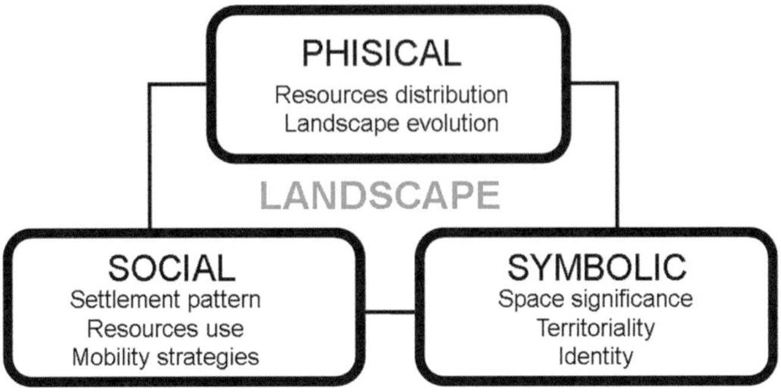

Fig. 4.1 Multidimensional landscape. From an actor's perspective, the landscape is multidimensional, and human decisions are a function of its various dimensions

> which have ever existed were hunter-gatherers or peasant farmers_ there are also powerful social, religious and psychological bonds (Roberts, 1987:79).

Thus, the landscape is not merely a static physical framework to which society adapts. Rather, societies act on the landscape, often making changes that transform it. In this sense, the landscape is a socio-cultural construct. According to Criado (1999), landscapes have physical, social, and symbolic dimensions. However, these dimensions cannot be conceived separately. This division is only a legitimate analytical tool for studying this complex subject (Parcero, 2002). This perspective seeks to deconstruct the dichotomous separation between nature and culture, a tradition of Western thought (Ingold, 1993). Nevertheless, it differs from other valid Western and non-Western perspectives (Ucko & Layton, 1999, cited in Engelmark & Larsson, 2005). Hunter-gatherer societies perceive themselves as part of nature, not separate from it (Miotti, 1997).

Consistent with this, the study posits that the distribution of archaeological evidence is influenced not only by economic factors, but also by sociopolitical and symbolic motivations. Additionally, it is assumed that archaeological evidence can partially reveal the various ways humans have interacted with landscapes, such as acquiring survival resources, marking territories, and the symbolism associated with locations (Parcero, 2002). Within landscape studies, these theoretical guidelines provide a humanized perspective of space by avoiding the implicit determinism of approaches that consider only environmental variables when explaining location.

In short, our focus is on analyzing the decisions of societies in their use of the territory (Vicent, 1991). As Fernández Cacho (2006) puts it, the environment does not determine human behavioral patterns. Rather, it presents a variety of conditions, and humans establish their preferences, which may or may not be directly related to these conditions or other intangible factors (psychic, ideological, etc.). This position moves toward a possibilist vision of the relationship between human groups and the environment.

4.2 Theoretical Proposal

Theoretical-methodological frameworks are not rigid dogmas. On the contrary, they interact with each other to find answers to research questions. Therefore, this work incorporates aspects of Cultural Ecology, Spatial Archaeology, and Symbolic Archaeology. Often, these three approaches are presented in the literature as mutually incompatible because each focuses on a relatively narrow set of problems (García Sanjuán, 2005). Consequently, the explanatory variables chosen differ in each case. Cultural Ecology focuses on the study of human adaptation to the environment and the exploitation of natural resources. It privileges only the physical-economic dimension of the use of space and pays little attention to symbols, despite the fact that symbols are one of the ways human societies relate to the environment. Symbolic Archaeology is more concerned with understanding how past societies perceived their environment and how that perception was integrated into their religious, mythical, and symbolic constructions (García Sanjuán, 2005). Spatial analysis is defined as recovering information related to archaeological spatial relationships, as well as studying the spatial consequences of patterns of past human activity within and between contexts and structures. This includes examining how these patterns are articulated within settlements, settlement systems, and their natural environments (Clarke, 1977). This approach is not restricted to a particular body of theory, but rather is exploratory and based on spatial patterns found in the data.

Though there are specific incompatibilities, the aforementioned perspectives can be viewed as mutually complementary and greatly benefit the discipline of archaeology (García Sanjuán, 2005; Van Pool & Van Pool, 1999). Cognitive-processual archaeology is a theoretical perspective that supports this idea and offers an appropriate framework in which the specifically human quality of symbolism will play a central role (Renfrew & Bahn, 1993).

4.2.1 Landscape Archaeology

In our discipline, different approaches are referred to as Landscape Archaeology. Some of these approaches oppose the post-processual perspective (Rossignol, 1992), which is an archaeological analysis of the symbolic aspects of group-environment relations (Criado, 1993, 1999; Garcia Sanjuan, 2006; Tilley, 1994; Tilley & Bennett, 2001). However, as mentioned above, some positions reconcile aspects of both perspectives.

Some Landscape Archaeologies focus on prospecting on a regional scale, incorporating surface archaeological evidence, which is more accessible than stratigraphic sites for continuously analyzing regional space. For instance, settlement archaeology centers on sites and areas with significant archaeological remains. Others emphasize registries based not on sites but on artifacts and artifact assemblages. Examples include Distributional Archaeology, Off-site Archaeology (Foley, 1981), non-site sampling (Thomas, 1975), and site-less survey (Dunnell & Dancey, 1983). While the relationship between these density calculations and the human behavior that created them is not always explicitly defined, the objective of accounting for the regional archaeological structure is often clear (Ebert & Kohler, 1988). In other words, the goal is to measure different artifactual densities to define variations in space-use intensity over time. In Patagonian archaeology, this is called landscape hierarchy (Belardi, 2003; Borrero et al., 1991).

This theoretical-methodological perspective has proven particularly useful in cases where archaeological distributions are diffuse and lack clear boundaries defining discrete sites. Examples include the western United States (Thomas, 1975) and eastern Africa (Foley, 1981). This methodology uses sampling units to record artifact counts and register low-spatial-resolution data, which is useful for comparing different surface contexts.

Conversely, some authors use a methodology that records individual artifacts (typically with total stations or differential GPS) and employs intensive prospecting or full-coverage surveys. This approach provides high spatial resolution for locating and recording artifacts, plotting their attributes, and analyzing their deposit conditions. This type of survey completely covers areas smaller than 2500 m^2. This approach has been termed Distributional Archaeology (Connolly & Lake, 2006; Ebert & Kohler, 1988; Fanning & Holdaway, 2001; Wandsnider & Camilli, 1992). These investigations reveal a shift from studying human behavior to meticulously controlling record formation processes, invariably carried out in sectors with high artifact density.

Finally, it is worth mentioning the differences between this approach and Dunnell and Dancey's site-less survey. While this approach also uses individual artifacts as its unit of analysis, artifacts are not always the unit of record. Rather than controlling site formation processes, the foundation of this approach is the search for coherence between methodology and theory (Dunnell, 1980). Because this approach is theoretically associated with Evolutionary Archaeology or Selectionism, it takes artifacts as units of analysis, considers them part of the human phenotype, and applies the mechanisms of biological evolution.

Regarding the use of sites, artifact densities, or isolated artifacts as units of analysis, we believe they are all useful at different scales for addressing specific research questions. For this reason, we do not consider the concept of a site to be inadequate (see Rossignol and Wandsnider 1992, for a discussion). In this study, we adopt a perspective that differentiates recurrences in locational decisions for the differential use of places in mobile hunter-gatherer societies in this area. To that end, we calculate artifact densities and classify evidence as sites, artifact dispersions, or isolated finds because they could all be informative. To interpret and compare different surface contexts, we include information on formation processes. Our interpretations consider the economic, social, and symbolic dimensions underlying the decisions that led to the current locational patterns.

The Landscape Archaeology approach proposed in this work follows the explanatory scheme shown in the next section.

4.3 The Explanation Process

Figure 4.2 illustrates the process used to explain the locations of past hunter-gatherer settlements in the study area. This process is common to all archaeological explanations (Ebert & Kohler, 1988:97).

As the diagram shows, the explanation attempts to link the broad-scope theory of the organization of subsistence, settlement, and mobility of past human groups with

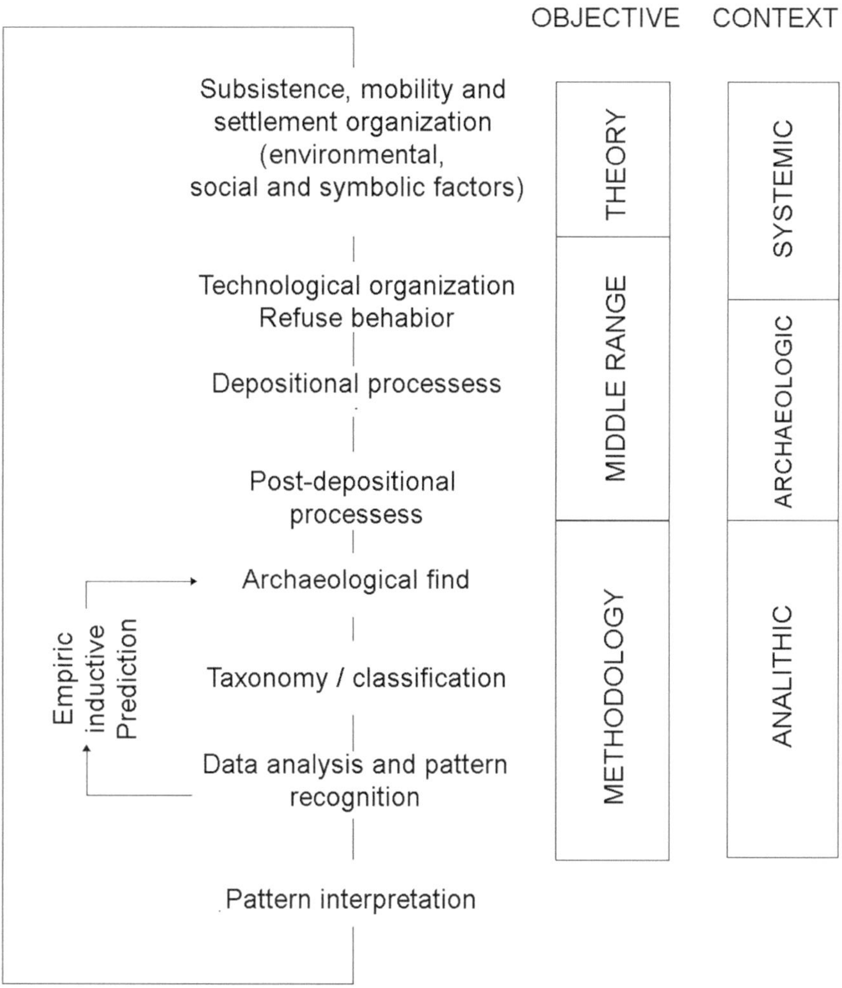

Fig. 4.2 The explanatory framework of archaeology. This diagram is adapted from Ebert and Kohler (1988: 103). According to the diagram, research can proceed using an inductive or deductive analytical methodology. In the first approach, researchers start with the data and use theory to explain it. The second approach begins with questions arising from theory, which are then contrasted with existing data

patterns in the archaeological record through a middle-range theory and methodology (Binford, 1977). This theory incorporates models of mobility, settlement, technological organization, and artifact disposal, as well as experimental studies and depositional and post-depositional processes.

The explanation can begin with either the theory or the data. The middle-range theory serves as a bridge that enables the generation of empirically falsifiable results from the general theory.

In this research, we first extract settlement patterns inductively. Then, we investigate these patterns through a deductive procedure by testing specific hypotheses to contribute to higher levels of knowledge.

The explanation process involves integrating archaeological information with ethnographic, historical, and environmental data to create models connecting the archaeological record to potential past events (Ebert & Kohler, 1988).

1. Specifically, models derived from ethnography and archaeological research are used to classify and interpret archaeological evidence in terms of different potential activities or functions that could have occurred at various locations.
2. A locational model is developed for each type of evidence, characterizing its distribution by spatial variables (environmental or otherwise). Each model is statistically tested against control data or negative evidence.
3. To delve deeper into the patterns generated, hypothesis tests are designed.
4. Given the amount and variety of spatial data handled throughout the process, Geographic Information Systems (GIS) are the most appropriate methodological tool for testing these ideas and modeling the configuration of archaeological landscapes.
5. The ability to model changes in these configurations throughout human occupation history depends on having adequate data to expand its temporal scope. In future, we hope to incorporate them through absolute and relative chronologies.

4.3.1 High-Level Theory, Systemic Context

Systems theory has proposed a useful model for analyzing complex interrelationships. Some of its basic concepts are essential for integrating the environmental dimension with social phenomena (Butzer, 1982). This systemic vision enables us to view human groups as dynamic systems that interact with natural environments and other social systems. Understanding these factors at play allow us to model the processes under study, such as the location of human settlements in the landscape (Clarke, 1972).

4.3.1.1 The Relationships Between Environment and Culture

This systemic vision is based on the idea that human groups are part of a changing ecosystem or natural environment. According to Dincauze (2000), any environmental change produces a response. This includes changes to the natural environment, as well as to societies that interact with it and are part of the social environment.

> The catalog of potential human responses is extremely varied, from biological evolution, phenotypic plasticity, adjustments in population density and distribution, behavioral changes at the scale of individuals or cultures, even ritual and technological innovations (Dincauze, 2000:73).

The author distinguishes genetic responses or adaptations from behavioral responses, which are defined as adjustments or strategies (Dincauze, 2000). Genetic responses are irrelevant on the short time scale of this archaeological study. Conversely, changes in behavior are the primary means by which humans adjust to their environment. When faced with a change in the natural or social environment, human groups can choose from various strategies regarding subsistence organization, technology, settlement, mobility, ritual, and other factors. Short-term environmental changes produce temporary responses in human behavior that are difficult to detect archaeologically. These responses do not involve new strategies but rather changes in the intensity of existing ones. For example, changes may occur in the seasonal organization of tasks, the provision of rituals, the resources sought, the foods exchanged, seasonal settlement patterns, or the size of social groups (Dincauze, 2000). In these cases, no changes occur in the social structure because individuals absorb the effects. Conversely, long-term environmental changes support and intensify new strategies involving adjustments to the cultural system through positive feedback mechanisms. Once the system has adjusted to several changes, it acquires a new state that is difficult to reverse. The main cultural subsystems undergo major reorganization. Changes in resources and technology can alter the niche. On a regional scale, settlement patterns can alter both the niche and the habitat (Dincauze, 2000). In archaeological research, understanding the natural environment of the past improves our ability to comprehend past societies. However, there is no simple, direct relationship between the environment and adaptive behavior (Dincauze, 2000). Societies are conditioned by the environment and, at the same time, affect it. Thus, some generalizations can be made regarding economic organization. Other aspects, such as social organization, beliefs, and ideology, are more indirectly related to the physical environment. These aspects have characteristics that cannot be explained by ecology alone. This position, called environmental/cultural possibilism, is widely accepted in contemporary archaeology and seeks to avoid interpretations close to any determinism.

While ecological relationships constrain the range of possible economic behavior, they do not dictate all aspects of economic institutions. Similarly, economics may constrain the scope of variation in social and political behavior, and social and political relations may limit the scope of religious practices and aesthetics. However, neither level is fully determined by the more basic level. Conversely, as material constraints become less direct, progressively higher levels possess a greater number of specific traits that cannot be explained ecologically (Trigger, 1992). Accepting these postulates means acknowledging that the location of human settlements depends partly on internal societal factors, including available technologies and intangible aspects such as customs and beliefs. Each case study requires a specific historical explanation (Pauketat, 2001).

Before presenting mid-range models or theories that bridge the gap between the data and its implications for location and mobility studies, an important point must be discussed. Can human mobility and location strategies be studied in all their dimensions using archaeological evidence? Correlating material culture with human behavior can be difficult, particularly at the social and symbolic levels (Trigger, 1992:364), but it is possible when adequate indicators are identified. Since material culture, its spatial relations, and its ecofactual associations are the only phenomena available to archaeologists, behavior—whether thought or action—must be inferred from these proxies (Dincauze, 2000). The challenge lies in the fact that the material evidence is one-dimensional while the behavior that gave rise to it is multidimensional. The problem lies in determining the extent to which material culture allows us to study the different spheres of human behavior. As Hawkes (1954) demonstrated, interpreting archaeological data about human activities becomes increasingly difficult. Studying the organization of subsistence and technology through material culture is simpler than studying social organization, beliefs, and ideology. Trigger (1992) presents differing views on whether this scale of difficulty is inherent to the nature of archaeological data or if it results from the ability of archaeologists to raise relevant interpretive questions. We contend that both perspectives contain some truth. How, then, can we study the more difficult human strategies according to the Hawkes scale? Formulating new questions is essential. This generates the need to review existing inferential tools and promotes advances in theory. It is also important to interpret archaeological indicators using models derived mainly from ethnographic analogy and ethnoarchaeology rather than ecology (Binford, 1972). In this regard, it is important to adopt methodological advances that allow us to approach aspects related to perception (e.g., using GIS to study visual basins, intervisibility, predominance, and accessibility).

Social and symbolic aspects have undoubtedly been part of past cultural systems, so they should be integrated into our research (Binford, 1962; Gamble, 1990). Beliefs, rituals, and customs influence people's activities and have material consequences. They are not separate from subsistence activities; rather, they are part of them. However, the fact that beliefs and symbolism are not universally generalizable limits the usefulness of middle-range theory in interpreting findings. One possible approach is to use direct historical analogies supported by adequate ethnographic documentation or ethnoarchaeological studies (Trigger, 1992). When such documentation is unavailable, the study of the sociosymbolic aspects of settlement location can propose alternative interpretations using available evidence. These interpretations can then be compared with other organizational aspects of the cultural system.

The difficulties observed by Hawkes (1954) also apply to the study of the economic and technological aspects of human organizations (Magnin, 2009b). These aspects are not exempt from symbolic determinations. For instance, food taboos, as documented in ethnoarchaeological studies, are an example of this phenomenon. Ideological factors influence patterns of fauna exploitation more than material factors, affecting bone assemblages (Politis & Saunders, 2002). Studies of hunter-gatherer societies demonstrate that worldviews based on daily experiences influence how the natural environment is exploited (Bonomo, 2006). Conversely, studying the difficult end of the spectrum requires formulating new questions, generating theoretical interpretations, and searching for adequate archaeological indicators (Bonomo, 2006; Carden, 2009; Carden et al., 2009; Gaffney et al., 1996; Hermo, 2009; Miotti, 2006; Magnin, 2009b; Van Pool & Van Pool, 1999).

In summary, we are interested in the following behaviors: spatial organization, use of space, and human mobility. The search for an explanation directs us to not only the physical and economic dimensions of the landscape but also its social and symbolic dimensions.

4.3.2 Middle-Range Theory, Archaeological Context

According to Merton (1968), as cited in Clarke et al. (2002), middle-range theories are empirically grounded theories involving sets of confirmed hypotheses rather than organized descriptive data or empirical generalizations that remain logically disparate and disconnected. Middle-range theories connect observations, inferences, hypotheses, and empirically supported research, and are tools for interpreting patterns in archaeological evidence. Based on the principles of actualism and uniformitarianism, these theories imply that past processes can be understood through present-day processes (Binford, 1981; Schiffer, 1983). These principles form the foundation of all disciplines that study the past because the subject of study is not directly accessible to the researcher.

4.3.2.1 Functional Classification Scheme of Archaeological Evidence Based on Middle-Range Theories

This work uses the following to interpret the archaeological evidence recorded in field surveys:

(a) General models of hunter-gatherer settlement and mobility (Binford, 1980, 1982; Borrero, 1985; Foley, 1981; Miotti, 1989; Miotti & Salemme, 1999; Sivertsen, 1980);
(b) The ethnographic record the locations of various activities in the landscape by human groups that inhabited the region or similar areas of Patagonia (Claraz, 1865–1866, 1988; Martinić, 1995; Musters, 1871, 1997; among others); and archaeological studies of hunter-gatherers (see below);
(c) Archaeological studies of hunter-gatherers (see below); and
(d) Relevant taphonomic and geoarchaeological considerations (Borrero, 2001a; Borrazzo, 2004; Butzer, 1982; Favier-Dubois, 2001; Foley, 1981; Rossignol & Wandsnider, 1992; Sivertsen, 1980; Waters, 1992; Waters & Kuehn, 1996, among others).

The expected variability in archaeological assemblages is associated with patterns of site use and movement that correspond to logistical organization in a heterogeneous environment (Borrero, 1985, 1994–95; Borrero et al., 1998; Franco & Carballo Marina, 1993; Goñi et al., 2004; Miotti, 1989, 2003; Miotti & Salemme, 1999; Salemme & Miotti, 2008). According to Binford (1980, 1981), it is expected that the archaeological remains will show evidence of residential base camps, where general activities are carried out; and logistical sites, where specific tasks are performed. This logistical organization allows us to assume that we will find differences in artifact assemblage composition that are useful for understanding the use of space. However, the high mobility of these groups may have prevented exact overlap between occupations (Binford, 1982). Furthermore, the visibility and preservation of the evidence depend on specific taphonomic factors. Some variability in the archaeological evidence is due to site formation history (Camilli, 1989; Dewar & McBride, 1992), a factor that must be considered when addressing the wide spectrum of activities carried out by past human groups.

The following paragraphs present twelve types of hunter-gatherer activities, their archaeological indicators, and the landscapes in which they occurred. To define these types, we searched for indicators within the ethnographic, ethnoarchaeological, and archaeological bibliographies, with a particular focus on surface archaeological research in Patagonia.

Negative evidence. It is a place of transit. The absence of evidence in conditions of high visibility is the archaeological indicator. These areas are far from places visited frequently or where hunter-gatherers concentrated resources (Magnin, 2013).

Place with rock art. It is a place where people engage in activities related to communicating ideas. Some interpretations of rock art production include symbolizing territorial boundaries, claiming places with resource concentrations, and expressing sacred values in ritual contexts with varying degrees of connection to the domestic sphere (Miotti et al., 1999; Carden, 2009). The presence of rock paintings or engravings is an archaeological indicator. These paintings and engravings are located where natural supports are available. Depending on their function, they are found on mobility routes, on the borders of territories, or in areas with a high concentration of resources. Analyzing rock motifs has sometimes revealed places where different social units aggregated seasonally (Aschero, 1996) and where different population segments converged for the seasonal use of space (Belardi & Goñi, 2006).

Quarry. This is a place where lithic raw materials are extracted for knapping. Archaeological indicators include flake negatives on rock outcrops, large cores, hammers, and knapping debris with a high proportion of cortex. This represents the initial stage of the production process for lithic artifacts (Andrefsky, 2005). Quarries are located where useful rock outcrops or on secondary deposits (Borrero & Nami, 1996; Pérez de Micou et al., 1992). Hermo (2008, 2009) previously identified and examined several quarries in the area.

Lithic workshop. This is a place where lithic artifacts were manufactured. Archaeological indicators include dense assemblages of varied knapping debitage, ranging from large cortex flakes to small flakes without cortex, as well as artifacts discarded due to manufacturing errors. These activities took place in the quarry itself or in locations used for waiting, stalking, or planning, such as stations (Binford, 1980).

Occasional knapping site. These are places where opportunistic knapping was performed (see Nelson, 1991: Binford, 1979). Some nodules were knapped to test the quality of the raw material and to obtain base shapes or cores, which were then transported elsewhere. Archaeological indicators include low-density artifact assemblages formed by cores with few extractions, flakes with cortex, and hammers. This type of evidence is expected to appear in areas where nodules are dispersed or where there are very small seams of rock containing less raw material than in quarries.

Extraction of plant resources site. It is a place where specific activities were carried out, such as extracting wood to make various artifacts and roots to make dyes. Archaeological indicators include heavy lithic artifacts with edges capable of processing hard substances. Based on the artifacts' morphological and functional characteristics, it has been proposed that denticulate knives could have been used for sawing wood (Aschero, 1975), as well as large scrapers (Álvarez, 2004; Espinosa, 2000; Franco & Carballo Marina, 1993), wedges (C. Bayón, pers. comm.), bifaces (Hermo & Miotti, 2003), chopping tools (Binford, 1980), and large scrapers (Borrazzo, 2005).

> …The smoothing brush, to work the molle stick for the looms or the calafate sticks for the awnings, was also used for the sash when there was no molle. The tall sticks were those of the molles, some that are straight, and tall, the molle on the slopes of the canyons, to smooth it well and leave it well polished, they put ash from the fire so that the wood looked good and they scraped it… (excerpt from the interview with Silvana Chapalala extracted from Aguerre, 2000:99).

This type of indicator is expected in sectors where woody species were available in the past.

Primary prey processing site. This is a place where limited post-hunt processing activities occurred. Archaeological indicators include assemblages with low artifact density. The artifacts have long edges with signs of use. For example, knives among the artifacts suggest the primary processing of prey (Espinosa, 2000). There may also be unformatted artifacts, such as flakes with long, potentially useful blades, as well as cores that were intensively reduced, from which these blades could have been extracted. There is also blade reactivation, knapping debris, and scrapers. Other artifacts include projectile points with impact fractures discarded after extraction from hunted animals, as well as heavy artifacts used to dismantle game (Miotti & Salemme, 1999; Silvertsen, 1980). It is unlikely that the same location will be reoccupied for processing hunted animals, as these activities are expected to occur near the point of slaughter. The point of slaughter differs for each hunting event but is usually close to sectors commonly used for hunting.

Hunting ground. This is a place where animals were slaughtered. Archaeological indicators include isolated finds, primarily projectile points and intact boleadora balls (possibly lost), as well as fragments of points and balls that were discarded due to breakage during hunting (Cassiodoro, 2011; Sivertsen, 1980). This activity could have occurred in various locations throughout the landscape. Natural traps could have been used to ambush prey. Ethnographic examples include the Aonikenk provoking rheas into falling by using steep areas, as noted by Musters ([1871] 1997). Conversely, Claraz (1865–1866, 1988) describes collective hunting in the Yamnagoo region of Río Negro. He points out that the hunters used shallow areas with brackish water, which gave them visibility of the prey without being seen by it. Miotti et al. (1999) used this scenario as an analogy to interpret the archaeological assemblages recovered in Santa Cruz locales. At Piedra Museo, the Pleistocene shallow lake attracted fauna while the nearby rock formation concealed the hunters. In another locality, Los Toldos, Miotti (1989) discusses the potential use of a natural geoform—a canyon—for confining and hunting guanacos. In the La Puna region, places frequented by pray animals such as the bottoms of ravines were used for opportunistic hunting techniques involving observation posts, waiting stations, and herding animals (Martinez, 2003). In this scenario, hunters took advantage of natural topographies and built stone structures or hunting blinds near hunting grounds. In Patagonia, hunting could take place in the open pampas, possibly with the use of horses in the post-contact period.

The chronicles indicate that the Aonikenk hunted while moving through their territory, using caravans of women, children, and the elderly as part of a strategy that included using fire to direct animals into traps within reach of their weapons (Martinić, 1995; Musters, [1871] 1997).

These examples suggest that hunting occurred in various topographies. Martinez (2003) argues that hunting weapons are only lost in long-distance hunting strategies, such as those involving a bow and arrow, dart shuttle, or throwing spear. We can add boleadoras and lost balls to this list (Martinić, 1995; Musters, [1871] 1997). Therefore, hunting weapons found alone in the landscape indicate a place where such weapons were lost. Projectile points may be intact or fractured due to use or trampling.

Hunting stands. According to Binford (1980), a station is a location where information about animals can be obtained, hunts can be planned, and the movement of human groups can be controlled. While stalking activities during hunting can occur in any area of the landscape where hunters can hide, the construction of artificial shelters is another option in areas where these conditions do not occur naturally. Only hunting-related activities are carried out at hunting stations. Artifacts are manufactured and activated in these shelters while waiting for animals that typically use these passageways (Binford, 1980; Cassiodoro, 2011; Claraz, 1865–1866, 1988). As previously mentioned, these stations may coincide with workshops. Archaeological indicators include straight or semilunar stone hunting blinds that can be oriented according to the wind direction. These structures provide greater shelter and prevent animals from detecting hunters by smell (Gradín, 1962). Structures measuring 1.50–2 m in length and 50–60 cm in height, which are semicircular in shape, have been interpreted as hunting blinds at Lake Strobel (Gradín, 1962) and in an area of the Central Plateau (Hermo & Vázquez, 1999). As will be seen later, these structures are smaller than the residential parapets proposed in various sectors of Santa Cruz and the Somuncurá Plateau in Río Negro (see Chap. 6).

In hunting areas, knapping debris is expected to be found on the windward side (Gradín, 1959–1960). Associated artifacts indicate the completion of final shaping tasks and the reactivation of tools while waiting for prey in these areas. These artifacts are small and very small internal flakes (Aschero et al., 2005). There may also be basal fragments of projectile points, indicating that the points were replaced on the shafts (Bamforth, 1986). Stone ball fragments may also indicate the recycling of these weapons. Currently, for example, the inhabitants of the plateau gather these artifact fragments and reuse them to assemble boleadoras by tying them with leather (L. Miotti, personal communication).

The variety of lichens on the exposed surfaces of the rock blocks forming the hunting blind indicates the blind's relatively old construction and allows for detection of modifications to the rocks' original positions (Magnin, 2009a). Hunting stations may be associated with hunting fields that lack ideal natural conditions for hunter concealment. They can also coincide with field camps (Binford, 1980).

Chenques. Chenques are funerary cairns formed by piles of rocks. They are similar to those described by travelers and in ethnographic accounts (Deodat, 1946; Viganti, 1934). These structures were usually built directly on the ground above the body of the deceased. Based on their location, one can infer that ritual activities were carried out (Musters, [1871] 1997). Indicators include oval, elliptical, or circular stone structures, which are sometimes associated with knapped artifacts. To distinguish them from other registered structures, which are assigned a different function, the shapes and sizes of the chenques are analyzed and compared with those of chenques with proven funerary functions in Chap. 6 (De Aparicio, 1933–1935; Goñi & Barrientos, 2000; Goñi et al., 2004).

Residential camps. They were multiple activity places where general activities took place, such as sleeping, taking caring of children, preparink of food, knapping and repairing lithic artifacts, and performing many other tasks (Binford, 1980). Evidence of these daily activities is expected to be found in these areas, which are usually organized around a hearth (Binford, 1988; Manzi & Spikins, 2008; Surovell & Waguespack, 2007). However, combustion structures are likely not preserved on the surface. There may be transported rocks that delimit them (Silvertsen, 1980; Casamiquela, 2000), or signs of thermoalteration may be observed in knapped materials that were in contact with the fire. Other indications include hot stones, which could have been used for cooking, as observed by Musters (1871), as well as ceramics, whose functionality is related to cooking and storage—activities primarily associated with domestic contexts (Cassiodoro, 2008b).

Regarding housing, Gradín (1976) described stone structures and residential parapets on the plateau of Lake Buenos Aires. These structures likely supported leather awnings, providing greater strength and protection against the wind (see other examples in Chap. 6). These structures, along with similar ones, are associated with bone remains, primarily from Lama guanicoe, as well as dense lithic assemblages; hands, mills, and ceramics (Cassiodoro et al., 2004; Espinosa & Goñi, 1999; Molinari &

Ferraro, 2004). Hearth structures, stones for awnings, mills, and grinding stones can be considered evidence of places equipment (Binford, 1979; Cassiodoro, 2008). Regarding the analysis of lithic assemblages, residues from all manufacturing stages are expected in places of residence, including primary reduction and core preparation. Assuming that greater variability in shapes indicates a greater variety of activities carried out (Andrefky, 2005), one would expect a high artifactual richness, as well as different morphological types in the knapping debitage. These are all indicators of the multiple activities developed in a long- or re-occupied base camp (Binford, 1982; Camilli, 1989; Kelly, 1992). Additionally, the presence of specific technological types may also suggest tasks associated with residential areas. Gamble (1990) points out that these are places where most of the time is spent working with leather. In the case study, this activity could be related to the manufacture of awnings, quillangos, and other items. A high percentage of artifact types associated with this activity would be expected, particularly scrapers (Aschero, 1975, 1983). Previous archaeological cases have observed high proportions of these types, for example, at the Cueva Maripe site (Hermo, 2008).

Basal fragments of projectile points, which may indicate the replacement or repair of weapons, are found at campsites. This activity occurs in residential areas (Bamforth, 1986; Camilli, 1989; Nelson, 1991). This knapping activity produces a significant amount of debris (Camilli, 1989). Therefore, the assemblage is expected to exhibit a high artifactual density (Foley, 1981). Finally, the presence of distal projectile point fragments may reveal where hunted prey was consumed (Cassiodoro, 2011).

Archaeological indicators tend to be concentrated in caves, rock shelters, and open-air sites with topographic protection from inclement weather and near springs (Vignati, 1934). In unsheltered areas, stone structures may have served as bases for awnings (Gradín, 1976). Cassiodoro (2011) mentions that stone awning bases are found in Pampa del Asador in areas close to bodies of water and in relatively high topographic positions. In those cases, they generally form groupings.

Transient camps. These are places where people carried out domestic activities such as eating, sleeping, knapping, and repairing artifacts, while staying outside of a residential base. Their characteristics depend on the logistical tasks associated with them (Binford, 1980). The presence of artifact assemblages of intermediate density is an archaeological indicator. Since the activities carried out in these places vary, it is necessary to rule out a greater density of buried materials, which could indicate activities performed over a longer period of time or successive reoccupations. This may be consistent with a residential base camp. Indicators may include artifacts knapped primarily without cortex and of smaller sizes. Cores are not expected to be discarded at these camps unless they are exhausted. The bases of replaced projectile points, as well as medial and distal fragments with use-related fractures, may also be found in these places. This could indicate their entry into the complex during the hunting of prey. Evidence of a hearth is also expected.

Evidence of transitory camps is expected to be found in areas more than ten kilometers (the equivalent of a two-hour trip) away from a residential base camp (Binford, 1982).

The description of the activities carried out by hunter-gatherers in this scheme is a simplification because it is restricted to the possibilities offered by the archaeological record. In Chap. 6, we use this schematization to provide a functional interpretation of the archaeological evidence, incorporating taphonomic considerations. In this description, we were careful not to use environmental indicators, as contrasting the expected and observed spatial locations of environmental variables constitutes an independent instance of the initial classification. This allows us to extract patterns in locations.

We have systematically explored this line of analysis, recognizing the effects of equifinality and acknowledging that only some aspects of complex spatial behaviors from the past can be differentiated—those with the most visible archaeological indicators. Similarly, behaviors that have been established over time and repeated by successive generations will be addressed more satisfactorily.

Throughout the history of the sector's occupation, the way the space has been used has likely changed for various reasons. However, due to the nature of the archaeological evidence, it is currently impossible to distinguish between different chronologies. Therefore, this must be considered a limitation of the data. The analysis applied in this work enables us to identify general trends in the distribution of evidence from successive settlements (Dewar & McBride, 1992; Foley, 1981).

4.3.3 Methodology, Analytical Context

The work methodology involves a series of steps that apply various analytical techniques. Environmental and cultural data serve as inputs from which spatial relationships are extracted. First, an exploratory location model is generated for locational analysis, followed by hypothesis testing assisted by GIS. Briefly, the first procedure detects general patterns in the data, which are then investigated using other tests. Each step is detailed in the following chapter.

4.3.3.1 Locational Models

Some archaeological models aim to predict the probability of discovering sites in unexplored areas (Maschner, 1996a). However, the objective of this study is to describe the precise locations of archaeological data in order to formulate hypotheses and guide further research seeking to explain observed patterns (Van Leusen, 1996). The result is a locational model of the different types of archaeological and environmental evidence in the studied area, which makes it a useful research tool.

Inductive vs. deductive approach

According to their argumentative logic, models are inductive when patterns extracted from archaeological distributions in a particular area or chronology are extrapolated to unknown regions (Woodman, 2000b, among others). Conversely, models are considered deductive when these patterns are theoretically defined and contrasted with archaeological evidence (see, for example, Kohler and Parker 1986; Stančič and Kvamme 1999, and their application to a Patagonian archaeology case study in Matteucci and Scheinsohn 2004).

While many authors have debated which procedure is correct, both approaches are useful and complementary.

In theory, a predictive model can be developed through pure deduction from a theory or through pure induction from empirical observations. In practice, however, most models use both. The deductive approach requires selecting independent variables, a process guided by theoretical knowledge. Likewise, an inductively generated model must be tested by contrasting it with the data (Warren, 1990a). Which procedure is used depends on the objectives and the adequacy of the available data for analysis.

Cultural Resources Management vs. Academic Approach

The ability to predict site locations makes models valuable tools for government agencies when deciding about managing and protecting cultural resources. This application is most prevalent in the United States and Canada, where significant economic resources are allocated to these plans and models have a long history (Fernández Cacho, 2006). Consequently, these models are primarily used in territorial planning programs (Kuna, 2000; Wheatley & Gillings, 2002). The goal is to find archaeological evidence and protect it. However, determining the location of archaeological materials may also be important. For example, when planning work that could potentially destroy sites. In archaeological research, this tool is important for planning prior to fieldwork. It directs prospecting toward areas most likely to yield positive results in discovering new sites, which considerably reduces economic costs (Fernández Cacho, 2006; Kuna, 2000; Scheinsohn, 2001).

Here, the focus is on evaluating the possibilities that these models offer as research tools (Savage, 1990). In approaches oriented toward the management of cultural resources, models are primarily created to predict where archaeological evidence will appear or be absent. In academic approaches, however, the aim is to understand the location factors. From a perspective more in line with the theoretical guidelines of this work, the aim is to understand the factors involved in past location choices (Van Leusen, 1996).

The approach applied in this work: Description of patterns and testing of hypotheses

This work seeks to contribute to an ongoing debate in which numerous authors from various disciplines that use these models have participated, not only from archaeology. The current status of this topic indicates that archaeologists widely recognize the use of these tools and understand that relationships between variables in the form of patterns should not be confused with causal explanations. For instance, in our case study, identifying a strong correlation between the analyzed independent variables and the location of the archaeological evidence in the landscape does not explain the latter. Causality must be demonstrated, not assumed (Köhler & Parker, 1986; Wescott, 2000; Wheatley, 1996), because the found pattern found is merely an objective, a methodological one, in addition to high- and medium-range theory in the explanatory process (Fig. 4.2) (Ebert & Köhler, 1988). Explanations are achieved by combining other lines of evidence that provide information and theoretical tools that allow us to interpret the obtained patterns and understand why people located their camps, burial sites, or task areas in specific sectors of the landscape.

In summary, we acknowledge that generating a locational model reveals existing trends in the data, and agree with Kvamme (1995:5) that:

> Good archaeological models…can provide insight into past land use and settlement patterns by portraying principal trends and eliminating some of the statistical noise in the data. By making the essence of a spatial pattern clearer, such models offer great heuristic potential in regional studies.

The next step is to more intensively analyze the locational patterns that emerge from testing particular hypotheses in more detail.

4.3.3.2 Types of Variables Considered

Roper (1979) classified locational studies as either man-land or man-man relations-based. Important factors structuring settlement patterns in the former are distances to resources, while important factors in the latter are social variables. Although this work assumes that past settlement decisions were strongly influenced by natural environmental characteristics (Warren, 1990b), the theoretical framework adopted considers other possible social and symbolic factors. In archaeological research, it is commonly accepted that factors other than economic ones have influenced human decision-making regarding the location of settlements in the landscape or the organization within a particular settlement. Nevertheless, such factors are generally not used as predictive variables in models. This is either because these factors are archaeologically invisible or difficult to quantify (Woodman, 2000a, 2000b), or because archaeologists do not know how to use them due to an unknown influence on location (Köhler, 1988; Van Leusen, 1996). Some argue that environmental factors influence settlement more in hunter-gatherer societies than in "complex" societies (Kvamme, 1988:332). Other authors note a scaling problem in the sense of Altschul (1988). The environmental aspect can be modeled more easily than the symbolic aspect, the latter of which is only approximated in specific cases. Consequently, most existing models have not included cultural variables in their analysis (Fernández Cacho, 2006). Models that use the deductive procedure have particularly been criticized for deriving environmental deterministic inferences (Van Leusen, 1996; Wheatley & Gillings, 2002).

Furthermore, the application of Geographic Information Systems (GIS) technologies in geography has been criticized for introducing an inherent bias. However, these criticisms have sparked debates leading to innovative responses that seek to overcome these restrictions (Lock & Harris, 2000). Based on numerous examples of archaeological research, it is possible to include purely cultural variables depending on the particular case. Nevertheless, we agree with authors such as Van Leusen (1996), who argue that cultural factors involved in location can be studied even when only environmental variables are used. In the words of Hasenstab (1996) both, settlement systems and ecosystems are complex, so we should not expect to find simple correlations between them. The purpose of locational modeling is to identify environmental factors that influence settlement behavior and consider them alongside non-environmental factors that also impact settlement behavior.

The selection of location analysis variables was based on a review of general literature about hunter-gatherers in Patagonia. The inclusion of variables related to social and cognitive aspects was possible when material correlates of these constructs were identified (Aldenderfer, 1998). This study analyzed some variables for possible implementation in the analysis, including distance to permanent waterways, distance to temporary tributaries, location in a sheltered sector of the landscape, distance to the landscape's main mobility routes, location in an area with high environmental visibility, availability of lithic raw materials, and the presence hunting prey (guanacos), firewood, and plant resources (see Chap. 6).

4.4 Conclusion

This chapter introduced theoretical concepts that enable us to study how human groups interacted with the landscape in the study area. The ethnoarchaeological and archaeological information presented here, together with visibility controls and site formation processes (see Chap. 6), constitute medium-range theoretical constructions that allow interpretation of the archaeological evidence from field surveys. Ethnographic information and archaeological knowledge from other regions complement this interpretation.

Theoretical aspects of the methodology were also introduced (Ebert & Kohler, 1988). Geographic Information Systems (GIS) and other spatial technologies are particularly important to this work, so their theoretical and methodological aspects are discussed throughout different chapters (Chaps. 5, 6, and 7).

Chapter 5
Methodology

Abstract One of the most original contributions to this research is the collection of surface data using an off-site sampling design. Though not intensive, this design systematically covers the area of interest. This prospecting strategy yields fundamental baseline data necessary for generating digital models to analyze the archaeological landscape. In addition to describing the general methodological procedure, this chapter begins by defining the units of analysis and discussing the available data and their suitability for addressing the problem.

Keywords Archaeological and environmental survey · Geographic Information Systems (GIS) · Locational analysis · Hypothesis testing · Model formalization · Pattern detection

5.1 GIS and Other Spatial Technologies

An information system can be defined as a set of useful tools that allows one to obtain information from data (González Pérez et al., 1997). In the case of Geographic Information Systems (from now on GIS), this data has a spatial component, which distinguishes them from other database management systems. GIS are, then, spatially referenced databases that allow users to display data graphically. However, they are more than computerized mapping because they allow the storage, mathematical manipulation, rapid retrieval, and flexible presentation of spatially referenced data (Allen et al., 1990; Green, 1990; Kvamme, 1990). GIS can be described as systems that allow superimposing a series of *layers* of data that are of interest for a particular study problem and relating them to each other in search of patterns (Chapa et al., 1998). This information can be represented in two different ways, and through a vector or raster formats combined in the same project. Data presented in the first format are suitable for representing discrete variables, with well-defined limits, such

Supplementary Information The online version contains supplementary material available at https://doi.org/10.1007/978-3-032-08127-8_5.

L. A. Magnin, *The Landscape and Mobility of Hunter-Gatherer Societies*, The Latin American Studies Book Series, https://doi.org/10.1007/978-3-032-08127-8_5

as a river channel, or the units defined on a geological map. Conversely, raster formats facilitate the representation of continuous variables, such as the height above sea level in a digital elevation model of the terrain. In this case, the data is organized into a large number of cells of the same size. Each cell contains individual values for the variable represented by the coverage. These cells constitute the unit of quantitative analysis when addressing spatial problems, as they allow the information from various layers of evidence to be compared cell-by-cell according to an anchor given by the coordinates that determine their location.

The greatest advantage of raster-based GIS is the ability to make reliable inferences using statistical tests. Before quantitative methods were introduced, many analyses of spatial associations were based on visual inspections and a limited range of data categories. However, several authors have warned about the limitations of such analyses at both the intra-site (Blankholm, 1991) and inter-site (Hodder & Orton, 1990; Kvamme, 1990) scales. Using specific statistical techniques is essential for spatial analysis, and GIS facilitates generating the necessary data. For instance, predictive models can be generated from manual measurements of map variables. Nevertheless, the high cost of time and labor restricts analysis to relatively small samples of the total available data (Kvamme, 1990). The introduction of GIS, however, allows us to expand our analysis of big data using statistical techniques in modules integrated into GIS or in external statistical software packages.

5.2 Applied Methodology

5.2.1 Objective, Scale, and Scope of the Analysis

The analyses proposed below allow us to study the variability of human use of space by addressing the distribution of archaeological and environmental evidence and evaluating it through an explanatory framework (Chap. 4).

The study aims to obtain results at different scales. First, a lower detail information macro/scale feeds from the field surveys that record the distribution of archaeological evidence throughout the La Primavera Canyon and adjacent geomorphic features. At this scale, the formation processes are analyzed through field observations, which are complemented by studies using remote sensing images. On the other hand, at a micro-scale, the analysis generates useful higher detail information to characterize the landscape at different locations. This is achieved by analyzing the objects collected and the field records of sites, surface artifact assemblages, isolated artifacts, stone structures, and rock art. Each specific place (Binford, 1982) provides evidence that allows us to propose possible human activities that could have left their mark in evidence capable of lasting on the land surface. At this scale, formation processes are approached by lithic taphonomy studies (Borrazzo, 2004). We expect that the results contribute to the research conducted at the macro or regional scale (García Sanjuán, 2005; see Chaps. 2 and 9).

About the time scale, it is broad, as the analyzed surface assemblages can span from the Early to Late Holocene. This estimate is based on the existing radiocarbon sequence in Cueva Maripe, Cueva Mora, and La Quinta, as well as the technological indicators of the assemblages (see Chap. 7).

5.2.2 Methodological Procedure

The methodological proposal developed in this work (Fig. 5.1) consists of four successive methodological steps.

The process begins with identifying the problem, for which a data collection protocol is then designed. Prior to conducting fieldwork, we gathered information on the area of study, including aerial photographs, satellite images, radar images, and

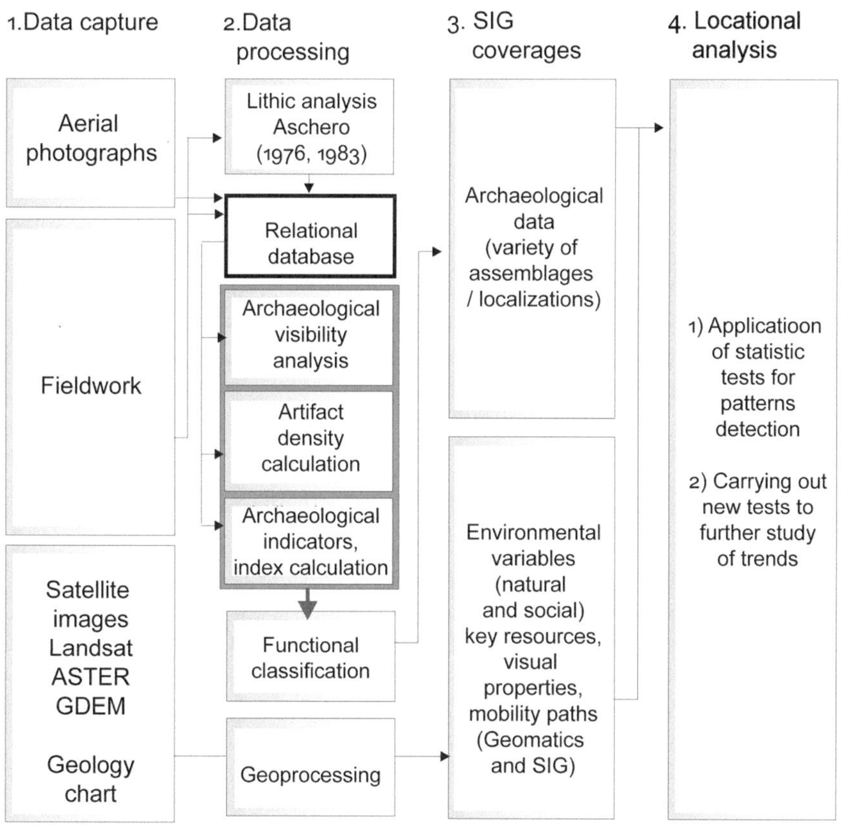

Fig. 5.1 Methodological procedure

maps. During fieldwork, we collected surface archaeological materials and recorded environmental data.

The second step is data processing. The collected materials undergo lithic analysis and the information on other types of archaeological evidence registered (e.g., rock art, stone structures) and environmental data, are entered into a relational database and incorporated into the GIS. This allowed us to analyze ground stability conditions affecting the differential visibility and preservation of archaeological evidence, calculate artifact densities, evaluate indicators of functionality, and calculate indices. These steps allow for the functional classification of archaeological evidence.

This classification is used to obtain a series of vector-format point data coverages. Each coverage represents locations in the landscape where different types of activities were carried out (step 3). Similarly, based on the available environmental data, a series of geoprocessing steps are carried out to generate coverages representing variables that could affect the location of archaeological evidence. These coverages and the previously classified archaeological data are integrated into a location analysis (Step 4). This exploratory analysis allows different hypotheses to be tested statistically, deepening the study of the patterns of interest.

5.3 Data Capture

Spatial data that is analyzed using GIS consists of primary and secondary data (Connolly & Lake, 2006). Primary sources include information collected in the field and from remote sensors, such as aerial photographs and raw satellite images. Secondary sources consist of processed and interpreted information available in paper or digital format and processed satellite and radar images. Table 5.1 details the various data sources used as baseline information at different stages of the project.

5.3.1 Aerial Photographs, Satellite Images and Cartography

Some of the contextual information comes from remote sensors. Remote sensing refers to the collection and interpretation of information about phenomena with which physical contact is not made. Since the launch of the first digital sensors in the 1970s, a wide range of products have been used in archaeology to study the environment and locate archaeological sites (Conolly & Lake, 2006). For this study, an intensive data search was conducted. Data available through institutional agreements that served the study's objectives and data that were low-cost or freely available online were used. While these data are important for establishing the environmental context of the archaeological evidence, direct knowledge of the terrain is also crucial because it provides the experience necessary for generating criteria for using data from remote sensors. The rest of this chapter describes the process of acquiring and managing field data.

Table 5.1 Various data sources used as baseline information at different stages of the project

Description	Identification and year of data collection	Origin	Format	Georreference	Band and spectral region/ coverage	Wavelength (μm)	Scale/ resolution	Horizontal/ vertical error RMSE (95% confidence interval)
GPS*	2008	Prospection	Vectorial, shp	Geographical coordinates Reference geoid: WGS84			4–12 m	
Field form*	2008	Prospection	Microsoft access	Related GPS data table			Observations made within 15 m. Estimation of data detail	
Aerial photographs*	1M_3049 1M_3050 1M_3051 (1968)	Geographic National Institute	Digitized at 20 microns (1270 DPI), Raster				1:60,000 Spatial resolution: 2 m	
Multispectral satellite image	S_19_45	Landsat 4/5 GeoCover TM Circa 1990	MrSIDTM	Geographical coordinates Reference geoid: WGS84	7 Mid-infrared 4 Near infrared 2 Green visible		Spatial resolution: 28.5 m	RMSE 50 m

(continued)

Table 5.1 (continued)

Description	Identification and year of data collection	Origin	Format	Georreference	Band and spectral region/ coverage	Wavelength (μm)	Scale/ resolution	Horizontal/ vertical error RMSE (95% confidence interval)
Multispectral Satellite Image * (1)		Landsat 5 TM	Raster	UTM S19	1 Visible blue 2 Green visible 3 Visible red 4 Near infrared 5 Mid-infrared 6 Infrared thermal 7 Mid-infrared	0.45–0.52 0.52–0.60 0.63–0.69 0.76–0.90 1.55–1.75 10.42–12.50 2.08–2.35 (spatial resolution)	30 m 30 m 30 m 30 m 30 m 60 m 30 m	45 m/(1 and a half pixel)
Multispectral satellite images		ASTER (Advanced Spaceborne Thermal Emission and Reflection Radiometer)	Raster		1 VNIR (nadir) 2 VNIR (nadir) 3 VNIR (nadir l) 3 VNIR (backw) 4 SWIR 5 SWIR 6 SWIR 7 SWIR 8 SWIR 9 SWIR 10 TIR 11 TIR 12 TIR 13 TIR 14 TIR	0.52–0.60 0.63–0.69 0.76–0.86 0.76–0.86 1.6–1.7 2.145–2.185 2.185–2.225 2.235–2.285 2.295–2.365 2.360–2.430 8.125–8.475 8.475–8.825 8.925–9.275 10.25–10.95 10.95–11.65	15 m 15 m 15 m 15 m 30 m 30 m 30 m 30 m 30 m 30 m 90 m 90 m 90 m 90 m 90 m	

(continued)

Table 5.1 (continued)

Description	Identification and year of data collection	Origin	Format	Georreference	Band and spectral region/ coverage	Wavelength (μm)	Scale/ resolution	Horizontal/ vertical error RMSE (95% confidence interval)
Digital elevation model* (2)	S48W070 S48W069 (2009)	Aster Gdem V.1	Raster	Geographical coordinates Reference geoid: WGS84/ EGM96	VNIR SWIR TIR	0.78–0.86	1 arc second ~ 30 m	Relative vertical precision 10 m Absolute vertical precision 7 m Relative horizontal precision 10 m Absolute horizontal precision 7 m

(continued)

Table 5.1 (continued)

Description	Identification and year of data collection	Origin	Format	Georreference	Band and spectral region/ coverage	Wavelength (µm)	Scale/ resolution	Horizontal/ vertical error RMSE (95% confidence interval)
Digital elevation model* (3)	s48_w070 s48_w069 (2000)	SRTM V.1	Raster	Coordinates Reference geoid: WGS84/ EGM96	SIR-C X-SAR	5.6 cm 3.1 cm	For the study area: 3 arc second ~ 90 m	Band C/X Relative vertical precision 10/ 6 m Absolute vertical precision 16/ 16 m Relative horizontal precision 15/ 15 m Absolute horizontal precision 20/ 20 m

(continued)

Table 5.1 (continued)

Description	Identification and year of data collection	Origin	Format	Georreference	Band and spectral region/ coverage	Wavelength (μm)	Scale/ resolution	Horizontal/ vertical error RMSE (95% confidence interval)
Digital elevation model* (3)	s48_w070_1arc_v3 s48_w069_1arc_v3 (2000)	SRTM V.3.0 (released 2015)	Raster	Geographical coordinates Reference geoid: WGS84/ EGM96	SIR-C X-SAR	5.6 cm 3.1 cm	Global 1 arc second ~ 30 m	Band C/X Relative vertical precision 10/ 6 m Absolute vertical precision 16/ 16 m Relative horizontal precision 15/ 15 m Absolute horizontal precision 20/ 20 m

(continued)

Table 5.1 (continued)

Description	Identification and year of data collection	Origin	Format	Georreference	Band and spectral region/ coverage	Wavelength (μm)	Scale/ resolution	Horizontal/ vertical error RMSE (95% confidence interval)
Geologic chart**	4769–IV. Monumento Natural Bosque Petrificado (1999); 4769-III. Destacamento la María (2001); 4969-II. Tres Cerros (1994)	Secretaría de Minería. Instituto de Minería y Recursos Minerales (SEGEMAR)	Digitized raster/vector	Projection: Gauss Krüger Faja 2 Datum: Campo Inchauspe			1:250.000 (40–45 m)	
Topographical charts**	4769–33. Cerro Virginia (1970–80) 4769–34. Gobernador Moyano (1970) 4769–35. Monumento Natural Bosque Petrificado (1970) 4769–36. Cerro del Noventa y Seis (1980)	Instituto Geográfico Militar (IGM)	Digitized vector	Projection: Gauss Krüger Faja 2 Datum: Campo Inchauspe			1:100,000	

(continued)

Table 5.1 (continued)

Description	Identification and year of data collection	Origin	Format	Georreference	Band and spectral region/ coverage	Wavelength (μm)	Scale/ resolution	Horizontal/ vertical error RMSE (95% confidence interval)
SIG 250 Topographic Digital chart**	Vectorial cartography	Instituto Geográfico Nacional (IGNM)	Topographic Digital IGN. Vector	Projection: Gauss Krüger Faja 2 Datum: Campo Inchauspe	– Tributaries – Roads – Establishments		1:250,000	
INTA coverages	Vectorial cartography INTA: FISIONOF. SUELOSF	INTA	Digitalized by INTA. Corrected georeference	Projection: Gauss Krüger Faja 2 Datum: WGS84	– Vegetation – Soil		1:250,000	
Geomorphological map and hydrography** (4)	La Primavera Archaeological locality	Modified from Gómez and Magnin (2008)	Vector	Projection: Gauss Krüger Faja 2 Datum: WGS84			1:50,000 (4–12 m)	

(*) Primary and (**) secondary GIS base data. References (1) NASA (2013); (2) ERSDAC (2003); (3) USGS, (2005); (4) Gómez and Magnin (2008)

5.3.2 *Archaeological and Environmental Data Collected Through Fieldwork*

The data collected in the field aimed to record the diversity of surface archaeological evidence. This included lithic assemblages and isolated finds, as well as lithic artifacts of different sizes. Other types of evidence were also documented, such as rock art and stone structures of various shapes. The absence of archaeological evidence was also recorded. This dataset is necessary for identifying landscape usage patterns.

Additionally, we recorded the most significant geological processes that affect the land surface and the preservation of archaeological sites. To accomplish this, we recorded the following variables: the presence of soil or type of sediment; approximate granulometry and humidity; geomorphology; slope; and indicators of assemblage disturbance, such as evidence of landslides; holes made by humans or animals; fences; vehicle routes; paths; animal wallows; gullies of surface runoff; and burials (see Chap. 6). We also observed the materials found on the surface, such as the stages of weathering in bones, and the presence of patinas, lichens, verdigris, ventifacts, and signs of thermal alteration on the surface of artifacts. We also recorded data on differential visibility conditions due to vegetation cover, as well as the distribution of resources such as water, wind protection, accessibility, rock outcrops suitable for knapping, ocher outcrops, salt deposits, and support availability for rock art.

Finally, we controlled for possible biases introduced in the data recording process, such as the time of data collection, through a set of variables.

5.3.2.1 Adequacy of Current Environmental Data as Proxies for the Environment of the Past

When presenting data sources for analyzing archaeological distributions in the landscape, it is important to note that this data represents the recent landscape. At present, numerous environmental data from aerospace technologies are accessible. The environmental data for the study area was collected between 1968 and 2008 and is presented in Table 5 (Supplementary Material). While these data are suitable for evaluating current formation processes, their scope and limitations as proxies for the distribution of useful resources for past populations must be evaluated. To advance in that direction, we use the available paleoenvironmental information for the region (see Chap. 2) and some geoarchaeological studies of the archaeological localities (Gómez & Magnin, 2008; Zárate et al., 2000). By the moment, there are no paleogeographic reconstructions in map format showing old geomorphology, or the past distribution of permanent watercourses, channels, shallow lakes and floristic species. In this work we recognize that the Central Plateau is a relatively stable region and this allows us to assume that the landscape known by early populations was similar to the current landscape (Zárate et al., 2000). However, when analyzing human interactions with the landscape, seasonal variations and oscillations in the availability of resources over time need to be considered. The hydrology dynamics of water availability in temporary shallow lakes and ditches are among the most important. Research in hydrology in extra-Andean Patagonia (Hernández et al., 2008; Mazzoni et al., 2000; Melendi, 2023) is useful for the area addressed in this work. Future interdisciplinary research on this topic will allow us to better understand the changes that occurred in the landscape and strengthen our explanations.

The methodology and theoretical framework guiding the research will be improved with better data, when available. GIS are particularly suitable for analysis because they facilitate restudy. Current data is an adequate starting point for proposing the distribution of past resources under similar inferred conditions. For example, in the case of water, a critical resource, it can be argued that, during much of the area's history of occupation (including arid periods, such as the current one), water was permanently available in springs. As mentioned in Chap. 2, during other periods, however, in addition to these water sources, there was greater relative availability of water in shallow lakes and streams fed by a more abundant rainfall regime than exists today. Nevertheless, these sources were always temporary. Based on paleoclimatic studies, the periods of greatest effective humidity occurred between 7000 and 3000 years before present (BP), when the climatic optimum occurred in the Central Plateau (Salemme & Miotti, 2008), as well as during the Little Ice Age, a cold and wet period that lasted approximately 400 years and occurred 500 years ago (Morales et al., 2009).

Generating detailed maps of environmental distributions in different periods, together with relative geochronology studies, will allow us to narrow the uncertainty surrounding the probable dates of surface archaeological materials.

5.3.2.2 The Survey Design

The sampling strategy was designed to systematically cover the archaeological locality. This strategy builds on previous archaeological research. It is suitable for theoretical reasons as well as for fieldwork logistics. The design centered on previously known and characterized archaeological sites that were multiple activities loci. These sites include Cueva Maripe (Miotti et al., 2007, 2009), La Playita (Magnin & Blanco, 2006), and La Quinta.

Previous studies have identified an archaeological bias toward prospecting the most accessible geoforms (Gómez & Magnin, 2008). To address this issue, the design provides continuous spatial coverage that is not directed toward particular geoforms. This differs from other prospecting designs, such as Foley's (1981) proposal or Borrero et al.'s (1992) work, where the transects were directed toward homogeneous geoforms or landscape units.

The surveyed sector extends over 10 by 10 km, equivalent to approximately two hours of walking from these sites. In the study of settlement systems, analyzing the exploitation territory (Higgs & Vita Finzi, 1972; Roper, 1979) or camp range (Binford, 1982) is considered the minimum area for spatial pattern analysis. This is equivalent to the space used during a day of activities carried out from a base camp where immediately available resources are located. Binford (1982) calls this area the "search radius" in his economic zoning model, which is complemented by broader scales: the logistic radius and the annual range (or action range, Bettinger, 1980).

The fieldwork covered the search radius and part of the logistic radius that could be traveled to and explored from centrally located residential sites. The spacing between transects is wide, so the surface area covered is relatively small. For budgetary reasons, this design was preferred to more intense coverage. This decision aligns with the research objective because it is more effective for analyzing the location patterns of mobile hunter-gatherer groups.

Studying landscape use in mobile groups poses additional challenges. As Binford (1982) points out, places may be reoccupied with different functions over time. Thus, the present archaeological record is a remnant settlement pattern (Dewar & McBride, 1992), where each recovered assemblage does not represent a single occupation event. Rather, it is a palimpsest in which activities that occurred more frequently over successive reoccupations are mostly represented.

The chosen survey design reduces patterns of spatial variation, or autocorrelation, that spatial data generally exhibit. However, this effect is undesirable because statistical inference procedures applied in modeling and locational analysis assume independence between observations (Kvamme, 1990). This design is also analytically superior for some GIS-based analyses (e.g., visibility, Fisher et al., 1997). Finally, the location of the roads in the studied sector provides easy access to all the diagrammed transects.

The fieldwork was carried out in two stages. First, in 2006, a prospecting and registration strategy was planned and tested in the field. In the GIS environment, the archaeological locality was covered by a 120-point survey grid with 1-km spacing. All points were transferred to a GPS unit for navigation and to search for reference points in the field. These reference points allowed for the planned transects of the fieldwalking to be carried out in a north–south direction (or south-north, as appropriate, based on the layout of each field day). Eleven transects were defined, each with ten sections, and they were named with letters and numbers (see Figs. 5.2 and 5.3). Thirty-six transect sections were covered in the field, and sampling and recordings were carried out on field forms (Magnin, 2006d). Based on the experience of the first stage, it was possible to calculate the approximate time required to complete the total survey of the locality using this methodology, estimate the amount and type of expected archaeological evidence for the total area, and improve system registration using field forms.

The second field stage was developed in 2008. To avoid repeating the same transect sections from 2006, the survey grid was moved 300 m southwest. The grid points to be traversed were entered into GPS navigators to guide the prospectors according to the transect layout design. Based on previous field experience, the field registration form was improved, and only the 2008 data were used for GIS analysis. Topographic profiles were generated in GIS using a DEM (Figs. 5.2 and 5.3) to plan the transects for fieldwork. These profiles were useful for planning daily activities and choosing where to set up camp. Table 5.2 presents information regarding the development of fieldwork each day.

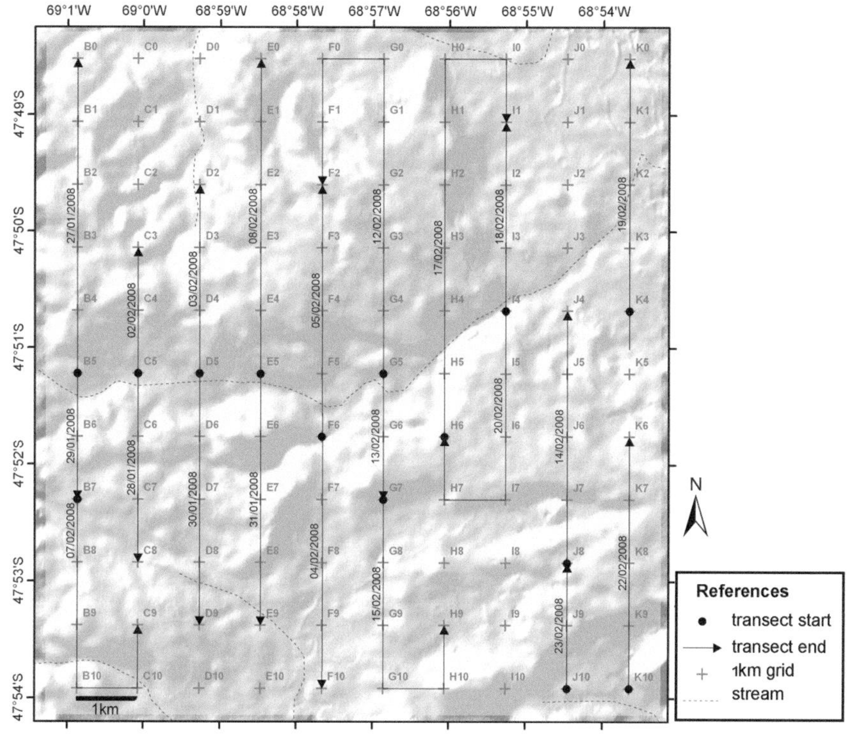

Fig. 5.2 The grid illustrates the systematic sampling design. The arrows show the route taken each day

The observation and discovery units are the 10-km-long transects. The survey team consisted of two pairs of fieldwalkers, each pair separated by 10 m. They toured the different transect sections guided by GPS navigation. Each prospector can see approximately 2.5 m on either side of the transect center, and 91 transect sections, each 1 km long, have been completed. Therefore, the minimum sampled area is estimated to be 182,000 m^2 or 0.018 km^2. When an archaeological discovery was made along the transect route, the sampled area was extended to cover a circular area of approximately 20 m in diameter. Counts of artifacts within this area allowed the team to characterize the archaeological evidence. If only one artifact was detected, it was defined as an isolated find, or a finding within the sampling unit. If two to 24 artifacts were found, it was defined as a concentration. If more than 24 artifacts were found, it was characterized as a site (Borrero et al., 1992; Table 5.3; Fig. 5.4). Sampling units with an absence of surface artifacts were also regularly recorded. This sampling strategy was useful because it represented the different types of archaeological evidence and allowed for environmental observations about the presence of resources and site formation processes.

Form, frequency, and detail in data recording

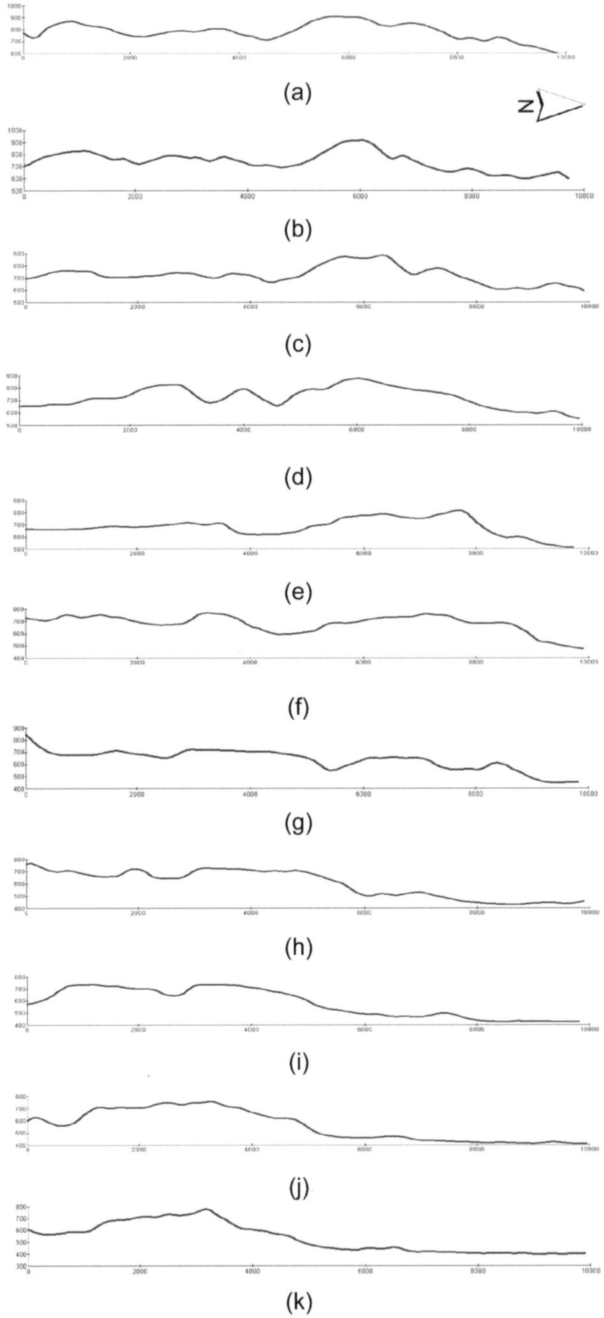

Fig. 5.3 Topographic profiles drawn on the 10 km transects from south to north were used for planning daily fieldwork. The letters indicate the name of each transect

Table 5.2 Results of the field work carried out in La Primavera in 2008

Date	Working hours	km walked	Completed forms	Site and concentration schemes	Photographs	Transect name	*N* collected artifacts
01/27/2008	8:45	5	20	8	86	B	203
01/28/2008	8:35	3	22	4	73	C	262
01/29/2008	0:00	2	4	1	138	B	93
01/30/2008	9:25	4	22	3	164	D	318
01/31/2008	11:00	4	13	2	170	E	92
02/02/2008	11:00	2	6	3	152	C	700
02/03/2008	5:20	3	8	0	121	D	72
02/04/2008	5:30	4	10	4	34	F	187
02/05/2008	5:30	4	12	1	188	F	209
02/07/2008	7:50	5	14	3	153	B–C	280
02/08/2008	6:50	5	18	1	105	E	273
02/12/2008	10:25	8	24	2	80	G–F	315
02/13/2008	6:14	2	13	4	81	G	312
02/14/2008	6:50	4	17	1	50	J	278
02/15/2008	8:20	5	23	0	119	G–H	387
02/17/2008	8:35	8	29	0	77	H–I	322
02/18/2008	6:35	3	17	2	50	I	291
02/19/2008	6:50	4.5	17	2	58	K	334
02/20/2008	6:40	5	22	1	75	H–I	302
02/22/2008	8:45	4	15	2	77	K	321

(continued)

Table 5.2 (continued)

Date	Working hours	km walked	Completed forms	Site and concentration schemes	Photographs	Transect name	*N* collected artifacts
02/23/ 2008	8:35	2	11	1	111	J	470
21 days	$X =$ 7:22	86.5	337	45	2162		6021

Table 5.3 Initial classification according to the number of artifacts counted within the sampling units

Evidence class	No. of artifacts within a 20 m diameter circle
Negative evidence	0
Isolated find	1
Concentration	2–24
Sitie	≥ 25

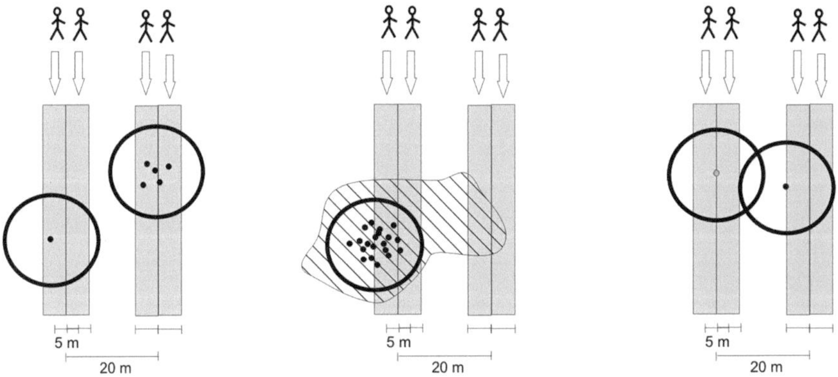

Fig. 5.4 Example of transects used as a unit of discovery of the archaeological evidence. When artifacts were found on the surface, they were defined as isolated finds, concentrations, or sites, based on the density of artifacts within a 20-m-diameter circle. The figure on the left illustrates the registration of an isolated find and a concentration. The central figure illustrates a site whose perimeter was later delineated using GPS. In some cases, as shown in the figure on the right, the areas of the circles around a record overlap. For laboratory analysis, hexagons of a similar area were used instead of circles, since this geometric shape allows the entire space to be covered without overlap (see Chap. 6)

Field registration was carried out using printed forms, digital photographs, field notebook notes, and GPS points, tracks, and areas (Fig. 5.5). At each point of the survey grid—that is, at the beginning and end of each transect section—a field form was filled out, where both environmental and archaeological data were recorded, and at least two photographs were taken. This record was made regardless of whether

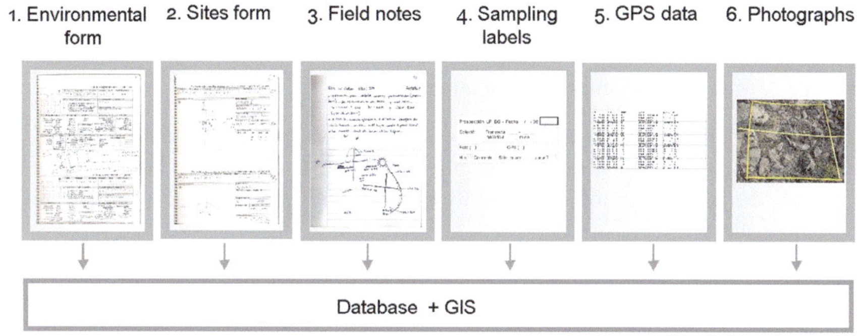

Fig. 5.5 Registration forms that were completed in the field and entered into the database (see the database design in the Supplementary Material)

archaeological evidence existed at that point on the land surface. Along each transect section, a GPS position was recorded, photos were taken, and a field form was filled out each time an environmental change occurred or a concentration or site was recorded. An example of environmental change, is the presence of resources considered important or a change in geomorphology was registered as a "ground truth" control point to improve existing cartography. With respect to the cultural evidence, It was not possible to do this with all the isolated findings because it exceeded the possibilities of recording in the field.

In many cases, the recorded lithic assemblages exceeded the areas covered by the transects. Their limits were defined by considering a buffer area of approximately 5 m without new findings. A section of the form intended for making schematic maps was used to record them. The procedure was to take a GPS reading at the approximate center of the artifact dispersion and record its approximate limits by plotting area polygons with the GPS. In some cases, this was impossible given the large extent and dispersed distribution of artifacts in large landscape units. For example on deposits covering pediment levels were registered using a different strategy: isolated findings and negative evidence were recorded using a GPS point and notes in the field notebook.

Sampling and collection of archaeological materials

Samples of materials, especially lithic artifacts, were collected for laboratory analysis. According to a conservative approach, the collected samples included the fewest possible artifacts representative of the total set recorded in the field. The size of the samples collected in each case was influenced by how much weight the operators could carry in their backpacks during a workday, during which they traveled about 15 km cross-country, as well as by various conditions. These conditions included climatic factors, sunlight conditions, and the remaining time in the workday to conduct a survey, particularly at extensive sites with a high density of artefacts.

Due to the variety of evidence and different circumstances, it was impossible to use a single criterion for sampling. In each case, however, one of the following techniques was applied and recorded on the forms.

(1) Isolated findings were collected and recorded using the form and labels.
(2) Concentrations and sites were recorded using the form, the labels, and, if necessary, the field notebook. The totality of small and sparse artifact concentrations was collected.
(3) At the sites, materials were collected along a north–south transect crossed by an east–west transect, with sampling at 1-m intervals. When sufficient time was available, materials visible on the surface were also collected in 1 by 1-m quadrats. When the artifact density was too high, approximately half of the visible artifacts were collected (operators collected one artifact for every two seen).
(4) In sites and structures with sufficient sediment, 1 by 1 m quadrats were delimited. Within these quadrats, subsurface sediment samples (the first 5 cm) were taken and sifted through a 4 mm mesh. These samples were taken at the intersection points of cross transects. These sieve samples record the presence or absence of small lithic artifacts indicative of re-knapping or retouching activities.

In sectors of the landscape where the archaeological evidence was so continuous that it made registration and collection intense and prevented progress of the work, as in several quarried terraces, another strategy was necessary. In these cases, all visible artifacts were collected within a 2-m-diameter unit located every 20 m along the transect section.

Documenting the sampling techniques is essential for calculating artifact densities (see Chap. 6).

5.4 Conclusion

This chapter details the methodological scheme followed and develops the first step within the scheme: obtaining the data for analysis. Since the quality and adequacy of the base data significantly impact the results of analyses and inferences, this chapter details the various aspects of data collection. The chosen methodology aims to detect patterns in the data to facilitate discussions about record formation processes and the context of human behavior. One of our goals is to apply it to other areas and places (Köhler, 1988).

Chapter 6
Analysis of Archaeological Evidence

Abstract This chapter processes the archaeological and environmental data collected during fieldwork and remote sensing. As a result, GIS coverages modeling the relevant variables for studying the settlement pattern in the La Primavera locality are obtained. Finally, the data are subjected to statistical analysis to determine the location of the different types of evidence.

Keywords Archaeological visibility · GIS data · Informed Surface data classification · Locational pattern detection · Site formation processes · Univariate statistical analysis

Throughout this chapter, the steps 2, 3, and 4.1, proposed in the methodological design, are developed (see Fig. 5.1). Archaeological data recorded in the field, as well as environmental data obtained through fieldwork and remote sensing imagery, are introduced in a Geographic Information System (GIS) environment. Statistical tests are then used to explore these datasets for patterns and singularities of the contexts. These patterns will enable us to begin analyzing how hunter-gatherers interacted with the landscape in the La Primavera locality.

6.1 The Archaeological Data

Since our focus is on spatial analysis, we use a spatial unit of analysis here. Each unit is georeferenced and related to archaeological and environmental data. Our goal is to identify patterns in the variables modeled in the GIS coverages that characterize the location of different types of archaeological evidence in the landscape. In these patterns, the dependent variables are archaeological and the independent variables

---

Supplementary Information The online version contains supplementary material available at https://doi.org/10.1007/978-3-032-08127-8_6.

L. A. Magnin, *The Landscape and Mobility of Hunter-Gatherer Societies*, The Latin American Studies Book Series, https://doi.org/10.1007/978-3-032-08127-8_6

are environmental. However, this does not mean that environmental variables determine or explain the detected patterns. Additionally, some cultural variables, such as distance to access roads and distance to the location of other types of archaeological evidence, are considered to test their influence on the location of different types of archaeological evidence.

6.1.1 Definition of the Analysis Unit: The Assemblage/Location

It is important to introduce discussions about the concept of sites, which are relevant to the archaeology of surface distributions that have been studied over the past few decades. The use of this concept is inconsistent in the literature, and it seems that there are as many definitions as there are archaeological investigations (Rossignol & Wandsnider, 1992). This work uses both on-site and off-site information. The relationship between the unit of analysis, the methodology of data collection, and the methodology of data interpretation used in this study is explained in three points.

(a) The archaeological data analyzed includes types of evidence beyond that concentrated in sites, in order to understand the diversity of activities developed in the landscape (Gradín & Aguerre, 1992). Even the absence of archaeological data is studied.
(b) Artifact density indicates the duration of occupation and reoccupation of places in mobile hunter-gatherer societies (Belardi, 1992, 2003; Foley, 1981; Thomas, 1975; Zvelebil et al., 1987).
(c) Studying the settlement pattern requires information throughout the study area. The technique used is not an intensive survey, but rather, sampling at regular intervals generates useful data for applying spatial techniques to achieve the study's objectives (Orton, 2000).

Since there are sectors of the landscape with high concentrations of materials that are conceptualized as sites, the concept of site is not ruled out here. A site is a high-density artifacts assemblage, but its meaning does not derive from a direct analogy in the systemic context (Dewar & Mc Bride, 1992). The same would be true for an approach that uses artifacts rather than sites as the unit of analysis. We argue that measuring artifact densities alone does not resolve the difficulties of interpreting them in terms of human behavior. Whether the site concept is used or discarded and replaced by artifact densities, natural formation processes must be controlled (Schiffer, 1983) and artifact variability must be analyzed for interpretation (Schofield, 1991). For instance, the density of artifacts resulting from activities such as occasionally knapping a nodule may resemble the density of lithic artifacts derived from programmed activities in a place where a group of people stayed for a longer time and discarded only a few artifacts (Binford, 1979; Kuhn, 1994). Therefore, in this work, density is not taken as a direct measure of the intensity of occupation in the above-defined unit of analysis (in Chap. 5). However, we agree with authors who take this measure as a useful indicator for comparing occupation intensities on a regional scale (Belardi, 2003; Borrero, 2001). In addition to artifact density, assemblage structure and attributes considered in lithic analysis are important for inferring the behaviors that produced them (Schofield, 1991). The diversity of artifact classes present can provide insight into the types of activities and how the artifacts were used. Cortex percentages and artifact sizes indicate whether the artifacts were generated during the extraction of raw materials or later in operational chains during manufacturing or maintenance activities. Additionally, fractures on finished artifacts may indicate the reasons for discarding the piece, such as during its manufacture or use (Andrefsky, 2005). Interpretation also analyzes environmental information to correct possible biases due to differential visibility (see below).

In conclusion, although we found the notions of site, concentration, and isolated finding useful for recording evidence during fieldwork (Borrero et al., 1992), we reclassified the data in the laboratory. During this process, we defined sets of high, medium, and low density based on the specific data of this case study. The limits of the different classes were established based on a mathematical criterion according to the type of distribution, adjusting the number of artifacts per unit area. The unit of analysis is a place (Rossignol & Wandsnider, 1992) or a plot of land with an artificially defined area and cultural content derived from the analysis of lithic assemblages and other archaeological features recorded within it. The continuous space under study is divided into units whose shape and size are consistent with the survey and registration methodology. These units of analysis, which are grouped into density classes, do not directly reflect the human behavior that generated them or the intensity of use of a place over time. We found that the best way to assign meaning to them is to consider multiple variables arranged in a model (see the model approach in Chap. 4). To highlight this concept, we refer to the unit of analysis as an assemblage/location and assign it a unique numerical identifier (ID).[1]

6.1.2 *Model Formalization*

The first step in classifying the assemblages/locations was transforming the theoretical model (Chap. 4) into a scheme explaining the necessary information. This scheme's main antecedents are found in Silvertsen (1980) and Miotti and Salemme (1999). Throughout this chapter, the processing of data is developed based on fieldwork and laboratory work to obtain the necessary information.

The indicators used to classify each set are presented and provided in Table 6.1. The table shows how the various activities carried out in the past translate into archaeological expectations. As stated in Chap. 4, the relationship between the indicators and the inferred activities is a middle-range theory derived from ethnography and general models applicable to hunter-gatherer societies. This theory is also

[1] Throughout the text, in figures and tables, these IDs are used to refer to particular assemblages/ locations, since some of them have names (generally the most important sites), but others do not.

based on previous archaeological studies and the analysis of possible sampling and preservation biases.

Therefore, the assemblage/location's function is an inference about possible past activities in a given place in the landscape. The archaeological indicators used are: (1) artifact density, considering the visibility conditions given by surface dynamics at a particular landscape location; (2) the proportion of lithic artifacts at different stages of the operational chain; and (3) the presence or absence of particular types of lithic evidence, rock art, or stone structures. The resulting classes identify the sectors used most frequently and refer to the activity(ies) that could have been carried out in the analyzed locations.

Although Microscopic functional analysis was attempted it did not yield results because of a difficulty recognizing use traces in artifacts that have undergone significant post-depositional processes (Mansur, 1999), a problem that is exacerbated in surface artifacts. Dr. Marcela Leipus and Dr. Virginia Lynch carried out test observations using a Nikon Epiphot 200 microscope with eyepieces that allow magnification from 500× to 50×, and a Nikon SMZ 800 magnifying glass with a magnitude of 63× to 10×, on some artifacts from the collection (a red siliceous rock core, a formatted obsidian artifact, some scrapers, and unformatted siliceous rock artifacts). However, the abrasion beaches present on the edges observed under the microscope, due to the action of weathering and the sandy sediments in which they were found, prevented finding use traces. For this reason, this methodology of analysis is not applicable here. Thus, this study derives the possible uses of the lithic artifacts from macroscopic and descriptive analyses (Aschero, 1975, 1983). It is assumed that primary functional inferences of lithic artifacts can be made based on certain edges, points, and surfaces, which provide information about the type and mode of action of an instrument (Aschero, 1975). Although functional studies using microscopy have demonstrated that morphology cannot be unequivocally related to functionality (Alvarez, 2004), the proposed analysis is one possible approach that can be applied to surface assemblies. Therefore, we propose characterizing each assemblage/location by the diversity of activities represented, with the most specific definitions expressed only in the theoretical model. Assemblages with greater artifactual diversity represent loci of multiple activities. They are designated based on the density and variability of the recorded evidence, taking into account the effects of natural and cultural formation processes that could affect the archaeological record. One of the most evident natural processes in this case study is sedimentary dynamics. On terrain surfaces where erosion predominates and vegetation is scarce, the visibility of superficial and subsurface archaeological materials is high, resulting in palimpsests of lower resolution and integrity. The opposite occurs on surfaces where sediment deposition predominates. There, high burial and greater vegetation cover are expected, resulting in decreased archaeological visibility (Chap. 2). Certainly, many other formation processes are involved. As will be seen later, not all observation cases present the same information in this regard. The model allows these data to be integrated and weighted.

Table 6.1 Functional classification scheme of archaeological evidence columns 1 through 12 list the inferred activities

	Inferred activities	1	2	3	4	5	6	7	8	9	10	11	12
		NE	A	Q	OK	PP	HG	HS	Ch	MAL	TC	W	V
	Archaeological indicators												
1	Isolated artifacts/in high visibility sector	x	-	x	p	x	p	-	-	x	x	x	-
2	Isolated artifacts/in low visibility sector	-	-	x	-	-	-	p	_	x	x	x	-
3	Intermediate artifact density/ high visibility	x	-	-	-	p	-	p	-	x	p	x	-
4	Intermediate artifact density/low visibility	-	-	p	x	x	x	-	-	p	p	p	-
5	High artifact density	x	-	p	x	x	x	-	-	p	x	p	x
6	High diversity (H index)	x	-	x	x	x	-	x	-	p	p	x	x
7	Low diversity (H index)	x	-	p	p	p	p	-	-	x	x	p	p
8	Quarried rock outcrops	x	-	p	p	-	-	-	-	-	-	-	-
9	Tested nodule	-	-	p	p	-	-	-	-	-	-	-	-
10	Core	x	-	p	p	-	-	-	-	p	p	p	-
11	Striker	x	-	p	p	-	-	-	-	p	p	p	-
12	First stages of the reduction sequence (1)	x	-	p	p	x	x	-	-	-	-	-	-
13	Last stages of the reduction sequence (2)	x	-	x	-	x	x	p	-	p	p	p	x
14	Non-formatted artifacts	x	-	-	-	p	-	-	-	p	p	-	-
15	Formatted artifacts with long edges (side scrapers, knives, *raclettes*)	x	-	x	x	p	x	-	-	p	p	-	-
16	Projectile point (complete or with impact fracture)	x	-	-	x	-	p	p	-	-	-	-	x
17	Projectile point base	x	-	-	x	p	-	p	-	-	-	-	x

(continued)

Table 6.1 (continued)

	Inferred activities	1	2	3	4	5	6	7	8	9	10	11	12
		NE	A	Q	OK	PP	HG	HS	Ch	MAL	TC	W	V
18	Projectile point tip	x	-	x	x	-	-	-	-	p	p	-	x
19	Stone bola with impact fracture	x	-	-	x	-	p	p	-	p	p	-	x
20	Bola preform	x	-	x	x	x	x	p	-	-	-	p	-
21	Projectile point with manufacture fracture	x	-	x	x	-	x	p	-	p	p	p	x
22	Big scraper, wedges, big side scrapers	x	-	x	x	x	x	x	x	p	p	-	p
23	Bifaces	x	-	-	x	p	x	-	-	p	p	p	p
24	Scrapers, drillers, awls, notches	x	-	x	x	x	x	x	x	p	p	-	-
25	Grinding stone artifacts	x	-	x	x	x	x	x	x	p	p	x	x
26	Combustion features, boiling stones	x	-	-	x	-	x	x	-	p	p	-	-
27	Choppers	x	-	-	p	p	x	-	-	p	p	-	p
28	Cairn, circular/oval stone structure (chenque)	x	-	-	-	-	-	-	p	p	p	-	-
29	Stone bases for skin windbreaks	x	-	-	x	-	-	-	-	p	p	-	-
30	Semicircular or straight stone structure	x	-	-	-	-	-	p	-	-	-	-	-
31	Small stone structure	x	-	-	x	-	-	-	-	p	p	-	-
32	Rock art	-	p	-	-	-	-	-	-	-	-	-	-

The abbreviations are as follows: NE (negative evidence); A (rock art site, or place with rock art); Q (quarries for extracting lithic raw materials); OK (occasional knapping place); PP (hunt primary processing site); HG (hunting ground); HS (hunting stand); Ch (chenque, or possible funerary cairn); MAL (multiple activities locus, or residential base camp); TC (transient camp); W (lithic workshop); and V (site of extraction of vegetal resources). The archaeological indicators for these activities are detailed in rows 1 through 32. The essential indicators for defining an activity are symbolized by the letter p. If an indicator must necessarily be absent, it is symbolized by an x. A hyphen (-) means the flag is irrelevant to defining the class

6.1.2.1 Calculation of Artifact Densities

An initial density estimate was made in the field as defined by approximately circular areas with a radius of 10 m. This method allowed for the recording of isolated archaeological finds, concentrations, and sites within those circular units (Borrero et al., 1992). The locations of the isolated finds were registered with a GPS waypoint. For the concentrations and sites, their boundaries were outlined by tracing their area with a GPS. In the laboratory, the spatial relationships between the findings were calculated more precisely by entering the GPS data into GIS and covering the data with a mesh of hexagonal polygons of a similar surface area. The hexagonal polygons were chosen because they were similar to the observation units used in the field and allowed areas to be covered without overlap. Figure 6.1 presents two schemes showing how a concentration (case 1) and a site (case 2) are delineated. Case 1 shows that the field registry of an isolated find (gray dot) and a concentration (black dot), after superimposing a hexagonal grid in GIS, were confined to the same unit. Therefore, they were analyzed as part of the same concentration. Case 2 shows how GPS was used to define an area that included all the artifacts that were visible on the surface and spatially associated (hatched polygon). In the field, an area with no finds within a 5 m buffer defines the limit of a site. In the laboratory, a series of contiguous hexagons was arranged until the entire GPS-delimited area was covered. These hexagonal units are the units of analysis used to calculate the artifact densities. The materials collected within each hexagon belong to the same assemblage. As will be seen later, density is one of the functional classification variables. Materials collected within adjacent hexagons were analyzed as part of the same assemblage/location. The calculated densities are presented in Supplementary Material, Table 1.

A bar diagram shows the distribution of artifactual density values. As Fig. 6.2 shows, the distribution is highly skewed to the right. The distribution's non-normality could be verified using a Shapiro–Wilk W test at a 0.05 confidence level. This estimate of the approximate dimension of artifact dispersion in each plot is the first step

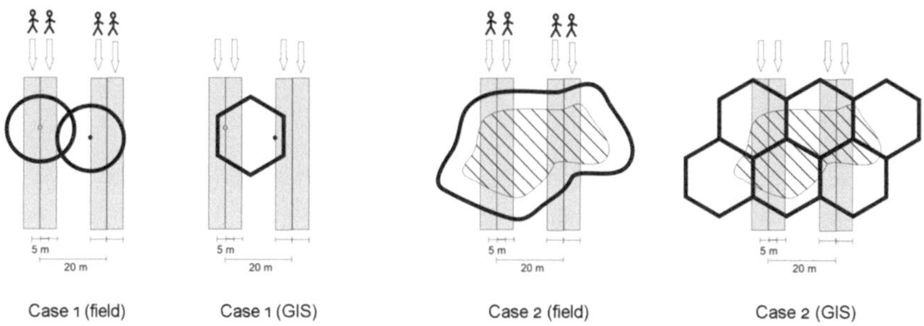

Fig. 6.1 Figure illustrates the two methods of registering archaeological evidence during fieldwork: GPS points (Case 1) and polygons (Case 2). In GIS, the data are superimposed on hexagonal area units to calculate artifact densities. These hexagonal areas are similar to the 10-radius circular areas used in fieldwork (344 m^2)

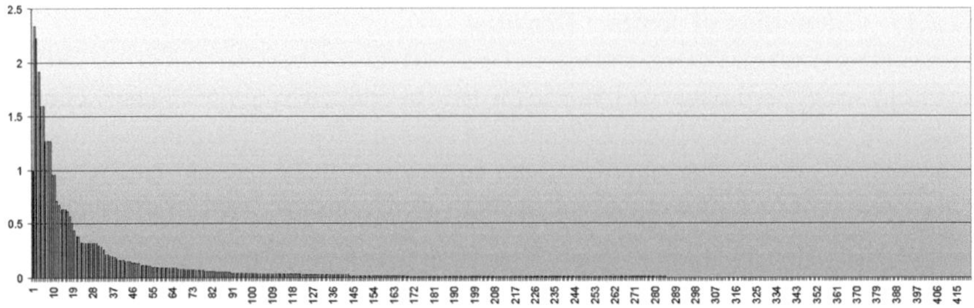

Fig. 6.2 Bar graph shows the distribution of artifact density values calculated for each assemblage/location. The data show strong positive asymmetry

in analyzing occupation patterns. The non-normality of the distribution indicates significant differences in the densities of the analyzed sets.

Following Borrero et al.'s (1992) proposal, the density values were separated into three categories. However, the intervals were not established in advance but rather defined for this case study. The data were reclassified using intervals that characterize the existing variability. A univariate classification was performed to generate three intervals in the density data for this purpose. Geometric intervals were chosen as the most appropriate method for the type of distribution obtained (Conolly & Lake, 2006).

The upper limits of each class were defined using the following formula: (Eq. 6.1):

$$\frac{L \times X_1, L \times X_2, \ldots L\times}{X_n} \tag{6.1}$$

The multiplier was defined by the formula (Eq. 6.2):

$$X_n = H/L \tag{6.2}$$

where $n =$ number of classes, $H =$ the value of the highest interval, $L =$ the lowest, $X_3 = 2.331131078/0.002906977$, $X = 3\sqrt{2.331131078} = 9.29055615$.

Thus, the distribution was divided into three classes to categorize the 419 artifact assemblages according to their density level: high, medium, or low (see Table 6.2 and Fig. 6.3).

The maps in Figs. 6.4, 6.5, and 6.6 show the spatial distribution of the assemblages with high, medium, and low density values, respectively.

The graphs show that the densities of the assemblages in class 1 are more variable and that this variability decreases in classes 2 and 3. This initial approach enabled us to distinguish and order sets of different densities into classes. However, this distinction alone does not account for the formation processes that affect artifact density in various locations. The following section provides an assessment of each set based on indicators that identify some of the natural factors affecting the obtained density values.

Table 6.2 Intervals of the classes defined for the variable density

Density class intervals	Upper limit of the interval (*N* artifacts)	Upper limit of the interval (density)	Distribution percentage	Intervals (artifacts/unit area)	*N* of assemblages in each class
1-High	801.9090	2.331106833	1	92–802	32
2-Medium	86.3144	0.216394057	10	9–92	99
3-Low	9.29	0.023255814	89	0–9	288
	Total		100		419

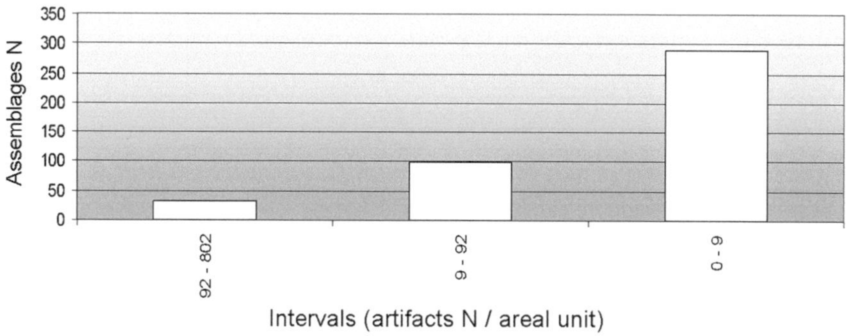

Fig. 6.3 Frequency distribution of assemblages within defined artifact density intervals

6.1.2.2 Archaeological Visibility Analysis

The assemblages defined in the previous section can be better understood in terms of human group occupation redundancy when considered alongside an assessment of the general visibility conditions at each location. As established in Chap. 2, null-to-low sediment burial potential conditions predominate in the study area, and 35% vegetation coverage results in very good to excellent general visibility conditions. However, these conditions vary at larger scales. For this reason, this section proposes a method for establishing particular visibility values for each archaeological assemblage/location. The following section proposes a method for evaluating archaeological assemblages that considers multiple factors. We included three types of data indicative of archaeological visibility conditions: environmental data collected in the field (see Supplementary Material, Table 2), a geomorphological map generated from aerial photographs, and a lithic taphonomy analysis of the collected materials. These indicators allow us to assign a value between "4" and "– 4" to each assemblage/location. This value is obtained through the algebraic sum of "+ 1" for each indicator of processes that increase archaeological visibility and "– 1" for each indicator of processes that reduce it. This value provides an estimate of the archaeological visibility at the discovery location of each assemblage.

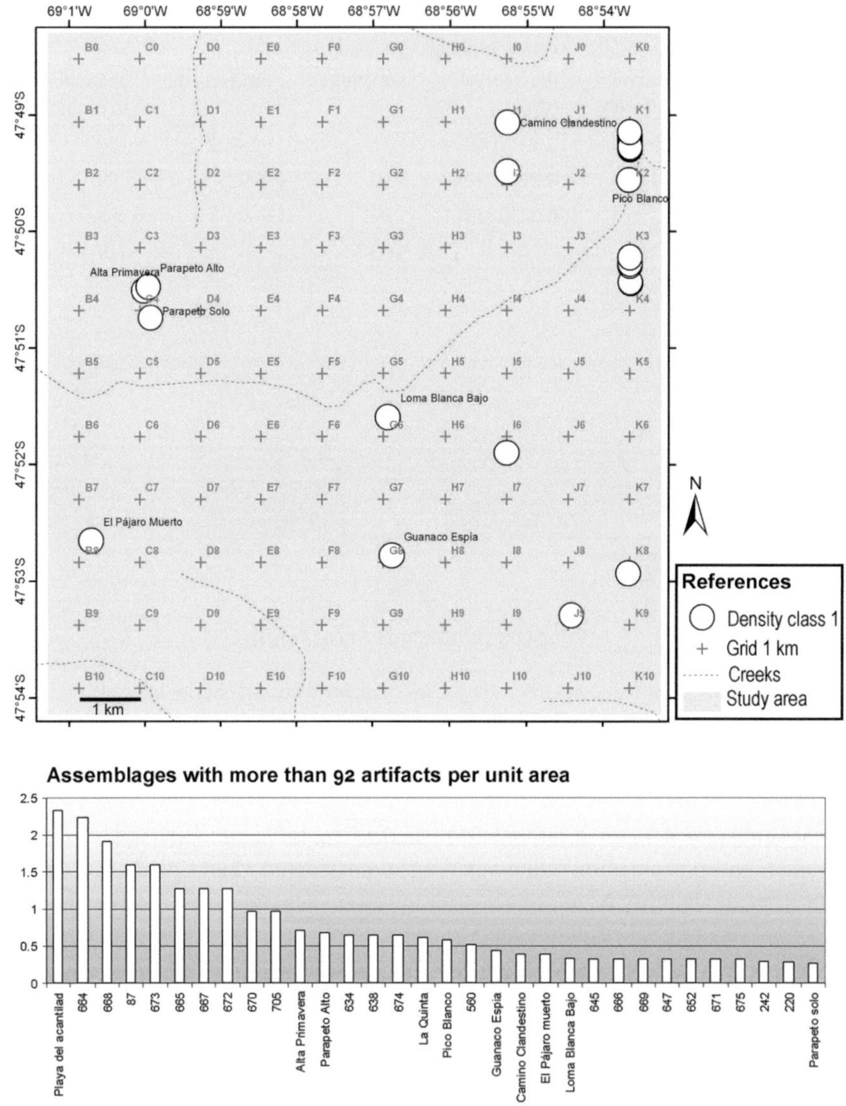

Fig. 6.4 Distribution map of high-density assemblages and bar graph showing the frequency distribution of densities (unit area 344 m^2) for all assemblages within class 1. This data constitute 1% of the analyzed assemblages. The *x*-axis of the graph presents the name or identifying number of each site

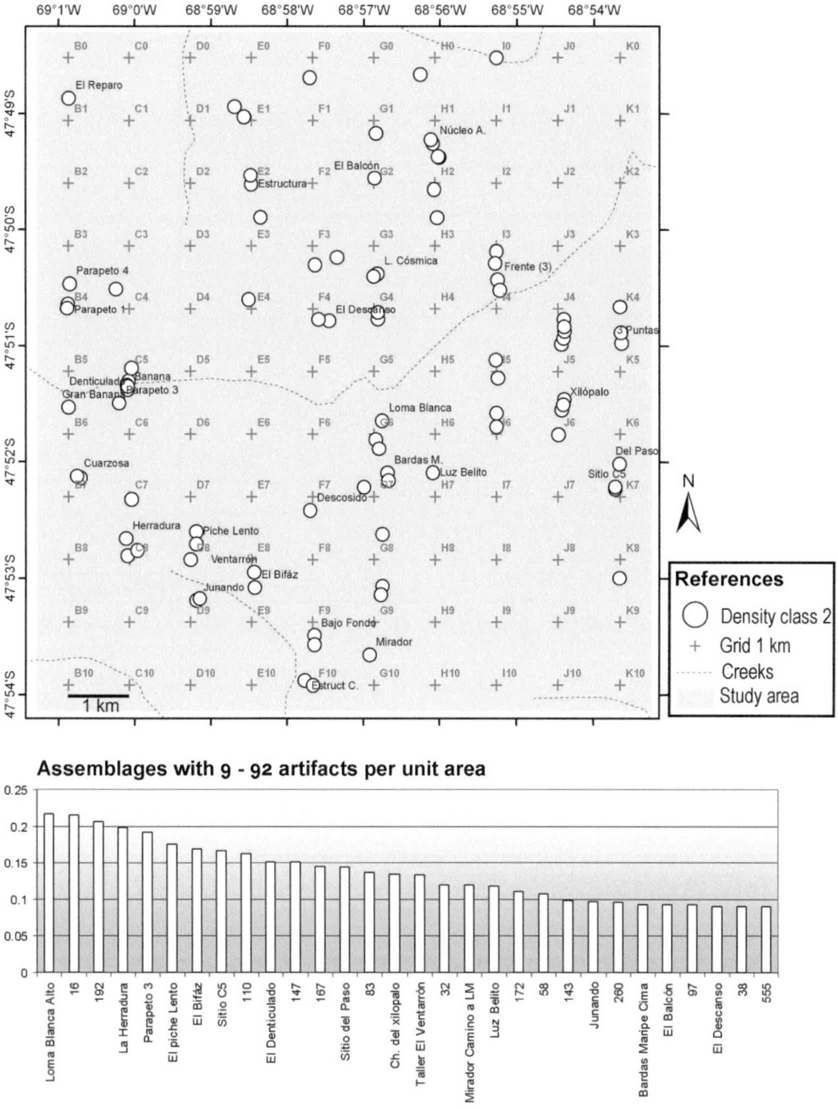

Fig. 6.5 Distribution map and bar graph for the first 30 assemblages within class 2 assemblages. The base grid is 1 km. These data constitute 10% of the analyzed assemblages. The *x*-axis of the graph presents the name or identifying number of each site

Characterizing archaeological visibility based on field data. Taphonomy at the scale of the observation unit

A total of 561 environmental forms were completed during fieldwork. These forms were filled out at the beginning and end of each 100-m transect section, at every

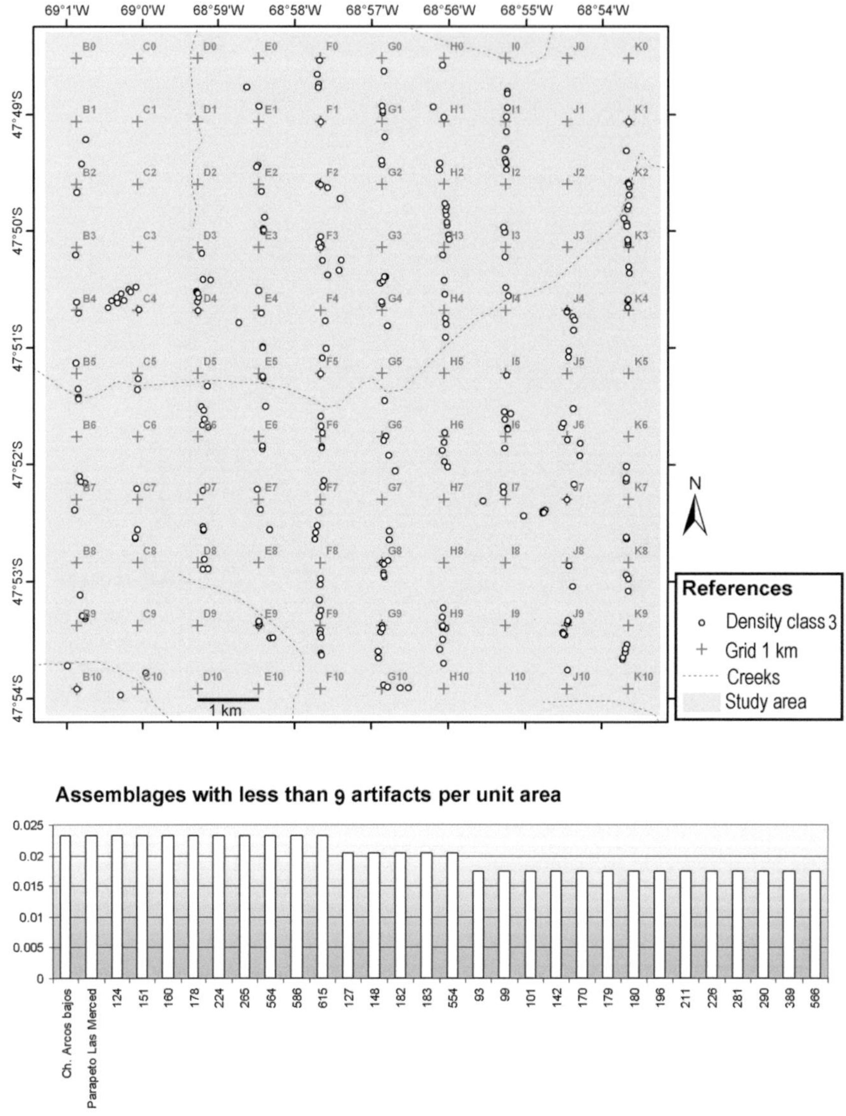

Fig. 6.6 Distribution map and bar graph for the first 30 assemblages within class 3. The base grid is 1 km. This data constitutes 89% of the analyzed sets. The *x*-axis of the graph presents the name or identifying number of each site

archaeological find, and at locations between them (see Chap. 5). Nine of the variables included in the form provide information on archaeological visibility. These include the type of geoform, the presence of rocky outcrops, ditches, minor gullies, meadows, trails, wallows, or animal pits; the percentage of vegetation cover; and the presence of buried materials (see Table 6.3).

Table 6.3 Indicators of archaeological visibility inferred the predominant process and value of archaeological visibility

	Observations	Predominant process	Visibility
Geomorphology	Plateau edge	Erosion	1
	Plateau top	Erosion	1
	Hill top	Erosion	1
	Pass	Erosion	1
	Low hill top	Erosion	1
	Pediment	Erosion	1
	Rocky outcrop	Erosion	1
	Temporary stream	Erosion and transport	1
	Gully	Erosion	1
	Canyon slope	Transport	– 1
	Hillside	Transport	– 1
	Low hillside	Transport	– 1
	Valley slope	Transport	– 1
	Canyon bottom	Deposit	– 1
	Valley bottom	Deposit	– 1
	Cave	Deposit	– 1
	Rock shelter	Deposit	– 1
	Wet meadow/vegas	Deposit	– 1
	Dune	Differential	0
	Closed depression	Differential	0
	Open depression	Differential	0
	High pampa	Differential	0
	Alluvial cone	Differential	0
Fauna	Animal trails	Erosion	1
	Animal wallow	Erosion	1
	Animal burrow	Erosion	1
Plant cover	0–25%	Erosion	1
	26–50%	Erosion	1
	50–75%	Deposit	– 1
	76–100%	Deposit	– 1
Artifacts	Wind abrasion	Erosion	1
	Lichens	Stability	1

A single value is obtained from all the existing information for each assemblage/location. This value is positive in the case of high visibility and negative in the opposite case

Characterization of archaeological visibility using a geomorphological map. Taphonomy at the scale of the landscape

To contextualize the location of the analyzed assemblages, detailed thematic cartography of the studied sector was generated, including hydrography and geomorphology. IGN vertical aerial photographs from a 1968 flight were used as basic data, covering the locality with overlap between them and the vector coverage of tributaries from the 1:250,000 scale SIG-IGN topographic map. Although the photographs were taken 40 years before the fieldwork, they are the only ones available for this area. The photographs have a spatial resolution of 2 m (1:50,000), which is sufficient for this study. They were orthorectified using a Landsat TM image as base data to correct their geometry. The different geomorphic units were defined based on the interpretation of textural features from the aerial photographs and supported by satellite image data. This was then completed with data taken in the field using non-differential GPS. The result is the map in Fig. 6.7, which has an approximate planimetric error of 35 m (Gómez & Magnin, 2008).

The first step in defining the units was interpreting the drainage network. There are ephemeral channels that carry water after scarce and evenly distributed rainfall throughout the year (see the climograph for Gobernador Gregores in Coronato et al.,

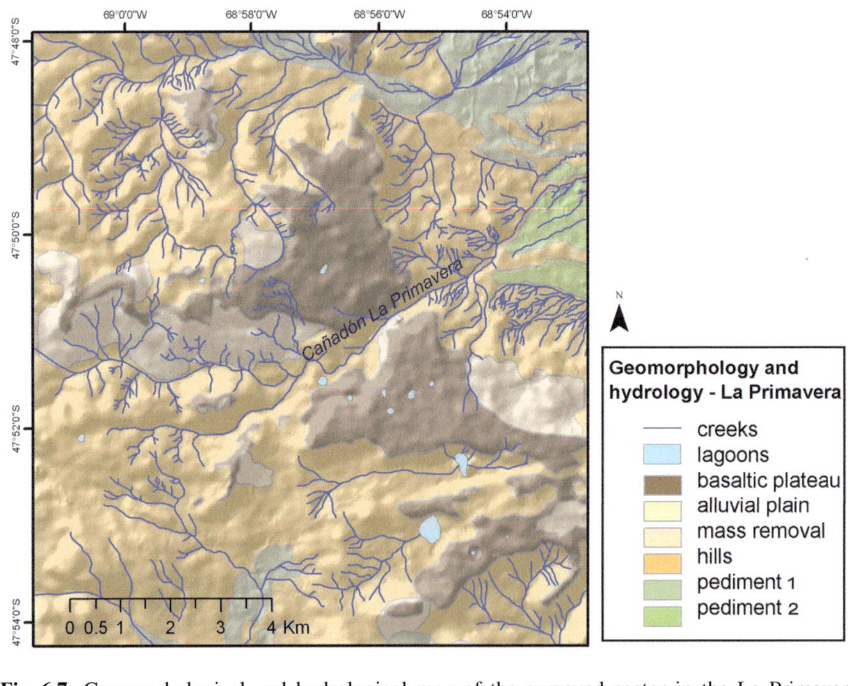

Fig. 6.7 Geomorphological and hydrological map of the surveyed sector in the La Primavera locality. The map is overlaid with 50% transparency on a digital terrain model (ASTER GDEM). The lines in the upper right corner of the map are GDEM errors (mole runs artifacts). Its effect on the processing carried out in this work is not significant

2008). Starting from the vector coverage of the tributaries on the 1:250,000 scale GIS-IGN topographic map, the network was completed through visual interpretation using aerial photographs and images. The outlined tributaries constitute the sources of the Zanjón Blanco and exhibit dendritic designs. The most conspicuous gully is the La Primavera Canyon, which runs in a direction different from the regional slope due to marked structural control, as it coincides with an SW-NE fault.

The identified geomorphic units are as follows: 1-Hills of rocky outcrops (a Jurassic volcanic sedimentary complex consisting of tuffs, ignimbrites, rhyolites, sandstones, and conglomerates); 2-Basaltic volcanic plateaus; 3-Areas of mass removal; 4-Aggradational plains (first and second levels, or terraces 1 and 2); 5-Shallow lake sediments; and 6-Alluvial plains.

These units are represented on Fig. 6.7. Table 6.4 analyzes the general geomorphological processes acting on each unit and the preservation and visibility expectations of the associated archaeological materials (Allen, 1991; Butzer, 1982).

Hills

The hills, the oldest geoforms in this landscape, extend over most of the study area. According to the geological chart, they correspond to rocky outcrops of the volcanic sedimentary type belonging to formations from the Mesozoic, Jurassic, and Cretaceous periods: The Chon Aike, Bajo Grande, and Baqueró formations (Panza, 2001).

1

Fig. 6.8 Geoforms recognized on the map 1) hills, 2) volcanic plateaus, 3) areas of mass removal, 4) aggradational plains 1 and 2, 5) shallows and shallow lakes, 6) gully in the current alluvial plain

2

3

Fig. 6.8 (continued)

4

5

Fig. 6.8 (continued)

6

7

Fig. 6.8 (continued)

Table 6.4 Geoforms, natural processes, and implications for the archaeological record

Geoform	Predominant geomorphic processes	Implications for the archaeological record
Hills	Fluvial and wind erosion. Movement by gravity on slopes. Sediment accumulation inside caves	Fluvial and wind erosion. Movement by gravity on slopes. Sediment accumulation inside caves. Depending on the degree of inclination of the terrain, archaeological materials are exposed, buried at the foot of the main elevations, or moved down the slope. Inside caves, materials are trapped in low-energy environments, though they may still move due to gravity
Volcanic plateau	Wind and fluvial erosion	The archaeological materials are exposed to weathering agents. Good visibility
Mass-removal areas	Accumulation: landslides and rockslides	High-energy environment. Archaeological materials buried or displaced downslope
Aggradational plains 1 and 2	Accumulation on Plain 1 (plains and alluvial cones) Fluvial and wind erosion on plain 2 (deposits covering levels of pediment)	Archaeological materials are found: (1) Buried in the sectors of alluvial plains and cones, though lighter materials may be exposed on the surface by digging animals, and (2) overlapping and mixing on erosion surfaces
Shallows and shallow lakes	Accumulation of sediment at the bottom of the shallow lakes due to hydro-wind action; deflation of the bottom and edges and accumulation by the wind from its eastern end to the east	The bottom and edges of the shallow lake contain highly fragmented, buried, or semiburied materials. Seasonal erosion and deposition processes act on these materials, favoring the formation of palimpsests. At the eastern end of the shallow lake, aeolian sediments accumulate, favoring the burial of materials when vegetation grows and acts as a fixer for these sediments
Floodplain	Erosion of the bottom of the channel. Accumulation of drift material in the floodplain and deposits	Archaeological materials associated with river channels are displaced and redeposited among the river sediments and at the bottom of the channels along the tributaries. Burrowing animals activity in loose sediments during dry seasons

The hills were formed primarily through fluvial action that intersected and undermined the rock units. As discussed in Chap. 2, mapping these hills in detail is important because they are potential sources of knapping raw materials and the primary context for cave and rock shelter formation. These sites were chosen for human habitation during the hunter-gatherer period and in later periods of occupation. For example, they were used as livestock ranch posts until the mid-twentieth century and as the Cueva Maripe site (Miotti et al., 2007). They are still used as shelters today.

According to the general model presented in Table 6.4, the predominant geomorphic processes in this area are fluvial erosion and wind erosion. However, gravity is another important factor in this environment, producing the downward movement of the heaviest and densest archaeological materials and burying them at the foot of the mountain ranges (Rick, 1976). Other factors, such as the action of large-to medium-sized animals (e.g., guanacos, rheas, sheep, and currently, a few horses), could have affected the distribution of evidence on slopes with an incline of 23–25° or greater (Allen, 1991; Butzer, 1982). As was observed in other regions, such as southeast England, the constant movement of livestock through the landscape causes the largest and roundest artifacts to displace when kicked by the animals. The smaller artifacts are less likely to be displaced.

Volcanic plateaus

The volcanic plateau unit was formed by effusions of the Las Mercedes basalt from the Upper Cretaceous and the Alma Gaucha basalt from the Tertiary. This unit originated as a lava flow that poured into the lowest areas of the Upper Cretaceous and Tertiary landscape. Later exogenous geomorphic processes, due to relief inversion, transformed these rocks into the current plateau reliefs. This same unit was dissected by fluvial processes responsible for the formation of the La Primavera ravine. Its perimeter suffered the action of exogenous processes that, due to the effect of gravity, generated mass-removal areas. In volcanic plateaus, wind and river erosion predominates, and there is a poor vegetation cover, so the archaeological materials present good visibility.

Mass-removal areas

The mass-removal unit forms through the sliding and falling of debris or rocks that have broken off from the unstable edges of basaltic plateaus due to gravity. In this high-energy environment, heavier archaeological materials are buried or displaced from their original locations. Compared to hill slopes, the rougher terrain will likely reduce the movement of materials downslope.

Aggradational plains

The aggradation plains are located east of the main elevations and run parallel to the regional slope. These plains contain diverse deposits, primarily Quaternary, resulting from the erosion of the slope's edges. They are composed of alluvial-colluvial deposits of silt, sand, and gravel. The two identified levels are associated with changes in the region's base level. The first level of the aggradation plains (Plain 1) is younger than the second level (Plain 2). These chronological differences are not significant in terms of the temporal extent of human occupation of the area because they formed before human groups arrived. Both plains are equally attractive sources of lithic raw materials. Their degrees of evolution determine different expectations of visibility and preservation. Plateau 2 probably has areas of greater visibility because it has a gentle but steeper slope that produces blanket runoff of rainwater on the surface. This process, along with deflation, removes fine sediments and accumulates larger rocks and artifacts (not by transfer, but by constant disappearance of the matrix). For this reason, the archaeological assemblages in Plain 2 have a higher probability of overlapping materials with different chronologies. In contrast, in Plain 1, the fan and alluvial plain deposits are characterized by the burial, transport, and redeposition of archaeological materials by a fluvial agent of variable energy (Butzer, 1982).

Floodplains

There are variable visibility conditions in alluvial plains. Some areas lack archaeological remains, which may be buried under fluvial-alluvial sediments associated with recent channels. Given that these are high-energy environments when the channels have water (see Table 6.4), any evidence found in this unit has a high likelihood of having moved from its original location.

Shallow lake sediments

The temporary shallow lakes found at the bottom of endorheic basins, hills, and basalt flows are small. The largest one, located southeast of the study area, measures 400 by 500 m. Their distinct coastlines are evidence of climatic variations that determine the seasonal presence of water and variations in its level. The sediments associated with these shallow lakes are sand and silt. During periods of greater humidity, the accumulation of fine sediments, such as sand and clay, due to streams and wind input favors the burial of archaeological materials in this environment. Furthermore, the edges of these bodies of water are subject to trampling by animals that come to drink, which facilitates the burial of materials when the ground is soft (Borrero et al., 1991; Espinosa, 2000).

On the other hand, in conditions of extreme aridity, artifacts tend to adhere to the substrate. They are often found in vertical positions alongside desiccation cracks, which increases the likelihood of fragmentation of superficial to subsurface lithic artifacts. Therefore, materials in these situations are highly fragmented and form palimpsests. During droughts, sediments are transported by wind, a process that is accentuated by other factors, such as overgrazing (Oliva et al., 1995). Fine- to medium-grain sediments are removed by the wind and deposited in aeolian plumes stretching from the shallow lake to the east due to the predominant west-to-east wind direction (Mazzoni et al., 1999, 2000). Archaeological materials deposited east of the water bodies are likely covered by these sediments. The opposite can be expected west of the shallow lakes, where deflation will predominate. These conditions vary and depend on vegetation, which acts as a sediment fixer.

Characterization of archaeological visibility based on observations of lithic artifacts that were collected

A third line of analysis comes from lithic taphonomy. Natural post-depositional processes can modify the surfaces of siliceous artifacts. These modifications constitute a record of the processes that the lithic assemblages underwent (Borrazzo, 2004; Burroni et al., 2002; Cassiodoro, 2011; Espinosa, 2000). Among these modifications, those indicating relatively long exposure times of the lithic pieces on the surface were considered (Favier Dubois, 1997). These include wind-abraded surfaces and lichen growth. The growth of lichens and the accentuated modification of artifacts' surfaces by wind indicate relatively long exposure times on land. However, the time it takes for each to form varies depending on local conditions. Therefore, experiments are needed to estimate these times. In the future, it will be necessary to extend this line of analysis to study the formation time of surfaces abraded by wind in this environment. The current research indicates that the presence of lichens and wind-abraded surfaces on artifacts is an indicator of their prolonged exposure to subaerial conditions. Therefore, stability in the surface conditions at the location of their discovery is inferred, indicating good archaeological visibility at that point. The assemblages were assigned visibility values using the indicators in Table 6.3, which are shown in Supplementary Material, Table 2. The results are summarized in Table 6.5. Positive values, which indicate greater visibility, predominate, comprising 61.26% of the sets. Values of medium and low visibility, on the other hand, are in the minority, accounting for 19.77% of sets with a value of zero and 18.95% of negative values. These results confirm the expected trends from the preliminary analysis in Chap. 2. At the same time, they allow us to confirm the validity of this procedure as a tool for estimating and comparing taphonomic processes affecting archaeological visibility among surface assemblages at the scale of the study area. However, this is not an exact measurement technique. Confirmation of these estimates will be made as new studies advance and greater geoarchaeological controls are implemented. Currently, this is a proposal to address the problem of differential visibility of archaeological evidence at the spatial scale under study. This is a crucial issue in surface archaeology (Allen, 1991).

Table 6.5 Distribution of archaeological visibility values in the analyzed locations

	Archaeological visibility value	Frequency	Percent frequency (%)	
Negative values (Low archaeological visibility)	– 4	1	0.16	18.95%
	– 3	9	1.47	
	– 2	33	5.39	
	– 1	73	11.93	
Values = 0 (Medium archaeological visibility)	0	121	19.77	19.77%
Positive values (High archaeological visibility)	1	266	43.46	61.26%
	2	66	10.78	
	3	36	5.88	
	4	7	1.14	
	Total	612	100	

6.1.2.3 Analysis of Lithic Assemblages

The lithic materials were analyzed based on Aschero's (1975, 1983) macroscopic and descriptive morphological proposal. The database design and variables are described in Supplementary Material-Chap. 5. All artifacts were assigned to artifact classes. The most numerous class was the debitage class, followed by non-formatted artifacts (NFA), formatted artifacts (FA), and finally, hammerstones (Fig. 6.9). Within the FA class, the largest group is scrapers, followed by artifacts with summary formatting, projectile points, and scrapers. Undifferentiated FA fragments occupy fifth place (Fig. 6.10). The graph in Fig. 6.11 and Table 6.6 show similar patterns for the distribution of different artifact classes in the assemblages. The most common classes, debris, NFA, scrapers, cores, and artifacts with summary formatting, are homogeneously distributed between the assemblages.

The required techno-morphological indicators were extracted from the total information obtained in the lithic analysis according to the model in Table 6.1. The

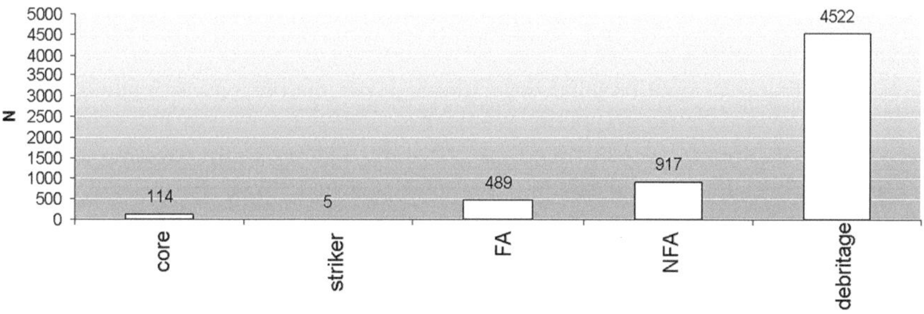

Fig. 6.9 Distribution of artifact classes in the total sample

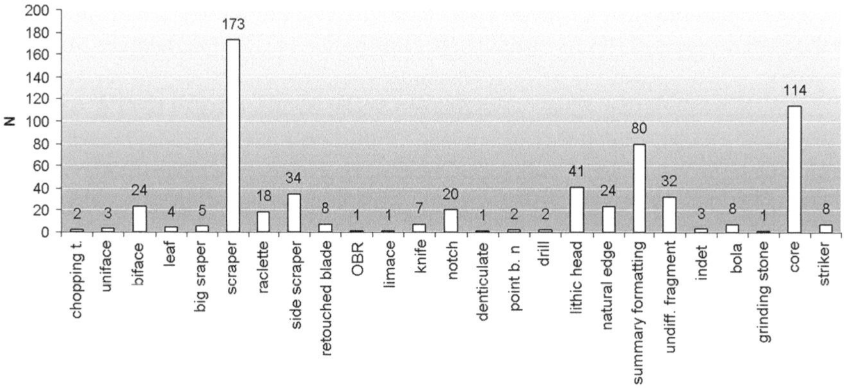

Fig. 6.10 Distribution of typological groups among the FA, cores, and strikers present in the total sample. The graph includes cores, strikers, and FA but excludes NFA ($n = 4522$) and debitage ($n = 917$)

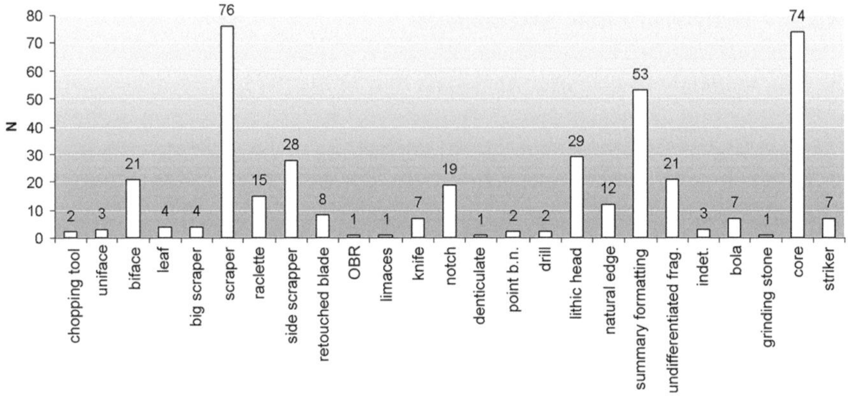

Fig. 6.11 Frequency distribution of assemblages that present different classes of artifacts. The graph excludes sets with NFA ($n = 221$) and debitage ($n = 301$)

observed frequency in each assemblage/location is presented in Supplementary Material, Tables 3 to 16.

Thirty-nine cores with isolated flake blanks (type A core, sensu Aschero, 1975, 1983) were recorded (Supplementary Material, Table 3); 75 cores of other types were recorded (Supplementary Material, Table 4); and eight strikers were recorded in seven different assemblages (Supplementary Material, Table 5).

As indicators of the initial stages of the operational chains, the following were recorded: NFA and debitage of sizes 3 and 4 (artifact sizes are defined as in Bagolini, 1968), with over 50% and 100% cortex; and all artifacts of sizes 5 and 6. Since values may vary in sets with different degrees of fragmentation, these data are presented as proportions of the total number of artifacts for comparison purposes. As indicators of the final stages of the lithic manufacturing chain, NFA without cortex and sizes 1 and 2 were included. As in the previous case, the data are presented as proportions of the total artifacts (236 sets have data for this indicator) (Supplementary Material, Table 6).

Artifacts with cutting edges were defined based on Aschero's (1975:6, Table 4) proposal of primary functional inferences. Useful artifacts for transverse-diagonal cutting were included: knives and other blades, such as side scrapers, raclettes, and retouched blades. NFAs are artifacts with edges longer than 3 cm and a shape that allows for gripping (Supplementary Material, Table 7).

A total of 173 scrapers (Supplementary Material, Table 8), 16 complete projectile points (Supplementary Material, Table 9), 13 projectile point bases (Supplementary Material, Table 10), and eight projectile point distal fragments (Supplementary Material, Table 11) were recorded. Among the weapons manufacturing indicators, four projectile point preforms and five projectile points with fractures likely resulting from errors during manufacturing were identified. Preliminary studies suggest two cases of perverse fractures and two cases of end-shock fractures (Crabtree, 1972;

Table 6.6 Hierarchy of artifact frequencies within each class

Classes	Frequency of assemblages that present each class	Artifact frequency within classes
Debitage	301	4522
NFA	221	917
Scraper	76	173
Core	74	114
Artifacts with summary formatting	53	80
Projectile point	29	41
Side scraper	28	34
Biface	21	32
Undifferentiated FA fragment	21	24
Notch	19	24
Raclette	15	20
Natural edge with complementary traces	12	18
Retouched blade	8	8
Knife	7	8
Bola	7	8
Striker	7	7
Leaf	4	5
Big scraper	4	4
Uniface	3	3
Chopping tool	2	2
Between notches point	2	2
Drill	2	2
OBR	1	1
Limaces	1	1
Denticulate	1	1
Grinding stone	1	1
Indeterminate	3	3
Total	397	6074

OBR: Oblique Bevel Retouch tools

Johnson, 1979; Weitzel & Colombo, 2006). One case of a boleadora (stone bola) preform was also included (Supplementary Material, Table 12).

Additionally, seven whole or fractured stone balls (Supplementary Material, Table 13), 24 bifaces (Supplementary Material, Table 14), two chopping tools (Supplementary Material, Table 15), five large scrapers, and one denticulate tool (Supplementary Material, Table 16) were recorded.

Evidence of quarried outcrops

Field surveys yielded new information that complements that previously recorded in the sector and nearby sectors (see Chap. 7). Two rocky outcrops of significant size (greater than 10 m long) with flaking blanks were found. These are the Platense and Gris quarries, with IDs 266 and 729, respectively.

Artifactual structure of assemblages

To analyze existing diversity within the total sample, Table 17 (Supplementary Material) presents the artifactual structure of assemblages.

We compared the assemblages to identify differences between them. As artifact classes are added, diversity increases; however, even in sets with fewer classes, differences exist since the types and proportions of artifacts present vary. As Fig. 6.12 shows, isolated finds and assemblages of two and three classes are the most numerous.

While this presentation allows us to observe existing differences between groups, quantitative comparisons require measures of diversity. These measures are calculated in the following section.

Artifact diversity of the assemblages

For two decades, the study of diversity in artifact assemblages has been of great importance and has been encouraged by contributions from other disciplines, such as biology and statistics (Lanata, 1996). Assemblage diversity has two dimensions: richness and homogeneity (Kintigh, 1988). Richness is the number of classes in a sample, while homogeneity is the distribution of artifacts across the different classes. In principle, diversity in groups provides information about the variety of activities carried out in a place. However, aspects such as sample size, typological classification methods, sampling design implementation, and bias control in the collection

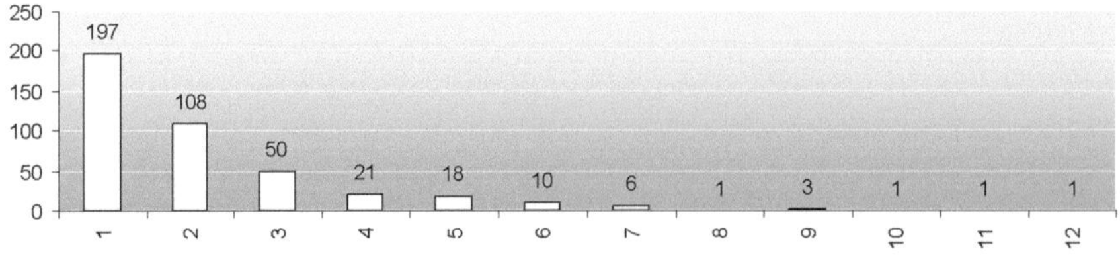

Fig. 6.12 Distribution of sets with *N* artifactual classes (strikers, cores, debris, NFA) and different typological groups within the formatted artifacts (Aschero, 1975 and its later modifications). Most of the sampled assemblages/locations present only one class of artifacts because the surveys recorded not only sites and concentrations, but also isolated artifacts

are important for reaching correct inferences (Lanata, 1996). Among researchers, the discussion is ongoing, as is the way of measuring diversity and interpreting the results. In addition to measuring the variety of activities carried out in a place, diversity measures have been interpreted within other frameworks, such as those concerning temporal stress (Franco & Borrero, 1996; Torrence, 1983) or greater or lesser residential mobility (Shott, 1986).

As seen in the previous section, the number of artifact types in each set provides an idea of the artifactual richness. However, it is inadequate as a direct indicator of diversity for comparing different groups since the number of classes present depends on the size of the sample (Lanata, 1996). To address this issue, the number of classes is not used as an indicator of diversity; rather, two indices are calculated: the artifactual richness index H of Shannon–Weaver (1949) and the homogeneity indices of Zar (1974) and Pielou (1977). The Shannon–Weaver H index expresses the probability that any randomly selected element from a set falls within a certain category. When all the elements of a set belong to the same category, H's value is 0; the maximum value of H depends on the number of categories present. Its formula is (Eq. 6.3):

$$H = \frac{n\log(n) - \sum_{i=1}^{k} f_i \log\log(f_i)}{n} \tag{6.3}$$

H Shannon–Weaver richness index (1949)
n Sample size
f_i Class frequency
k Class/category number.

The Zar and Pielou J index was used to measure homogeneity. This index varies between 0, indicating that the sample has only one category, and 1, indicating that the sample's items are distributed equally among all categories. The formula is (Eq. 6.4):

$$J = H - H\max \tag{6.4}$$

J homogeneity index
H Shannon–Weaver richness index
$H\max$ $\log(k)$
k class/category number.

These indices enable us to compare lithic assemblages of various sizes and determine their artifactual diversity. The application of a correlation test and the calculation of the regression coefficient between k and n values allow us to measure the degree of dependency and determine if there is a significant correlation between the two variables. These indices have been used in Patagonian archaeology, both in stratigraphic studies (Guráieb, 1999) and in surface assemblages (Bozzuto, 2007; Cassiodoro, 2008; Lanata, 1996). The consulted authors agree on calculating the indices of artifactual richness and homogeneity according to the number of typological groups added to strikers and cores in each collected lithic assemblage. In this study, we

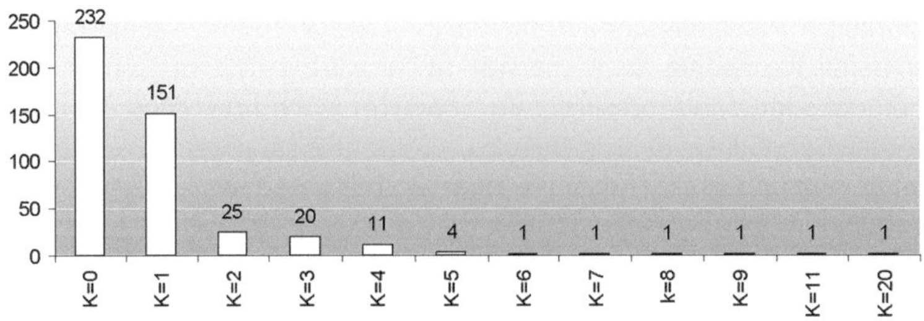

Fig. 6.13 Frequency of assemblages with k (class/category number) artifact classes. Assemblages and isolated findings ($n = 232$) presenting only debitage and NFA were excluded from the calculation of artifactual diversity indices. These are classified as "4_des" in the final result; see the last column of Table 18 in the Supplementary Material. The base 10 logarithm was used for its calculation, with the value of k corresponding to each sample

calculate diversity indices in the same manner as these authors to generate comparable results. As Fig. 6.13 shows, many assemblages ($n = 232$) contain debitage, NFA, or undifferentiated FA fragments. These classes are excluded from the analysis because their large numbers interfere with homogeneity measurements. Undifferentiated fragments of FA are also excluded because a class defined in this way would overstate the number of classes without representing greater diversity in past activities.

Table 6.24 shows the results of the techno-morphological diversity analysis. A Pearson correlation test was performed to evaluate whether assemblage diversity depends on sample size. At the 0.05 significance threshold, the observed value was 0.854, with a bilateral probability value of 0.0001, indicating a significant correlation between the variables. The regression coefficient yielded a value of $R^2 = 0.729$, meaning that 73% of the variation in the number of represented classes is explained by sample size. Even when the outliers "ID 45" and "ID 563" are excluded, the result indicates a correlation between the variables (see Fig. 6.14).

The Pearson correlation test and the regression test (R^2) indicate that much of the variability in the artifact assemblages depends on sample size. The above graph shows the dispersion and includes two outliers. One of them, ID 563, is the stratigraphic excavation in Cueva Maripe. This site was included to demonstrate its greater diversity of stratigraphic information. The graph below shows the distribution of values without the outliers. However, the correlation results are similar. These results suggest that diversity indices should not be used as the only explanatory variable. Although the result would have been a low or non-existent correlation between classes and sample size, we agree that, since diversity indices represent assemblage structure as a single value, they have the potential to confound explanatory factors (Shott, 1989). For this reason, we use them alongside other variables as part of a model that interprets archaeological evidence in terms of natural and cultural formation processes. To simplify interpretation and comparison, as well as to include the H index as another variable in the model, the continuous values obtained for the H index were transformed into categorical values. This was done by plotting the H and

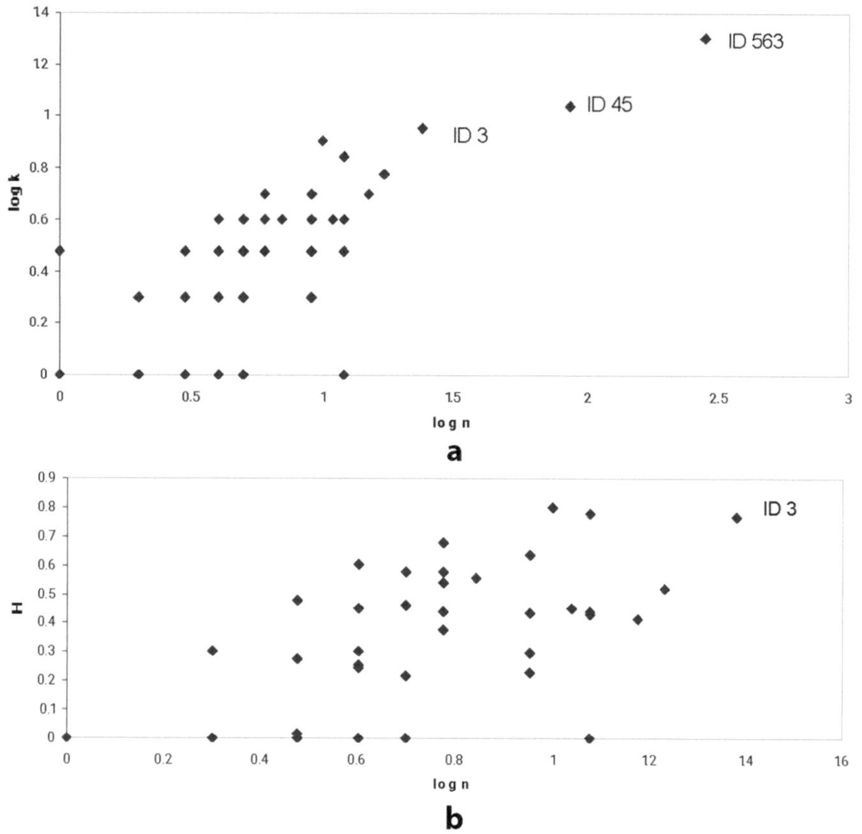

Fig. 6.14 Scatter plots of artifact richness versus sample size

J indices on a scatter plot (Fig. 6.15). Thus, the assemblages were divided into four diversity categories (see the last column in Table 24, Supplementary Material).

(1) Assemblages with high diversity (average $H = 0.76153124$; $k = 5$–8).
(2) Intermediate diversity (average $H = 0.49971535$; $k = 3$–11).
(3) Assemblages with 1 or 2 types of formatted artifacts (Average $H = 0.04806637$).
(4) Isolated finds or assemblages without FA, cores, or strikers.

As can be seen, this classification differs from one that uses only *k*.

Some authors have used *J* instead of *H* to characterize multiple activity sites, multiple activity sites with subseries of dominant activities, and limited activities sites. To include sets of very diverse *n* (between 451 and 2 in the surface sets, in addition to the isolated findings), it is more appropriate to interpret the *H* index within the model (see below).

The calculated indices are one of the most important tools for analyzing artifact assemblages. Technological diversity is defined as the number of artifact classes used

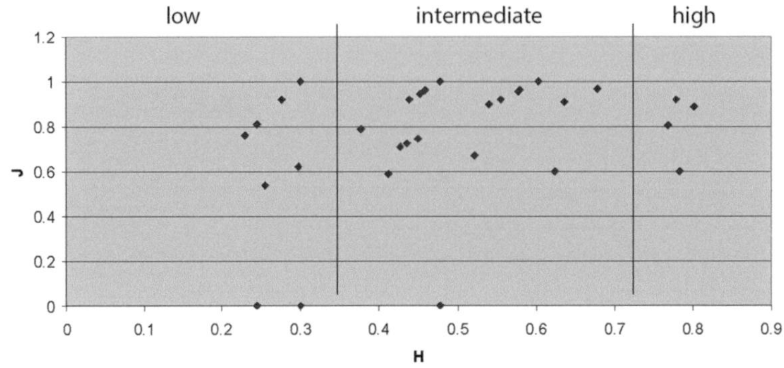

Fig. 6.15 Scatter plot of the *H* and *J* indices for assemblages with cores, strikers, and formatted artifacts of any typological group. The vertical lines indicate the sets' separation into richness categories

in everyday activities. This can vary based on factors such as the season of occupation, distribution and access to lithic resources in the landscape, and mobility strategies (Andrefsky, 2005). Our approach aligns with Andrefsky's (2005) idea that, at the local scale, greater artifactual diversity is related to longer permanence and greater reoccupation of sectors where more general activities were developed as opposed to sectors with more specific uses. General activities include food preparation and processing, artifact manufacturing, and shelter construction. Specific activities include hunting, collecting plant resources, slaughtering and processing animals, providing lithic raw materials in quarries, creating rock art, and burying the deceased, among others (Andrefsky, 2005).

As stated in the above model, the results of the diversity analysis are interpreted alongside other indicators that are not focused on formatted artifacts. This expands the range of activities that can be identified to characterize landscape use strategies. Evidence of lithic raw material quarrying and artifacts produced in the first and last stages of lithic artifact manufacturing indicates the stage represented in different assemblages. Together with other indicators, such as the presence of stone structures and rock art, this characterizes the archaeological evidence, which is then contrasted with a background of environmental information.

6.1.2.4 Stone Structures Variability

The model also includes information on the presence and type of stone structures. A total of 61 such structures were located in the macro-study area, in the localities of Bajo Grande, Piedra Museo, Jaramillo Petrified Forest National Park, Aguada del Cuero, and La Primavera. Some of these stone piles of various shapes and sizes are related to the hunter-gatherer way of life, while others are related to a more recent historical rural period. For example, there are cairns, which are stone piles that mark the location of estancia houses, corrals for livestock, posts, and shelters. The oldest archaeological structures can be identified by lichen coverage and associated hunter-gatherer materials (Albeck, 1995–1996). Lichens have basic ecological requirements of sunlight and nutrients, and they can be used as indicators of the stone structures' relative age and integrity. In this study, we use lichens to make comparisons between structures. The basic assumption is that greater lichen density and overlap in the rocky substrate, as well as changes in their coloring, indicate greater relative age of the structure. Based on these parameters and their association with archaeological evidence, structures most likely related to the hunter-gatherer lifestyle have been identified. These correspond to stacks of rocks—generally basalts or ignimbrites (see Chap. 2)—which have been interpreted as residential parapets, hunting blinds, burial sites, and smaller structures for other uses (Hermo & Vázquez, 1999; Magnin, 2009a; Miotti, 2006; Miotti et al., 1997).

The distribution of lichens on the rocks that form each structure indicates their degree of preservation. Lichens can only grow on rocky surfaces exposed to sunlight. Their presence on the illuminated surfaces of the rocks indicates good preservation of the original arrangement. Conversely, lichens growing on rock surfaces facing the dark interior of a stone structure suggest that the rocks were reassembled after lichens colonized the exterior surfaces of the original structure. The distinction between the internal and external sides of a structure is useful for closed structures or those built with walls wider than one rock.

Table 6.7 presents the archaeological stone structures in the macro-study area. They were classified based on their shape and dimensions. Open and closed forms were differentiated. Among the open structures, parapets 1 are smaller and parapets 2 are larger. Among the closed structures, small structures and chenques were differentiated.

Parapet 1 structures are relatively small, either semicircular or straight, and some are oriented north to south, functioning as wind barriers. Parapet 2 structures are larger, semicircular, or circular. Chenques are similar in size to parapet type 1, but they are closed stacks. However, a looted chenque may be semicircular or circular. Therefore, it is important to observe whether they show evidence of disturbance. Finally, there are small, circular, closed structures. One example is a looted cist burial in the Jaramillo Petrified Forest National Park. Although it would fall within the last class based on its measurements, it was discovered with human bones stained with ocher, confirming its function as a secondary burial site (Miotti, 2006).

This initial classification of stone structures in the region establishes a foundation for interpreting each structure found in the area under study based on its particular context.

Figures 6.16 and 6.17, as well as Table 19 (Supplementary Material), present the average measurements of the structure classes located in the Central Massif. Analysis of associated artifacts, location analysis using GIS, and stratigraphic excavations will complement our initial interpretations. Tables 20, 21, 22, and 23 (Supplementary Material) identify structures classified as "chenques" ($n = 14$), "parapets 2" ($n = 3$), "parapets 1" ($n = 15$), and "small structures" ($n = 14$) that were registered in the study area at La Primavera.

Table 6.7 Classification of the stone structures surveyed in the La Primavera locality and other structures in the macro-study area of the Central Massif

Nº	Reference	Locality	C_ ID	GPS_ ID	Height	Width	Length	Structure class	Name
1		LP	8	2211	67	210	300	Chenque	Chenque del viento
2		LP	232	2177	76	160	170	Chenque	Arcos Bajos—chenque N
3		LP	9	2178	110	130	150	Chenque	Arcos Bajos—chenque S
4		LP	10	2313	50	200	240	Chenque	Chenque del xilópalo
5		LP	11	1816	65	130	200	Chenque	Arcos Altos—chenque 2N
6		LP	28	1760	30	180	240	Chenque	Chenque Loma Pelada
7		LP	233	1815	55	135	260	Chenque	Arcos Altos—chenque 1S
8		LP	235	2153	35	160	180	Chenque	Chenque F2
9		LP	237	2156	40	240	330	Chenque	Chenque—Cabecera Cañadón Guanaco
10		LP	261	2359	90	200	210	Chenque	Co. de las estructuras—5
11		LP	261	2358	65	150	160	Chenque	Co. de las estructuras—2
12		LP	261	2358	60	180	180	Chenque	Co. de las estructuras—1
13		LP	712	2180	65	175	190	Chenque	Chenque B10
14		LP	736	294	85	140	160	Chenque	Ch. La Primavera 1
15		LP	100	1790	–	–	–	Chenque	Ch. La Primavera 2
16		LP	564	2589	40	100	200	Small structure	Cache
17		LP	9	2178	55	70	120	Small structure	Arcos Bajos—estructura del S
18		LP	23	2121	40	120	165	Small structure	Estructura de fogón—El Descosido
19		LP	27	2135	30	100	150	Small structure	Estructura cueva del círculo
20		LP	132	2123	28	110	140	Small structure	Estructura Zanjón de Aquilino
21		LP	236	2154	25	80	80	Small structure	Estructura F2
22		LP	249	1742	80	110	150	Small structure	Mojón

(continued)

Table 6.7 (continued)

Nº	Reference	Locality	C_ ID	GPS_ ID	Height	Width	Length	Structure class	Name
23		LP	250	1781	28	110	123	Small structure	Estructura Alta Primavera
24		LP	252	1841	60	90	110	Small structure	Meseta Sur yendo a Las Mercedes
25		LP	261	2358	90	100	170	Small structure	Co. de las estructuras—6
26		LP	261	2358	35	110	110	Small structure	Co. de las estructuras—3
27		LP	261	2359	50	100	100	Small structure	Co de las estructuras—4
28		LP	565	2594	30	60	60	Small structure	Estructura chica
29		LP	232	2177	65	90	110	Small structure	Arcos Bajos-ch del N
30		LP	48	1704	45	220	400	Parapet 1	Alta Prim. W.—Parapeto 1
31		LP	49	1705	70	170	470	Parapet 1	Alta Prim W.—Parapeto 2
32		LP	50	999992	65	200	280	Parapet 1	Alta Prim W.—Parapeto 3
33		LP	51	2548	57	70	330	Parapet 1	Las Mercedes Parapeto
34		LP	52	2090	72	210	290	Parapet 1	Alta Primavera Parapeto 1 (Solo)
35		LP	66	2098	80	200	200	Parapet 1	Alta Primavera—Parapeto 5
36		LP	67	1707	70	210	300	Parapet 1	Alta Primavera W Parapeto 4
37		LP	73	2204	35	200	200	Parapet 1	Parapeto (Estructura E2)
38		LP	96	1770	30	160	430	Parapet 1	Alta Primavera Parapeto 2
39		LP	242	2549	40	140	470	Parapet 1	Las Mercedes
40		LP	250	1781	28	110	123	Parapet 1	Estructura doble Alta Primavera—Parapeto 7

(continued)

Table 6.7 (continued)

N°	Reference	Locality	C_ ID	GPS_ ID	Height	Width	Length	Structure class	Name
41		LP	250	1781	50	230	420	Parapet 1	Estructura doble Alta Primavera—Parapeto 8
43		LP	262	2546	40	115	230	Parapet 1	Parapeto Las Mercedes
44		LP	682	2514	70	200	260	Parapet 1	Parapeto Luz Belito
45		LP	711	1848	60	80	340	Parapet 1	Parapeto Sitio del Paso
46		LP	–	999991	40	210	340	Parapet 1	Estructura 1
47		LP	583	1778	–	95	200	Parapet 1	Alta primavera—Parapeto 4
48		LP	238	2157	50	420	570	Parapet 2	Parapeto Cañadón Guanaco
49		LP	724	1777	40	350	600	Parapet 2	Alta Primavera—Parapeto 3
50		LP	725	1782	–	280	500	Parapet 2	Alta Primavera Parapeto 6
51		LL	737	2009	–	140	200	Parapet 2	La Lota estructura 2
52		LL	737	2009	–	–	320	Parapet 2	La Lota estructura 1
53	(1)	BG	–	259	50	140	150	Chenque	Chenque Tito del Valle
54	(1)	BG	–	264	80	210	250	Chenque	Chenque 1 del Moyano
55	(1)	BG	–	266	70	230	300	Chenque	Chenque 3 del Moyano
56	(1)	BPJNP	–	666	80	300	300	Chenque	Ch. camino Manantiales
57	(1)	AC	–	261	–	160	300	Chenque	Chenque 1 Co. Terciario
58	(1)	PM	–	1180	–	–	–	Chenque	Chenque El Sargento
59		LP	–	1806	–	–	–	Chenque	Chenque de la cruz
60	(1)	BG	–	265	–	–	–	Chenque	Chenque 2 del Moyano
61	(1), (2)	LD	–	300	–	–	–	Chenque	Chenque La Dorita
62	(3)	BPJNP	–	148	40	50	70	Cist burial	Entierro Madre e Hija

"N°": structure number; "Loc": archaeological locality (LP" La Primavera, LL: La Lotita, BG: Bajo Grande, LD: La Dorita, BPJNP: Bosque Petrificado de Jaramillo National Park); "C_ID": identification code of the assemblage/location; "GPS_ID": GPS point identifier code; "High": "Width" and "length": maximum measurements in centimeters from the external edge of the rock structure. If a measurement was not taken, it is indicated with "–"; "Class": type of structure; "Name": brief description. References: (1) Miotti (2006), (2) Hermo (2008), (3) Miotti et al. (1997), Hermo and Vázquez (1999)

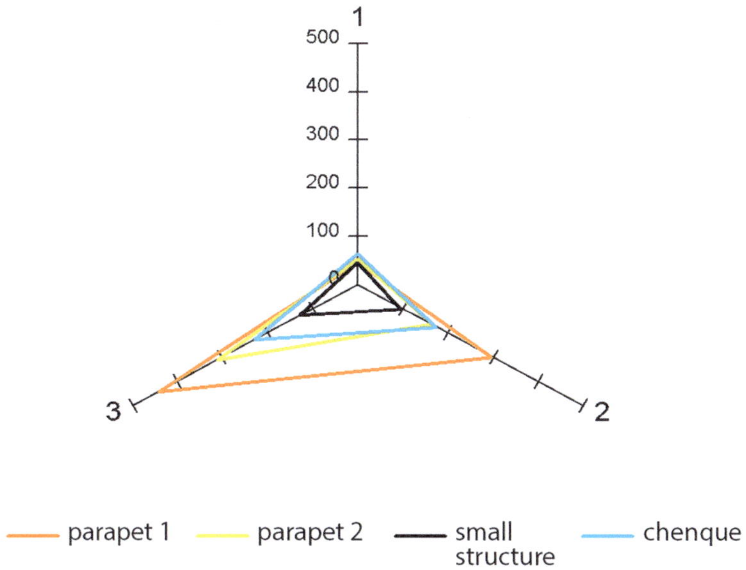

Fig. 6.16 Average measurements (in cm) of the different structure classes (1 = height; 2 = width; 3 = length)

Variability of chenques in the region

Based on the previous classification, 14 structures within the La Primavera locality are identified as possible burials or chenques. Since excavating burials was not part of the research agenda, the work methodology did not involve removing rocks from inside the structures. Instead, a low-impact strategy was employed, which involved describing and measuring the structures for comparison with verified burial sites in the region. As Table 6.8 shows, the chenques considered in this study are smaller on average than those described for the northern margin of the Deseado River (De Aparicio, 1933) and the Salitroso Lake Basin (Goñi & Barrientos, 2000). Conversely, their sizes are more similar to those recorded on the Atlantic coast, which average approximately two meters in diameter (Castro et al., 2000). Travelers also recorded variability in chenque dimensions. During his time in Chubut, Francisco Moreno mentioned that

> These cairns are formed with piled stones that surround and cover the human remains, apparently placed on an artificial floor of flat stones; The tallest one I know measures about three meters, and some of the stones that make them weigh from forty to fifty kilograms (Moreno [1879] 1997:96).

Others such as Pozzi (1936:30, cited in Berón et al., 2000) indicate that *the pile of stones ended 1 m or so from the ground.*

Variability of parapet stone structures in the region

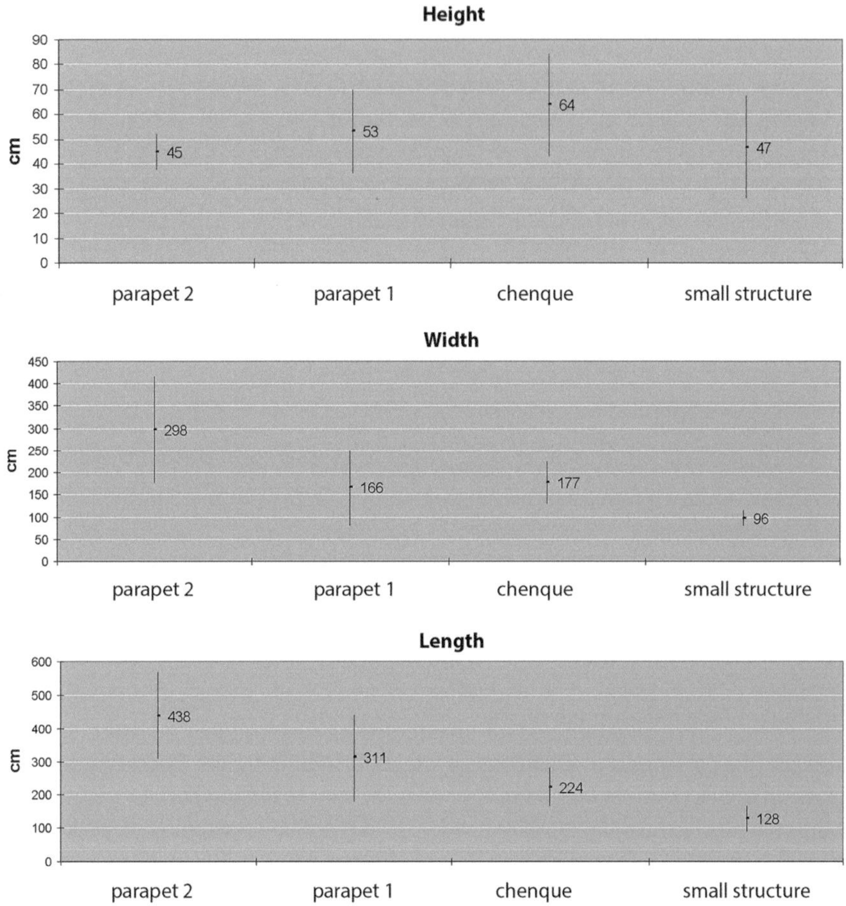

Fig. 6.17 Measures of central tendency and dispersion of stone structure dimensions of different types in the macro-study area ($n = 60$)[2]

Gradín first described the stone structures from a survey he conducted on the plateaus around Lake Cardiel in 1959. The author published this description in a report on Lake Strobel (Gradín, 1959–1960), in which he documented a significant number of similar structures, exceeding one hundred. Based on their shape in plan view, Gradín defined four groups: Group A, Group B, Group C, and Group D parapets. Group A is arranged in the shape of an arch; Group B, in a semicircle; Group C, at right angles; and Group D, with a circular layout interrupted on the east side. Most cases were registered on the basaltic plateau. Among them, 60 parapets have an arched plan and 30 have a semicircular plan. Sheltered by rocky ravines, the author found ten in a semicircle, four at right angles, and five in a circular shape. They are oriented south to north and have approximate dimensions of 5 m in length and 50 cm in height.

[2] The cist burial was not included in this calculation.

Table 6.8 Measurements of known chenques in the region, as well as those identified as chenques in the study area (La Primavera locality)

	Height (cm)	Width (cm)	Length (cm)
De Aparicio (1933), Deseado River north bank, close to Las Heras			
Chenque 2	100	225	350
Chenque 1	60	200	400
Mean	80	213	375
Goñi and Barrientos (2000), Salitroso Lake Basin			
SAC1-1	–	300	400
SAC1-2	–	370	380
SAC1-4	–	210	320
SAC1-6	–	250	500
SAC4-1	30	200	400
SAC4-2	–	240	650
Mean	–	262	442
SAC7, unexcavated chenques	–	160	350
	–	200	320
	–	250	300
Mean	–	203	323
Castro et al. (2000), Atlantic coast between Puerto Deseado and Bahía Laura			
Approximate mean	–	200	200
Santa Cruz Central Plateau (Table 6.25)			
Mean	64	177	224

> …given their orientation, they face the constant wind from the West. Generally, abundant stone chips appear on its east side… (Gradín, 1959–60: 146).

Furthermore, it mentions that the parapets of group "A" (in an arch)

> …almost always appear associated with lithic debitage, while those of group B) (in a semi-circle) are often found next to picaderos[3] where, in addition, I have collected splinters from scrapers, side scrapers, and arrowheads, in quantity (Gradín, 1959–1960: 146).

In general, they are found grouped in variable numbers ranging from three to ten, and there is a recurrence in terms of their location. Arched parapets are more frequently found on the plateau in windy, exposed areas, about 200 m apart. Semi-circular parapets, on the other hand, are found next to rocky ravines, closer together, and take advantage of the uneven terrain for shelter (Gradín, 1959–1960). Gradín proposes calling these sites "stations" and mentions that local residents attribute the purpose of these structures to hiding hunters from their prey.

[3] Picadero is the name given by local settlers to surface sites.

> …this interpretation is based on the orientation against the wind…which nullifies the animals' sense of smell, as well as on the circumstance that in a radius of approximately 100 m to the west of the parapets, arrowheads lost by ancient hunters are usually found… (Gradín, 1959–1960: 148).

The summary of the parapet characteristics defined by Gradín is presented in Table 6.9.

In a subsequent study of parapets and engraved art in Lake Buenos Aires, Gradín excavated a parapet at the top of Cerro Redondo, confirming its use as a temporary settlement or remains of tolderías (Gradín, 1976: 316). These settlements are generally located near water and sheltered by basalt walls, as in Laguna Honda. However, the parapets at Cerro Redondo are located atop this 100-m-high hill, over 1000 to 3000 m from a permanent water source. This position provides a dominant view of the Lake Buenos Aires Plateau. Since these locations are not sheltered, Gradín attributes the choice of this location as a campsite to its wide visibility. In other nearby complexes, such as the Cerro Chenque sector, a few housing units are grouped next to the walls, coinciding with scarce petroglyphs.

In Cerro Redondo, the parapets have an incomplete circular design and are oriented according to the wind. They are arranged in groups and aligned (five central parapets) as well as in isolation (Gradín, 1976). Its dimensions are shown in Table 6.10. Excavating three 5-cm layers allowed for the recovery of an assemblage containing knives, scrapers, and side scrapers, which could have been used to slaughter guanacos (Gradín, 1976). The assemblage also included cores, flakes, a hearth lens, bone remains presumably from a guanaco, and in the upper layer of the excavation, dart points and a ceramic fragment. This evidence, along with Spegazzini's 1884 text describing winter awnings near Río Gallegos, establishes their residential function. According to Spegazzini (1884), the awning was formed by

> …four rows of five or six larger poles, and two other lateral rows of as many shorter poles; four enormous quillangos covered everything, and their edges were stitched high, on land

Table 6.9 Parapet characteristics defined by Gradín (1959–1960b, 1976)

Group	Shape in plan view	Associated knapping products	Location	Use time (stay)	Function
A	Arch	Debitage	Plateau, separated by about 200 m	Relatively short	Hunting blind
B	Semicircular	Debitage and FA (scrapers, side scrapers, projectile points)	Next to ravines, on uneven terrain sheltered from the wind, closer to each other	Prolonged	Temporary dwellings or hunting posts
C	Straight	–	–	–	–
D	Circular	–	–	–	–

they were secured by thick stones…(Spegazzini, 1884:10) (see discussion about awnings in Casamiquela, 2000).

Gradín relates all the identified archaeological complexes in the different sectors of Lake Buenos Aires as evidence of the installation of temporary habitation sites, taking advantage of the hunting season and performing ritual ceremonies (Gradín,

Table 6.10 Habitational parapets and hunting blinds registered in the region

Source	Area	Locality	Structures registered (*n*)	Structures measured (*n*)	Height (cm)	With (cm)	Length (cm)
1	Strobel Plateau		More than 100 structures	–	50	–	500
2	Lake Buenos Aires Plateau	Cerro Redondo and other in nearby areas	17		65	200–500	200–500
3	Pinturas River	Circular structure occupied by Family Chapalala			–	500	500
4	Pampa del Asador	Co. Pampa 2, Sector A	8		–	–	–
5	Pampa del Asador	Co. Pampa 2, sector B	13		50	–	400
6	Del Águila Plateau		4		70	270	370
7	Asador Plateau		32	25	79	236	299
7	Pampa del Asador		34	29	59	172	404
7	Guitarra Plateau		84	55	74	266	351
7	Strobel Plateau		358	180	58	167	464
8	South Deseado Massif	La Gruta	6	6	62	238	510
		La Melisa	7	7	56	121	240
		La Barda	12	12	45	81	305

The measurements listed in the table are the average measurements or the range between the minimum and maximum dimensions, as stated in the publications. Sources: 1) Gradín (1959–1960), 2) Gradín (1976), 3) Gradín and Aguerre (1992), 4) Espinosa and Goñi (1999), 5) Rindel et al. (2007), 6) Molinari and Ferraro (2004), 7) Flores Coni et al. (2021), 8) Franco et al. (2021)

1976). They are formed by circular stone parapets, lithic workshops, and middens containing bone remains. There are also engravings on nearby rocks.

In Cerro Pampa 2 (Pampa del Asador), Espinosa and Goñi (1999) found thirteen structures built using arch-shaped pircas, which were studied by Cassiodoro et al. (2004). These structures are larger, allowing them to accommodate an awning. Some contain diverse lithic artifacts, including hearths and bone remains, which constitute evidence that they were used as dwellings for carrying out various activities (Cassiodoro, 2011; García & Pérez de Micou, 1980; Gradín, 1959–1960, 1976; Molinari & Ferraro, 2004).

A photograph published in Molinari and Ferraro (2004: 630) seems to confirm the function of the parapets as non-portable parts of the homes. In the photograph, an awning is assembled on a stone pirca in the Cerro Indice area (Santa Cruz Province), corresponding to a historical parapet. These authors located four parapets in the Ea. Menelik in the Vidalita stream canyon on the Águila plateau. They are all associated with bone remains, mainly from *Lama guanicoe*, as well as dense lithic assemblages, hands, mills, and ceramics.

Over the past decade, interest in studying stone structures has grown, shedding new light on the subject. The recent work of Flores Coni et al. (2021) summarizes 20 years of research on the high plateaus of the central-western Santa Cruz province. Their study of 535 structures in the Pampa del Asador and Asador, Guitarra, and Strobel plateaus revealed that their functionality varied according to changing needs.

Franco et al. (2021), in the southern Deseado Massif, also provided new information on structures that were interpreted as hunting blinds. These structures were located close to shallow lakes in the La Gruta, La Melisa, and La Barda Localities.

The data on the habitational parapets and hunting blinds studied in the region are presented in Table 6.10.

Returning to the data for this study, three large parapets were found in the surveyed area, which we refer to as parapets 2 (Supplementary material, Table 21). These are similar to the residential parapets discussed earlier (Table 6.10). These are parapets No. 3 and No. 6 of the Alta Primavera site (ID_724 and ID_725) and the Cañadón Guanaco parapet (ID_238). All of them have a semicircular plan. In contrast, two circular structures were located toward the west of the prospected sector (1 km outside its limits) in the La Lotita locality. Both structures are contiguous and located on an elevation from which a temporary shallow lake (690 m southwest) is within visibility. Regarding the smaller structures (structures 1) registered in the studied sector (Supplementary material, Table 6.22), their measurements coincide with those of the structures Gradín (1959–1960a) mentioned as probable hunting blinds. Based on ethnographic, ethnoarchaeological, and archaeological studies, visibility is important for defining the location of surveillance points where visual dominance is important for controlling herds of game animals (Binford, 1982; Claraz, 1988; Krist & Brown, 1994; Martínez, 2003). The parapets of Las Mercedes (ID_282 and 242), Luz Belito (ID_682), and Parapeto E2 (ID_73), in particular, can be assigned this function. Other archaeological studies (Martinez, 2003) discuss the potential uses of these structures and their implications for studying hunting strategies and technological changes.

The five parapets in the Alta Primavera sector and the four in Alta Primavera Oeste (see Fig. 7.17 in Chap. 7) may be related to hunting strategies or residential use.

In this regard, Casamiquela (2000) compiles ethnographic sources demonstrating variability in habitation structures among historical indigenous groups. We found the measurements of the base of the awnings of southern Tehuelche groups particularly interesting, as they coincide with those of the residential parapets described above. This was first observed by A. del Castillo in 1887 during his exploration of the Patagonian coast. *These tents have a general diameter of 5 to 7 m and 50 to 70 hides are used in them* (cited in Casamiquela, 2000: 31).

Martinić mentions the measures included in the portable home, or "kau," from his ethnographic and archaeological research. On average, the "kau" occupies 20 m^2 and houses one or two families, along with their belongings and dogs. Additionally, Casamiquela refers to a series of smaller windbreaks than those mentioned by Outes (1905). At the time of first contact with Europeans, the windbreak coexisted with the classic Tehuelche awning in the extreme south of Patagonia (Casamiquela, 2000). Martinić also mentions windbreaks that were possibly used during satellite hunting excursions by the Selk'nam people of Tierra del Fuego (Martinić, 1995). Another source, an interview with Silvana Chapalala, reports the existence of smaller single awnings inhabited by young men, as well as awnings where young and old women worked special tasks, such as silversmithing (Aguerre, 2000).

In sum, our classification of open structures into two groups, parapets 1 and 2, does not conclusively determine their function. However, this classification was analyzed in conjunction with surface evidence. This is the first step in analyzing this variability.

Small stone structures

A total of 14 small structures were identified in the study area. It is difficult to infer their possible functions, but for now, they can be considered constructions that are either spatially associated with sites or isolated. Examples of the latter are structures ID_27 and ID_132 near Zanjón de Aquilino and structure ID_564 near Cuevas de La U (Supplementary material, Table 6.23). The latter was proposed as a cache of lithic raw materials. Structures ID_27 and ID_711 are spatially associated with sites. Structure ID_27 is located near Cueva del Círculo, and structure ID_711 is located at the Paso Site. Their specific function is still unknown, but they could be equipping spaces. Gradín (1976) mentioned a similar structure in front of the central group of parapets of Co. Redondo (with measurements of 1 m in diameter and 60 cm in height). The author proposes that it could have been used to store and preserve food. According to Musters's account of the Manzaneros' customs, they used to leave bags of fat and provisions in different places, hoping to return to them in a not-too-distant time (Gradín, 1976; Musters, [1871] 1997). However, while the traveler states that these fat supply bags were hung on trees, Gradín does not rule out the possibility that the stone structures were used to contain these bags. A similar function has been proposed for the structure found in Piedra Museo, located in another sector of the macro-study area.

Other indicators

Signs of fire

Fire is fundamental to all domestic activities. It provides heat and is used for cooking. For this reason, it was included in the model as an indicator of these activities. However, as the chronicles mention, the Aonikek lit fires for other reasons and in places other than the camp. For instance, they used fire during residential movements, logistical parties, and hunts. They used it to communicate with others, request permission to enter hunting territories, and to corral animals (Musters, [1871] 1997). Furthermore, fire had great symbolic and ritual importance, as it was used during important ceremonies, such as those marking the onset of puberty in women and mortuary rituals (Manzi & Spikins, 2008; Musters, [1871] 1997).

Therefore, the presence of fire is not assignable only to domestic activities such as heating and cooking food in base and hunting camps, but it can also occur in camps established mainly for ritual reasons, such as Casa Bonita, where domestic activities were carried out simultaneously, such as preparing and consuming food (Musters, [1871] 1997). Fire may have even been used outside of camps to send smoke signals or for hunting (Musters, [1871] 1997).

Finally, taphonomic factors, such as naturally occurring fires in areas with archaeological remains, can be difficult to distinguish from intentional fire use. In this sense, only hearths that are structured in a way that denotes control can be interpreted as anthropogenic. This can be determined through stratigraphy. However, no charcoal remains have been found on the surface. Stone structures that controlled and delimited the fire area can then be proposed as indicators in these contexts. Among the small structures mentioned above, one associated with the El Descosido site (ID_23) is shaped like a closed circle, which could circumscribe a hearth (Fig. 6.18). This structure is associated with a set of 30 artifacts, including four scrapers, knapping debris, and NFA. This collection, along with the potential hearth structure, suggests the presence of domestic activities. Other small structures were mentioned in Cerro Redondo and are spatially associated with residential parapets that are interpreted as remains of tolderías (Gradín, 1976). Among these structures is a small circle of stones measuring 1.20 m in width, similar to the one presented here. However, the author does not specify its possible function.

For this study, other indicators of fire presence, such as thermal alteration of raw materials in lithic artifacts, were evaluated, and evidence was found in 60 surface lithic assemblages. Macroscopically recognizable indicators of thermal damage were taken into account for its recognition, such as peeling, dimples, roughness, and cracking (thermal action, according to Aschero, 1983). These indicators are useful for inferring the effects that a hearth can have on artifacts discarded there intentionally or accidentally. They can also be contrasted with gloss, color changes, and force waves that appear on raw materials when they are exposed to thermal action as part of rock treatment processes intended to improve their qualities for knapping. However, thermal damage can also occur during the intentional treatment of raw materials when temperatures exceed the thermal stress threshold (Cueto & Frank, 2008; Nami et al., 2000). This single indicator does not explain the exposure of rocks

Fig. 6.18 Small structure with rocks distributed in a circle. It is associated with a lithic assemblage containing scrapers, knapping debris, and NFA. The context suggests domestic activities (ID_23, El Descosido, $n = 30$)

to heat. Artifacts could have come into contact with fire: (1) in a controlled manner to improve their knapping properties (in this case, changes in the rocks are expected, but not thermal damage). This could have occurred in residential base camps or field camps far from them.

(2) On a site, without the intention of modifying their properties for knapping. This could have occurred by lighting a fire on artifacts discarded from a previous occupation, for example. It could also have occurred within sites due to the discarding of artifacts in hearths (Cueto & Frank, 2008). Alternatively, it could have occurred outside of places by lighting fires to send social communication signals (e.g., columns of smoke) or as an aid in hunting strategies, accidentally coinciding with locations where there were previously discarded artifacts. In this line, we conclude that it is not prudent to use thermal alteration alone as an indicator of surface site functionality, at least in the present case study. We will continue to evaluate its usefulness with other indicators once the functional context is established.

Rock art

Painted rock art was found in ten locations within the surveyed area. Three more sites are located 5 km away, one of which includes engravings (see Chap. 7). Rock art sites were studied in Carden's (2009) doctoral thesis and in other regional studies

(Miotti et al., 2007). Here, we focus on local patterns of art distribution (Table 24, Supplementary Material).

6.1.3 Classification of Archaeological Evidence

Due to the significant diversity of surface archaeological evidence and the intricate formation history of the assemblages, this classification serves as a framework for analyzing the diversity of the archaeological record and creating groupings to identify spatial patterns. This deductive classification (Ebert & Kohler, 1988) is based on ethnographic (Martinić, 1995), ethnoarchaeological (Binford, 1982), and archaeological (Borrero, 1989; Miotti, 1989) information summarized in the model presented in Chap. 4. Table 6.11 presents the steps followed to classify the evidence. The presence of rock art and stone structures interpreted as chenques allows the first two classes to be separated, as these structures are associated with artistic expression and burial rituals. The rest of the assemblages were classified using artifact density, diversity (*H* index), archaeological visibility, and the presence of certain artifact types. Lithic assemblages with high or medium artifact density form class 2.1, while those with low artifact density form class 2.2. Within class 2.1, those with high artifact diversity can be separated from those with lower diversity values (class 2.1.2). The former are the settlement loci of highly reoccupied residential camps (MAL). They exhibit high artifact density; however, sets with intermediate densities that coincide with low archaeological visibility can also be classified in this group. From the perspective of lithic analysis alone, it is impossible to determine whether these archaeological signs represent residential camps occupied for a prolonged period or multiple sporadic uses of the same location for various activities. Both possibilities result in the same archaeology, though they highlight the characteristics of a special place where prolonged residential occupations or successive reoccupations coincide (Binford, 1982).

The second group (group 2.2, low artifactual diversity) is interpreted as limited activity loci (LAL). The different activities that may have developed at these loci are suggested by the presence of diagnostic artifacts. In most cases, more than one artifact occurs at each locus. This group includes activities carried out in temporary camps (TC); resource extraction activities (lithic = Q and vegetable = V); workshops for preparing cores and other artifacts (W); the primary processing of prey, and other activities associated with hunting, such as knapping and restoring projectile points (hunting stations = HS), etc. In these cases, artifactual density tends to be lower due to the type of activity, time invested, and possibility of reoccupation. For instance, loci of primary processing of game prey are not expected to be reoccupied for the same activity. Therefore, the resulting archaeological imprint is subtle. The sets tend to be very small, and information loss from the final record tends to be high. The differentiations within the LAL group that we present in this scheme aim to recognize the diversity of assemblages. The existence of location patterns for these classes will

Table 6.11 Archaeological evidence classification scheme

1. Presence of a particular type of evidence		
Presence of chenque type structure		1.1. Chenque burial (**Ch**), $n = 14$
Presence of rock art		1.2. Art (**A**), $n = 9$
2. Density, visibility, and types of artifacts		
2.1. High and medium artifact density		
2.1.1. High artifactual diversity: **MAL**		
Locus frequently reoccupied as a residential camp, $n = 18$		
2.1.2. Medium or low artifactual diversity: **LAL**		
	Presence of edging in rock veins. Predominance of indicators of the first stages of extraction of lithic raw material	2.1.2.a Quarry (**Q**) $n = 4$ primary, 6 secondary
	Presence of scrapers. There may be stone structures associated. Least occupied residential sites	2.1.2.b Transitory campsite (**TC**) $n = 24$
	Presence of assemblages formed by cores and unformatted artifacts. There may be preforms. Projectile points may present manufacturing fractures	2.1.2.c Lithic workshop (**W**) $n = 5$
	Presence of hunting weapons. There may be hunting blinds associated	2.1.2.d Hunting stand (**HS**) $n = 16$
	Presence of big scrappers, chopping tools	2.1.2.e Vegetable supply *locus* (**V**) $n = 4$
	Presence of long edge artifacts. There may be stone structures associated	2.1.2.f Primary prey processing and other limited activities *locus* (**LAL**) $n = 130$
2.2. Low artifact density and diversity: **LAL**		
	Presence of isolated hunting weapons	2.2.a. Hunting ground (**HG**) $n = 10$
	Assemblages formed by cores with few flakes, unformatted artifacts and/or debris	2.2.b. Occasional knapping (**OK**) $n = 171$

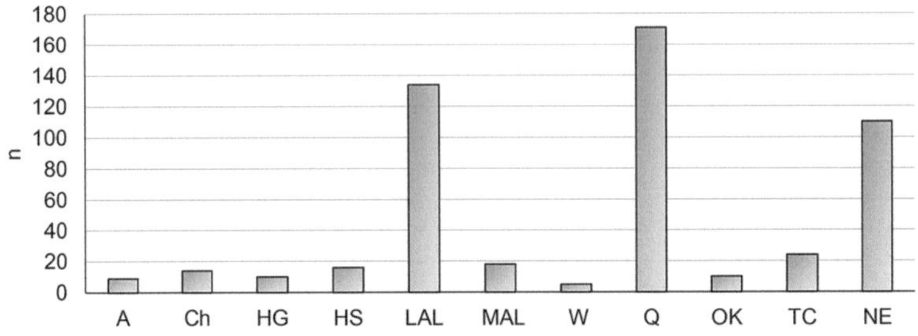

Fig. 6.19 Frequency distribution of the functional classes defined for the assemblages/locations. The acronyms on the *x*-axis correspond to the classes in Table 6.35. "NE": negative evidence; "A": rock art; "Q": extraction quarries in outcrops; "OK": occasional knapping; "HG": hunting ground; "HS": hunting stand; "Ch": chenque, possible burial; "MAL": multiple activity locus; "TC": temporary camp; "W": workshop; "V": locus of plant resource extraction; "LAL": other limited activities, such as primary processing of prey

inform us as to whether these classes are useful for analyzing the settlement patterning in the study area.

Within the group of low artifactual density and diversity (class 2.2), two types of activities are distinguished: hunting grounds (HG) and occasional knapping (OK) (Pérez de Micou et al., 1992). The first type is inferred from the isolated find of diagnostic artifacts, such as hunting weapons. The latter is inferred from sets of artifacts that are very limited in number, including cores with isolated flakes and unformatted artifacts.

Stone structures are sometimes spatially associated to contexts that suggest processing activities, such as lithic assemblages including scrapers, end scrapers, and notches. Occasionally, these structures are associated with evidences that indicate knapping activities, the replacement of projectile points, and activities involving the use of artifactis with long cutting edges. In other cases, they are associated with non-diagnostic artifacts, such as various types of debris, or they are not associated with any artifacts. Beyond the inferences that can be made about the activities they may have been associated with, such as hunting blinds or residential parapets, these structures denote an effort to prepare a space for future occupation. For this reason, they constitute evidence of the equipment of places (Cassiodoro, 2011; Kuhn, 1994).

The results of the classification based on the lithic analysis of the assemblages are shown in a bar graph (see Fig. 6.19). The classification results show the predominance of the OK, LAL, and NE classes.

6.1.4 Generation of GIS Coverages

The final step in classifying archaeological evidence is generating point vector coverages for each defined class. This information is entered into the GIS environment using the geographic reference provided by the GPS position associated with each assemblage/location. When there is more than one GPS position for the same unit, the position closest to the average center is selected. The following pages explore the distribution of different types of evidence in the landscape. It should be noted at this point that we used environmental evidence only to infer record formation processes (point 6.1.2) and not to classify archaeological evidence. The methodological justification for this approach can be found in the proposals of Kvamme (2006) and Stančič and Veljanovski (2000). These works avoid defining the different classes of assemblages in the initial classification using environmental characteristics. Using environmental characteristics to define site classes and then analyze their location to confirm the type of activities that took place would constitute circular reasoning. The functional classes defined from the archaeological record are then contrasted with environmental variables using independent datasets to define the classes and their locational patterns (see Methodological Step 4 in Fig. 5.1, Chap. 5).

Studies of locational patterns in archaeology employ a wide variety of approaches (Altschul, 1988). These approaches range from simpler applications aimed at identifying robust patterns to those that handle numerous environmental variables using complex statistical processes. GIS is suitable for obtaining and managing data in either case. For this study, we opted for simple statistical tests. More complex analyses are applied in Chap. 10. As mentioned in Chap. 3, this study employs both deductive and exploratory approaches to explore the data. We define the location variables and the location hypotheses of the different types of evidence based on the theoretical model (see the expected locations in Chap. 3). As discussed below, the information layers (maps) used to represent these variables are mainly derived from current topographic data. Therefore, we evaluated its use in each case based on the adequacy of the available information to model the relationships between hunter-gatherer groups and the landscape.

6.2 The Locational Variables

The environmental variables evaluated in the search for patterns are those that are frequently cited in hunter-gatherer literature as influential in settlement. Among these variables, distance to water sources is an undisputed critical factor in arid and semi-arid environments (Binford, 1980; Kelly, 1992). In the archaeology of extra-Andean Patagonia in general (e.g., Berón, 1994; Borrero et al., 1992; Cassiodoro et al., 2004; Goñi, 2000; Goñi, 2000–2002; Goñi et al., 2004; Morales et al., 2009; among others) and for the Santa Cruz Central Plateau in particular (Miotti, 1989), the importance of water sources is widely recognized. The availability of drinking water was the critical resource for the region's societies. This resource has had heterogeneous and scarce distribution throughout the Holocene. Based on the regional archaeological record, low, sheltered places with recurring drinking water, fauna, and plant resources—generally located inside ravines and shallows—are important settlement attractors and where the greatest number of small and large sites are located (Miotti, 1989, 1991, 2000; Miotti & Hermo, 2000). Accordingly, our location analysis anticipates that the settlement was influenced by this resource. On the other hand, the expected pattern in the archaeological record may vary. While some authors argue that water is such a vital resource that the distance to obtain it should be minimized (Jochim, 1976; Roper, 1979), others contend that, in situations where water is equally important, a different strategy may have been employed. For example, Yellen (1976) notes that the !Kung San of the northern Kalahari, whose movement is greatly limited by the distribution of water, settled at least half a kilometer from the water source to avoid disturbing the animals that also use it (Ebert & Kohler, 1988). Interestingly, although the most influential variables in settlement may be the same in various hunter-gatherer contexts, the specific pattern resulting from a recurrent strategy may vary from context to context. Another factor considered central to the location of hunter-gatherers is the existence of shelter (Jochim, 1976). In this region, residential areas are expected to be located near sheltered areas of the landscape that protect against strong westerly winds. These locations are also expected to be on relatively flat land and in low topographies. To analyze this, the land surface was divided into three categories: areas with gentle, relatively high slopes; areas with low slopes in relatively low positions; and areas with steep slopes. Regarding other variables, such as distance to lithic resources, they are not expected to strongly impact residential site locations given their abundance in this landscape (Hermo, 2009; Miotti & Salemme, 2004). However, in other types of assemblages/locations, such as W and OK, it may be important to describe trends regarding this resource. The distribution of the lithic resource reveals unique characteristics of this area, which are examined through an analysis aimed at understanding the movement of rocks on a local scale (Chap. 7). Location studies may include social factors in addition to environmental factors. There are different views on including these variables. Some authors claim that environmental factors are more important in hunter-gatherer societies or that they cannot be included in a model (see the discussion in Chap. 3). This work includes social factors, such as the visual properties of certain locations and proximity to mobility routes. The introduction of these factors enables us to examine aspects beyond the resolution of economic needs, potentially associated with the social environment and the sociocultural structuring of space. Some factors that would be highly relevant to the analysis, such as plant and animal community distribution, were restricted. This is because the available map data are not detailed enough to be representative of past conditions, as in the case of vegetation, and is difficult to model for mobile resources, such as fauna. All base data and their technical specifications are listed in Table 5.1. The most important data source is from surveys and a digital elevation model (GDEM Ev1.0), from which terrain slope and orientation coverages are derived. These coverages are fundamental for the analyses carried out. Generally, it is assumed that current relief data represent past relief and have not substantially changed throughout the entire period of human occupation (Zárate et al., 2000). The GDEM is an ASTER product generated from the analysis of stereoscopic pairs of radar images. The image presents terrain altitude data collected according to a 30-m grid (1 arc-second). This model serves as a base map on which archaeological data and relevant variables for pattern analysis are superimposed. These can be presented in vector or raster format depending on the most appropriate representation model for each variable type (qualitative or continuous). All raster coverages have a spatial resolution that coincides with that of the GDEM. In the GIS, all the used data were converted to the Gauss-Krüger Argentina strip 2 map projection with the WGS 84 datum.

Next, the locational variables are described, and their relationships with the types of archaeological evidence are analyzed.

6.2.1 Distance to Permanent Water Sources

Permanent water sources in this sector are located where the foothills of plateaus and wet meadow ecosystems meet (Mazzoni, 2007; Melendi, 2023). Through fieldwork, 21 meadows and 13 springs were recorded. Their positions were taken with a GPS with an accuracy of 6–4 m.

Within the recorded evidence, the state of the springs varies. There are: (a) active springs, where water currently flows with relative abundance; (b) wet meadows, where it is enough to dig a small well to activate them and obtain water based on their current use and the use by hunter-gatherers in the region (Aguerre, 2000); (c) currently dry meadows, which are active after rains and in the spring. Some are associated with historical constructions. We hypothesize that their desiccation may have intensified during historical periods. We acknowledge that these meadows and waterways may not have been available simultaneously in the past as they are not now, which does not contradict the long-term location systems study we propose here (Gamble, 1990, see Chap. 4).

In addition to the GPS-positioned springs and meadows in the field, the digital processing of two Landsat 5 TM multispectral images (scenes 229_93 and 228_93) was carried out in the laboratory. Using the known waterholes, another 23 wet meadows were identified in unsurveyed parts of the study area. The processing performed was the Abrams ratio combining bands 542. To maximize the procedure's effectiveness, we sought satellite images coinciding with the fieldwork date. This was possible for the first image, which had a coincident date (02/23/08). However, a cloud-free image of the second scene was not available. After consulting climate records from recent years, we decided to use an image from the previous year, from the same month (02/27/07). A mosaic was made with these images to cover the studied sector, and the aforementioned geoprocessing was carried out.

6.2.1.1 Permanent Water Sources and GIS

Distance measurements to this and other resources were carried out by calculating accumulated cost surfaces. This is done by applying a spread function to a cost-of-passage map. The function is designed to minimize the accumulated cost at the destination. Thus, the cost map models the minimum cost of moving from a given starting point to a specified destination. Generally, this destination is all cells on the map (Conolly & Lake, 2000). This technique generates a raster coverage in which the surface area represents the minimum incremental distance from any source feature to the edge of the map. The value of each cell indicates the cost of traversing the terrain from the origin. When generating this raster map, the calculation considers not only the Euclidean distance between two points (in a straight line), but also a friction map that represents the difficulty of traversing the terrain. As discussed in numerous works (e.g., Conolly & Lake, 2006; Van Leusen, 2002), these analyses have different methods that form the basis of accessibility and mobility studies. Therefore, it is important to specify the used algorithms and parameters. The distance (accumulated cost) maps generated here are products of various processing steps. They are based on the ASTER GDEM terrain elevation model. It has a spatial resolution of 30 m, which is inherited by all successive products generated. From this model, we generated a terrain slope map that we processed according to the friction model proposed by Tobler (1993). We adapted this model as used by Fábrega and Parcero (2007) to provide results in walking hours (0.001/Eq. 6.5).

$$W = 6\exp\{-3.5 * abs(D + 0.05)\} \tag{6.5}$$

where W is the walking speed (in km/h), S is the terrain slope.

There are numerous algorithms, but we chose this one because it provides results in walking time units, making it useful for interpreting distances to resources. The cumulative cost surface analysis, which measures the distance between the water resource and the surrounding landscape, provides a more accurate representation of human mobility than straight-line measurement (Euclidean distance). For the cumulative cost surface analysis, we used a buffer area of 6 km from the limits of the surveyed sector. The symbolization is a classification into ten regular distance intervals, each expressed in hours (Fig. 6.20). This thematic map allowed us to assign a distance value to each assemblage/location by intersecting both layers of information (see the "cdist_ag" column in Table 26, Supplementary Material). The central tendency and dispersion values were plotted separately for each class of evidence in Fig. 6.21.

The graph suggests that all classes are located within 9–23 min of a permanent water source. The types of evidence closest to these sources are rock art sites (A), followed by multiple activity loci (MAL). However, it is unclear whether the observed tendencies are significant. To address this, a K–S test is applied to statistically contrast them. The set of locations "NE" is the sample used for the contrast as part of the pattern recognition technique. Since no archaeological evidence was found at these locations, they serve as a control sample that describes the properties of the landscape itself. The presence of significant differences between the "NE" set and the classes of archaeological evidence indicates a reliable pattern (Kvamme, 1988; Woodman & Woodward, 2002).

In the case of sites with art (A) and workshops (W), it was not possible to apply the K–S test due to the low sample size. Therefore, they were not included in this test or in tests with the other variables. The proximity of art sites to water sources, as observed in Fig. 6.21, is a trend that must be considered in future studies. By adding

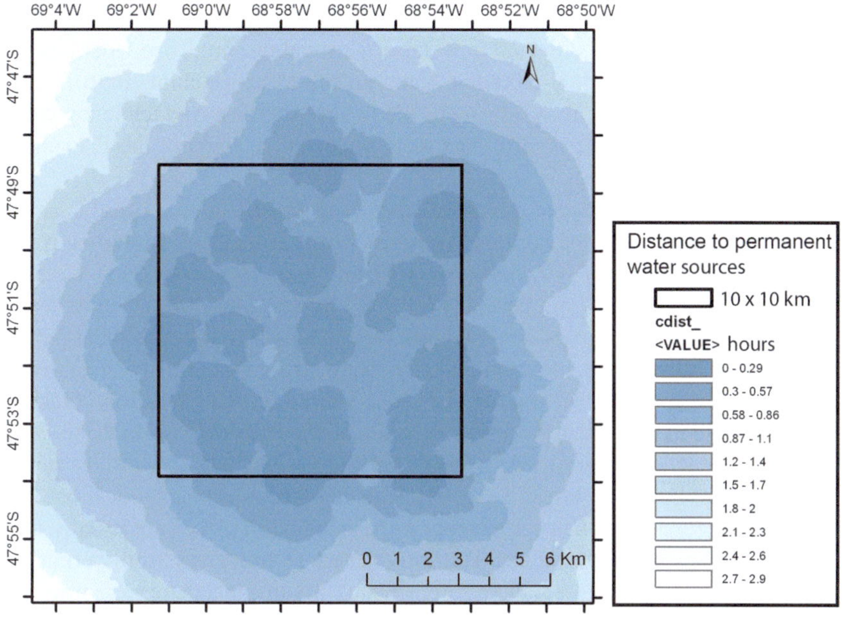

Fig. 6.20 Distance to permanent water sources

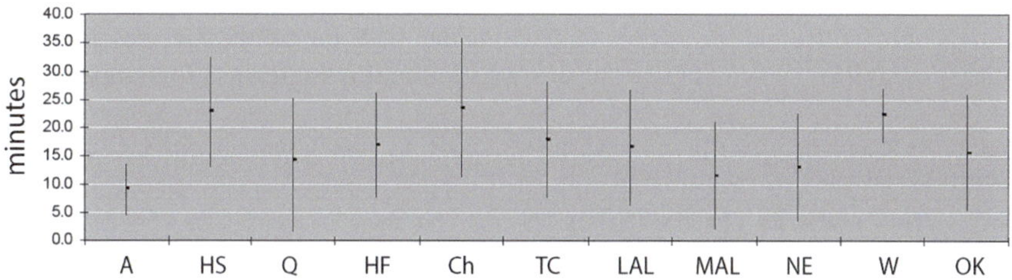

Fig. 6.21 Average distance values and standard deviations to permanent water sources

new data, the statistical significance of this trend can be determined. The association between springs and sites with art is probably related to the environment in which artistic practices are carried out, such as caves. Caves are naturally associated with the action of wind, which, on slopes of particular orientations, produces progressive perforation of rock formations (Miotti et al., 2007; Panza, 2001).

6.2.1.2 Pattern Observed in Relation to Permanent Water Sources

The results of the Kolmogorov–Smirnov (K–S) test show that although the distance to permanent water sources for the MAL class does not differ statistically from the control sample (EN), the MAL class is located closest to these water sources (Table 6.12 and Fig. 6.22). Given that water is a critical resource, one would expect springs to exert an important attractive force on the MAL (Jochim, 1976), which should impact the statistical significance. We believe the observed tendency is due to the manner in which the control data were collected, as eight of the EN sample points coincide with springs.

The other types of evidence are significantly separated from the control sample and located farther from the springs. These include the locus of limited activities (LAL), temporary camps (TC), chenques (Ch), and hunting stations (HS). Notably, the TCs are relatively far from permanent water sources. Since the TC class represents multiple activities carried out in camps, we expected it to show a similar pattern to the MAL. It is likely that some temporarily occupied camps (TC) were established near permanent springs over time in the area. Therefore, their location coincides with places also used as longer-occupancy residential camps (MAL). In these cases, TC and MAL cannot be identified as separate entities based only on a surface survey (Binford, 1982; Dewar & McBride, 1992). This class of assemblage/location (TC) can only be recognized when settlements were installed in places not usually used for MAL settlements. On average, TCs are located 18 min away from permanent water sources in our study area. Therefore, we must assume that there were means of transporting water from the springs. Chronicles and ethnographic works referencing Aónikenk groups mention transferring drinking water to campsites using leather bags and other organic containers, such as bladders and stomach bags (Martinić, 1995). Another observed pattern is that of class HS, which is also far from the springs. This information allows us to consider how the landscape was used for hunting activities. Since HSs are places where hunting strategies are planned, they

Table 6.12 Distance to springs

Set 1	*n*	*x* hours	*x* min	Set 2	*n*	*x* hours	*x* min	*D*	*P*
NE	127	0.216	13	MAL	18	0.1922	12	0.1934	0.550
NE	127	0.216	13	HS	16	0.3790	23	**0.5536**	0.000
NE	127	0.216	13	HG	10	0.2812	17	**0.3457**	0.173
NE	127	0.216	13	LAL	138	0.2754	17	**0.1866**	0.017
NE	127	0.216	13	OK	190	0.2608	16	**0.1407**	0.090
NE	127	0.216	13	Ch	14	0.3906	23	**0.4454**	0.009
NE	127	0.216	13	TC	24	0.2981	18	**0.3087**	0.033
NE	127	0.216	13	Q	12	0.2387	14	0.2171	0.538

K–S test results

"Set 1" = control sample; "x hours" = average cost distance between the class of assemblage/location and the resource "aguadas" expressed in hours; distance "x min" = the same value expressed in minutes; "Set 2" = sample that is compared with the control; "x hours" and "x min" = same as above referring to the different classes; "D" = maximum difference between the cumulative data distributions; "P" = critical value for an alpha = 0.05. Statistically significant differences are highlighted in bold font

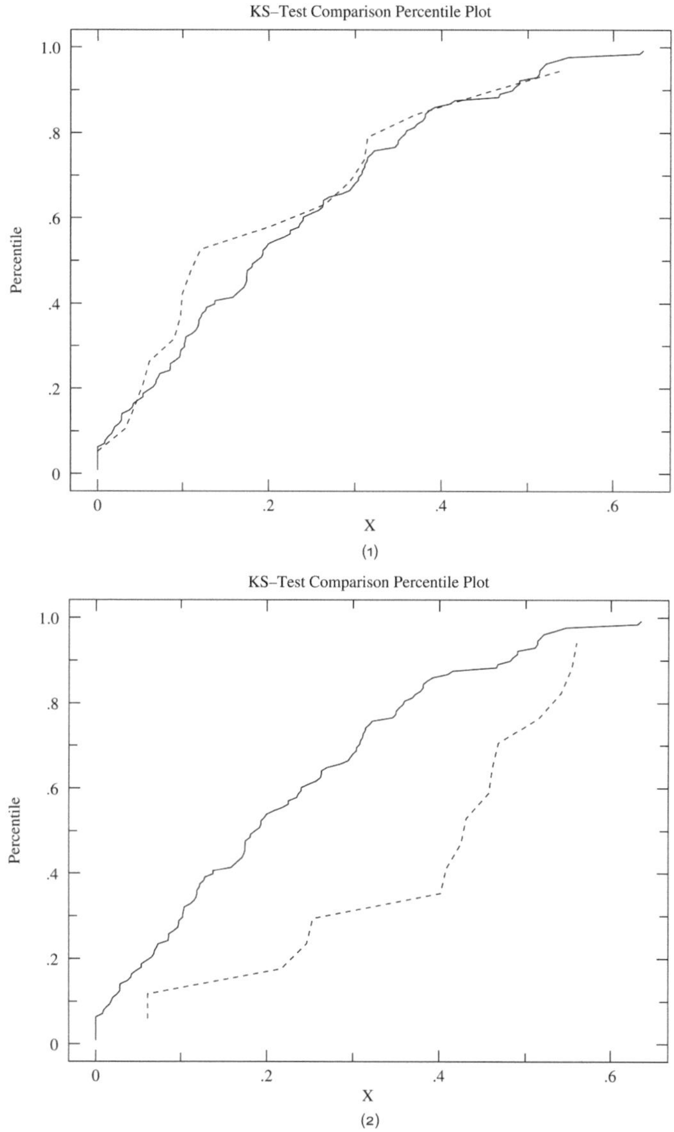

Fig. 6.22 Cumulative frequency diagrams showing the variable distance to permanent water sources. The continuous line shows the control class: NE; the dashed line shows the analyzed class: 1) MAL: multiple activities locus; 2) HS: hunting stations; 3) HG: hunting grounds; 4) LAL: limited activities loci; 5) Ok: occasional knapping; 6) Ch: chenques; 7) TC: temporary camps; 8) Q: quarries

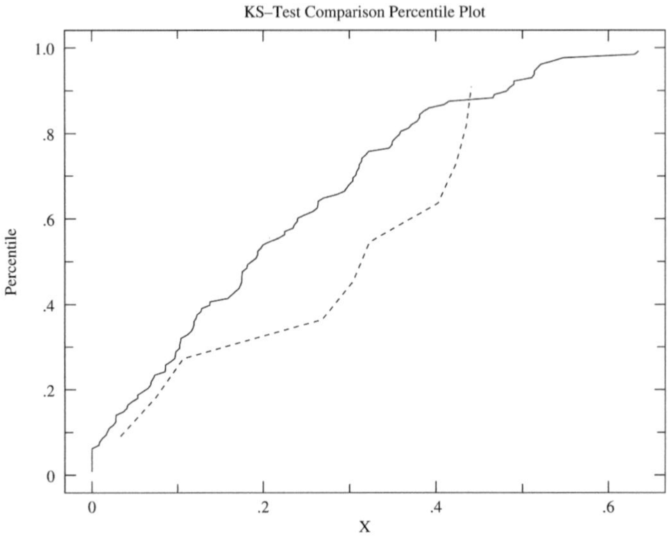

(3)

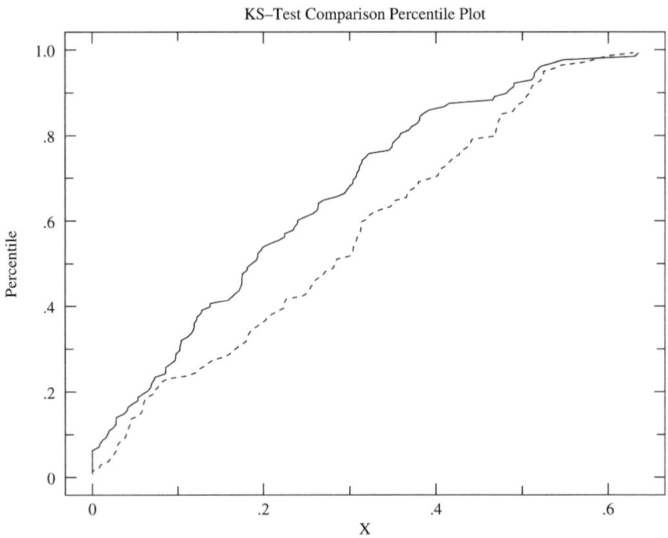

(4)

Fig. 6.22 (continued)

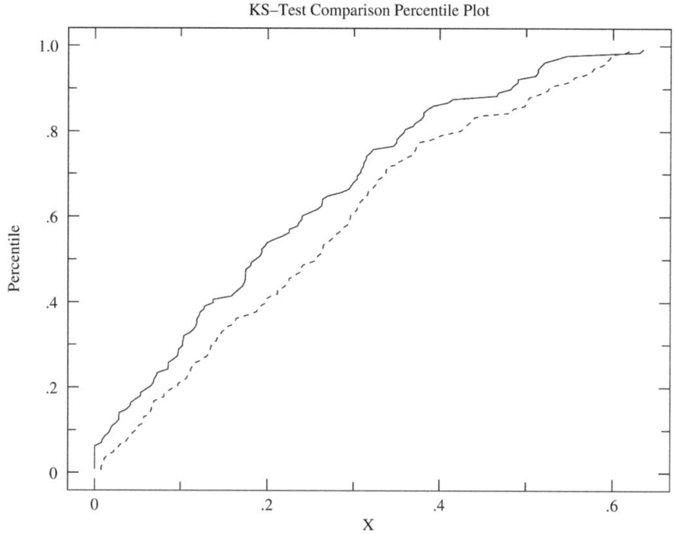

(5)

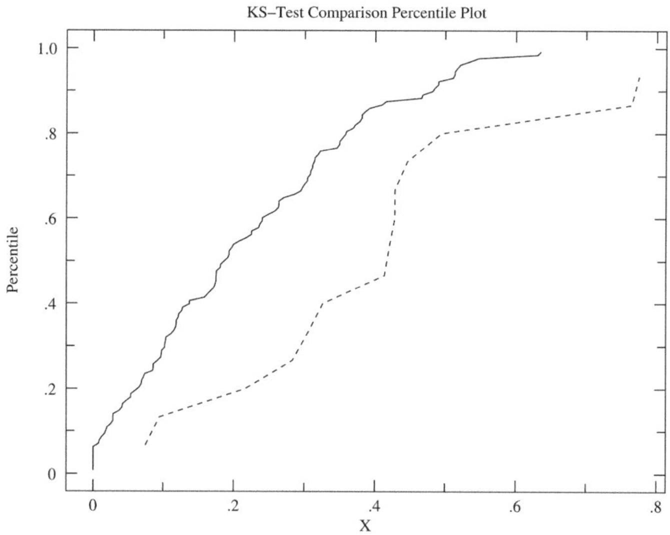

(6)

Fig. 6.22 (continued)

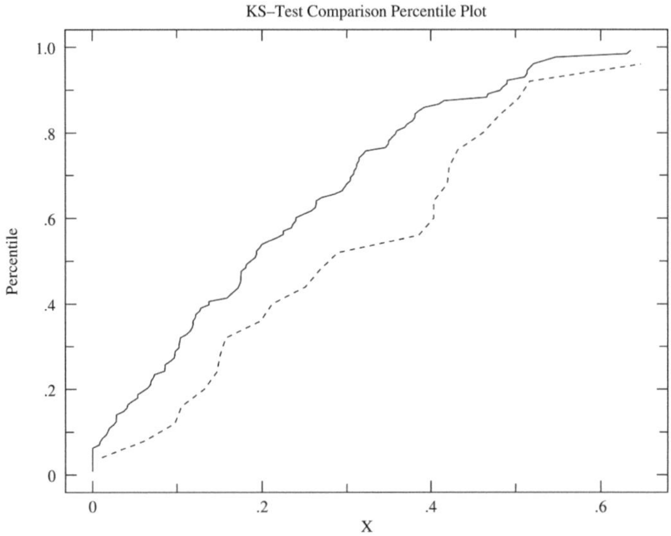

(7)

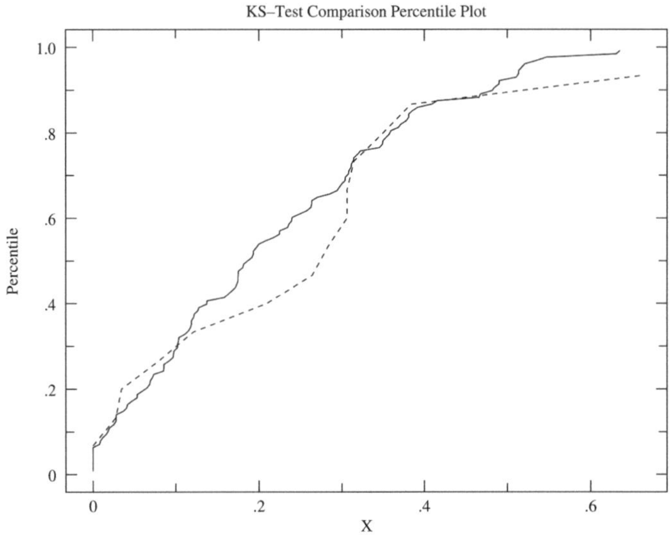

(8)

Fig. 6.22 (continued)

are likely far from domestic LAM sectors, which coincide with permanent water sources, since the normal human and dog activity in these areas would likely cause the animals to move away.

6.2.2 *Distance to Temporary Water Sources*

Other water sources in the study area are only available temporarily. These include shallow lakes and ephemeral tributaries. They only have water in them after sporadic rainfall throughout the year and after snow melts. Although there is no well-defined rainy season, the temporary water sources will most likely have water at the end of winter.

6.2.2.1 Temporary Water Sources and GIS

The information used to locate these resources in the studied sector was obtained by interpreting the drainage network on aerial photographs. Reference data came from the vector coverage of tributaries on the 1:250,000 scale GIS-IGN topographic map (Table 5.1). The tributaries were edited as a vector coverage of lines and the shallow lakes as a vector coverage of polygons in a GIS. Both coverages, along with the point coverage of springs, were processed for use in creating the map of accumulated costs (distances) (Fig. 6.23). Evaluating permanent and temporary water sources separately is important for considering aspects of settlement strategies concerning variations in water availability in the landscape.

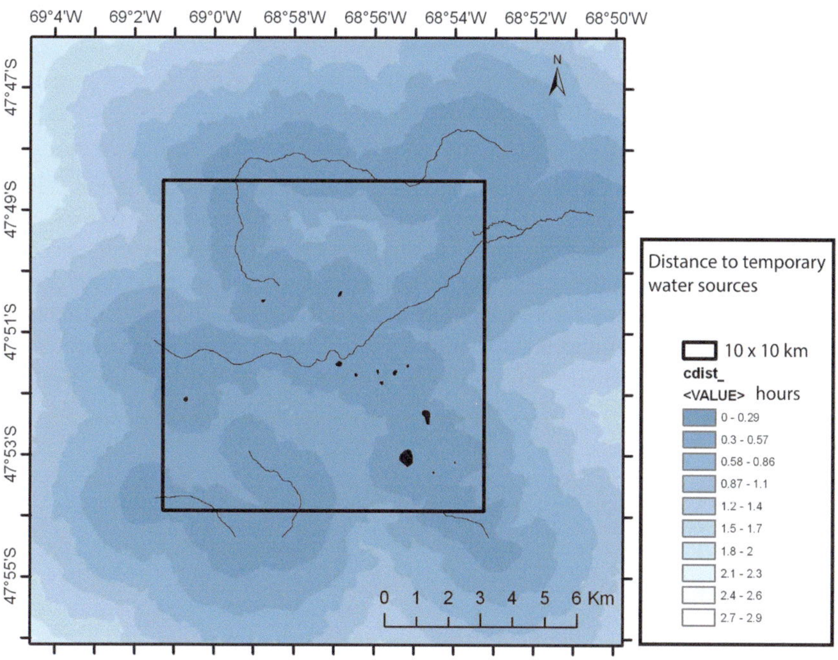

Fig. 6.23 Distance to temporary water sources

6.2.2.2 Observed Pattern of Distance to Temporary Water Sources

A visual analysis of Fig. 6.24 shows that, although the distances are generally smaller, all classes repeat the tendencies of the previous case. On average, all classes of evidence are located between four and 20 min from a temporary water source. The K–S test revealed significant differences in the distance value distributions for most of the classes. The art sites (A) class is the closest. The LAL class is also closer to the temporary water source than the control sample. Conversely, the OK, TC, W, Ch, and HS are farther away (Table 6.13 and Fig. 6.25). The chenques (Ch) class is located the farthest from water sources, both permanent springs, which are usually found at intermediate heights on plateau slopes, and temporary water sources in ravines and canyons. This shows a clear pattern for this class of evidence that, more than any other, indicates a kind of activity carried out far from this resource. On the contrary, the observed tendency in the LAL suggests that temporary water sources attracted longer-occupancy campsites and other limited activities during times of surface water availability. The only available rainfall records are for the Gobernador Gregores area. This area has similar continental conditions to the study area and is located 130 km southwest of it. The records indicate less than 200 mm of annual rainfall without distinct seasons of intense rain. Therefore, temporary water sources originating from rainfall are unpredictable and ephemeral in this area. There is a higher chance of water availability in temporary streams and shallow lakes after snow melts at the end of winter and in the spring (Melendi, 2023).

6.2.3 *Distance to Lithic Raw Materials*

The distance to sources of lithic raw materials is usually an important aspect in locational studies. However, the abundance of useful rocks in this landscape (see Chap. 2) suggests that the location of the site may not have been significantly influenced by this factor. Rather, the lithic resources could have been transported to residential areas and other locations where they were needed. However, analyzing distance to raw material sources is important for studying provisioning strategies, the circulation of rocks used to manufacture lithic artifacts, and hunter-gatherer mobility (Binford, 1980).

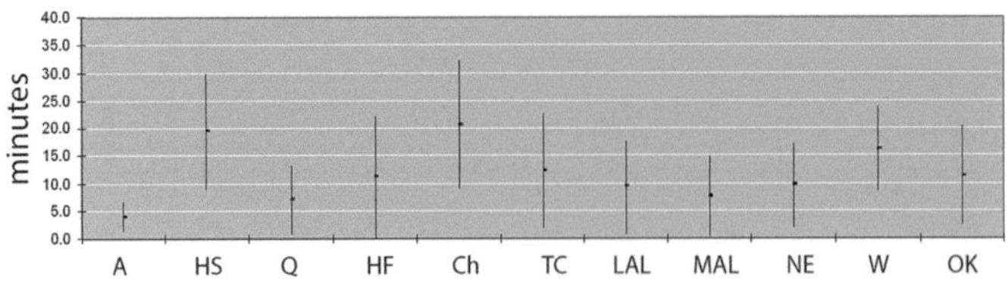

Fig. 6.24 Average distance values and standard deviations to temporary water sources

Table 6.13 Distance to temporary water sources

Set 1	*n*	*x* hours	*x* min	Set 2	*n*	*x* hours	*x* min	*D*	*P*
EN	127	0.1573	9.44	MAL	18	**0.1244**	7.46	**0.2957**	0.103
EN	127	0.1573	9.44	HS	16	**0.3246**	19.47	**0.5153**	0.001
EN	127	0.1573	9.44	HG	10	0.1843	11.06	0.2693	0.447
EN	127	0.1573	9.44	LAL	138	**0.1539**	9.24	**0.1403**	0.136
EN	127	0.1573	9.44	OK	190	0.1874	11.24	0.1218	0.195
EN	127	0.1573	9.44	Ch	14	**0.3407**	20.44	**0.5174**	0.001
EN	127	0.1573	9.44	TC	24	**0.2032**	12.19	**0.2444**	0.153
EN	127	0.1573	9.44	Q	12	0.1467	8.80	0.1530	0.905

K–S test results
"Set 1" = control sample; "x hours" = average cost distance between the class of assemblage/location and the resource "aguadas" expressed in hours; distance "x min" = the same value expressed in minutes; "Set 2" = sample that is compared with the control; "x hours" and "x min" = same as above referring to the different classes; "D" = maximum difference between the cumulative data distributions; "P" = critical value for an alpha = 0.05. Statistically significant differences are highlighted in bold font

6.2.3.1 Lithic Raw Materials and GIS

The known sources of raw materials for this sector include four exploited outcrops and eight secondary deposits (see Table 2.3). All of these sources are represented in GIS as a vector coverage of points. Secondary deposits, which are extensions of cobbles and blocks, could also be represented by polygons. However, because they are not homogeneous in raw materials spatial distrubution, they are represented by points located where larger nodules were registered. This topic will be discussed in more detail later. The exploited rock outcrops were also represented by points on the same coverage. In each case, a GPS point close to the average center of each quarry was selected. A cost-distance calculation was performed for each assemblage/location in relation to the deposits and exploited outcrops. In this case, the distance to any potential lithic resource in the landscape is not measured; only the distance to those with evidence of exploitation is measured. In other words, the variable is the distance to the class of quarries (Q). This decision is based on the widespread distribution of useful rocks in this landscape. Figures 6.26 and 6.27 show that, on average, all classes are located between 29 and 50 min from some Q. This is compared with the control sample (NE). This could indicate that, unlike the distance to water, the distance to this resource was not considered in the location of different classes of evidence. This coincides with expectations given that lithic raw materials are not vital resources and are widely and homogeneously distributed in the landscape (Cattaneo et al., 2006). Rocks and minerals can probably be transported more easily than water from sources to places of manufacture and use.

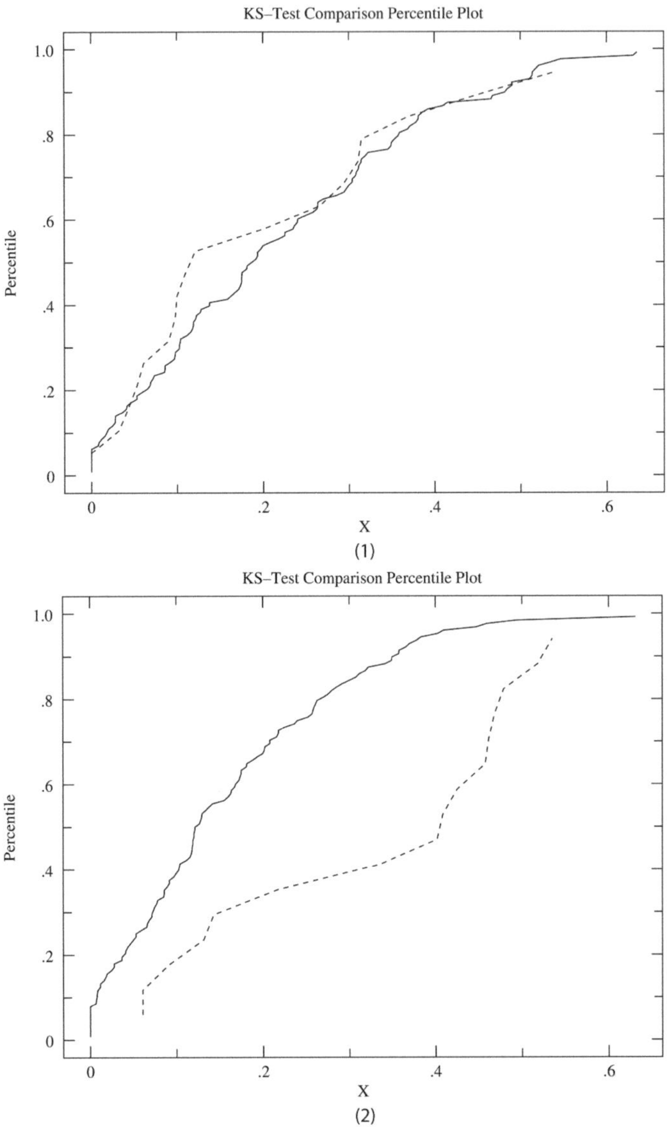

Fig. 6.25 Cumulative frequency diagrams showing the variable distance to temporary water sources. The continuous line shows the control class: NE; the dashed line shows the analyzed class: 1) MAL: multiple activities locus; 2) HS: hunting stations; 3) HG: hunting grounds; 4) LAL: limited activities loci; 5) Ok: occasional knapping; 6) Ch: chenques; 7) TC: temporary camps; 8) Q: quarries

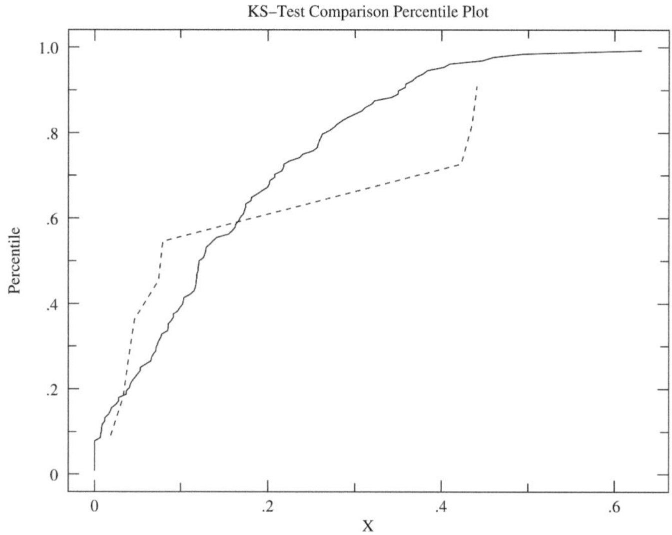

(3)

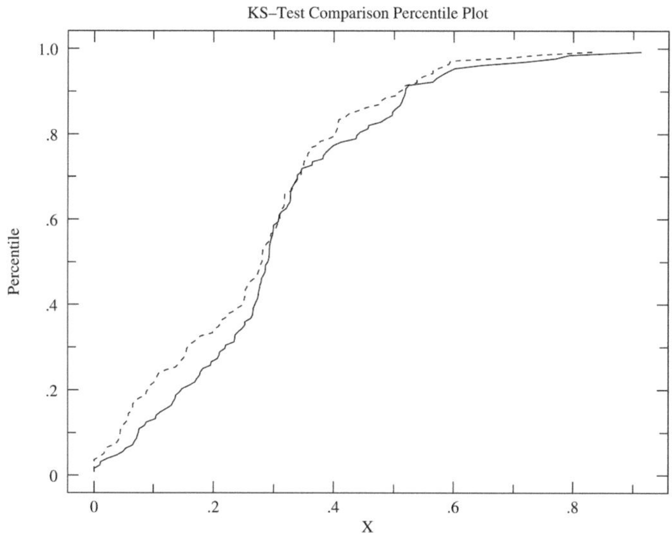

(4)

Fig. 6.25 (continued)

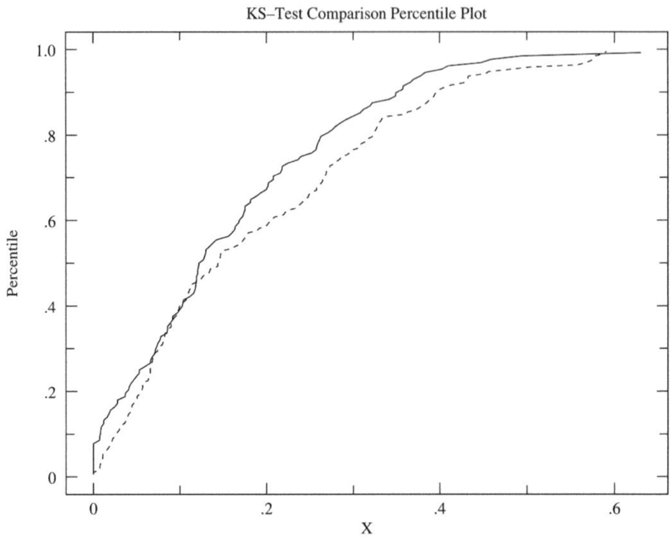

(5)

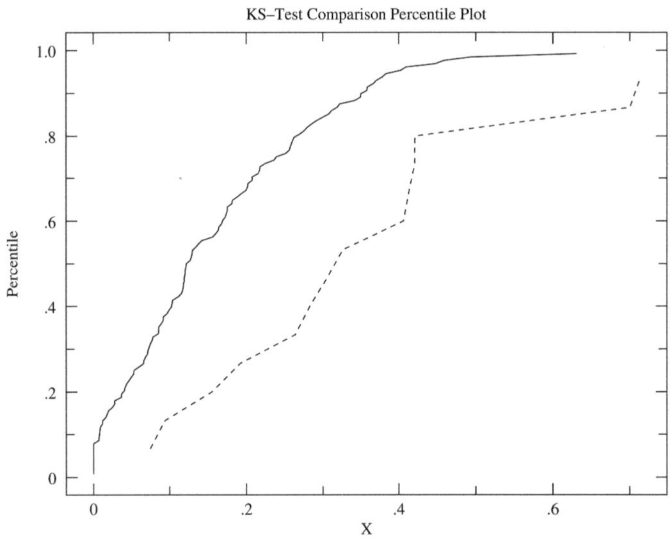

(6)

Fig. 6.25 (continued)

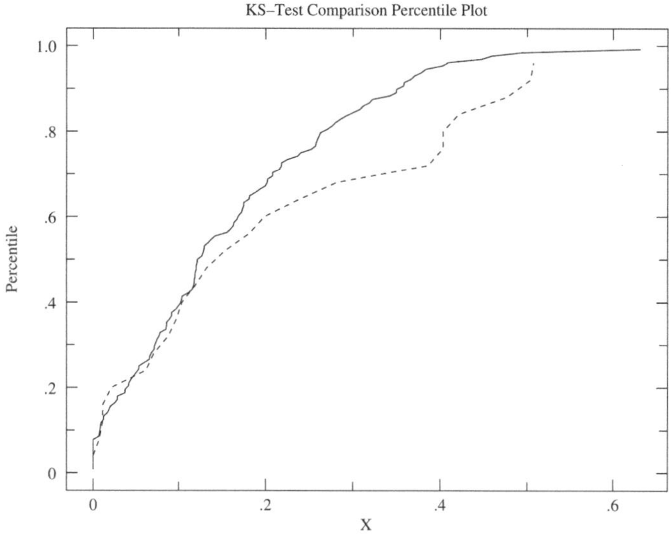

(7)

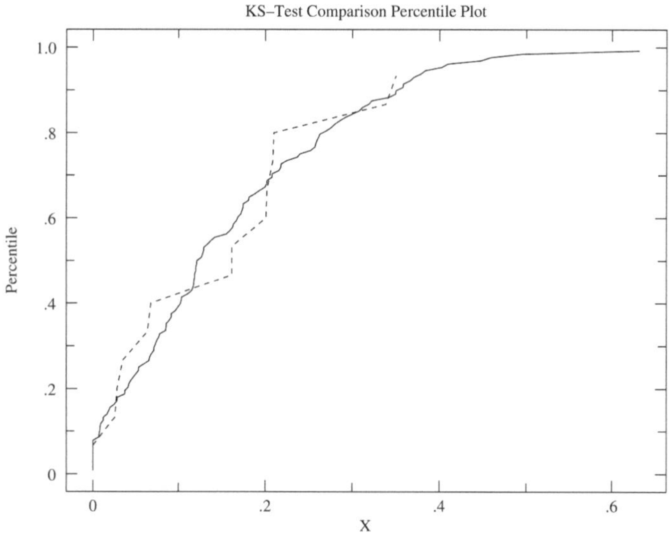

(8)

Fig. 6.25 (continued)

6.2.3.2 Observed Pattern of Distance to Sources of Lithic Raw Material

On average, all classes of evidence are located between 25 and 50 min from a lithic provisioning quarry (Fig. 6.27). These average distances are much greater than those for water sources, indicating that water was more important than lithic sources in the

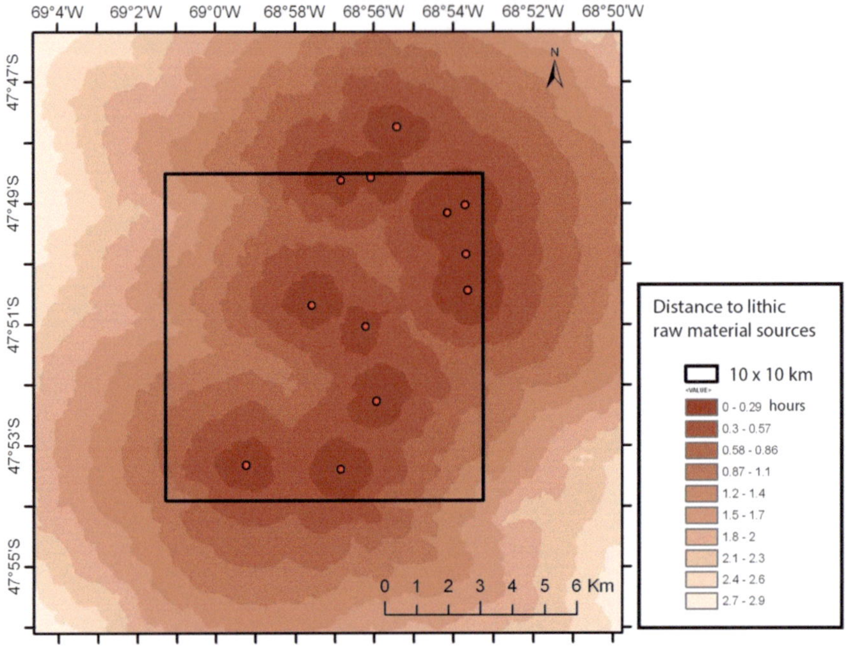

Fig. 6.26 Distance to sources of lithic raw materials

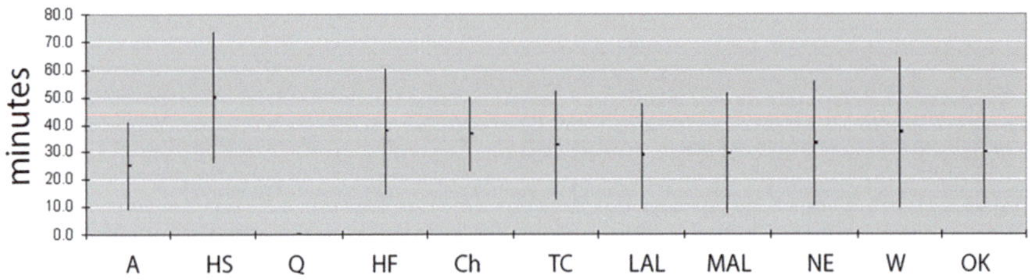

Fig. 6.27 Average distance values and standard deviations to the main sources of lithic raw material

settlement strategy. The patterns suggested by the graph were contrasted using the K–S test (see Table 6.14 and Fig. 6.28). The maximum differences in the cumulative distributions show that the MAL, LAL, and OK classes are significantly closer to the quarries than the control sample. Additionally, TCs are closer to the Q than the control sample, though no significant pattern was detected. Conversely, HS and Ch are significantly farther from lithic quarries than the control sample. MAL showed a pattern toward this resource. They are located close to sources of raw materials. This was unexpected given the abundant distribution of rock sources in the area. This led us to carry out a thorough analysis of the distribution of lithic resources in the sector and how they were used and discarded in MAL sites to learn about the use and circulation of lithic resources in this landscape (see Chap. 7). On average, the hunting stands

Table 6.14 Distance to sources of lithic raw material

Set 1	*n*	*x* hours	*x* min	Set 2	*n*	*x* horas	*x* min	*D*	*P*
EN	127	0.5496	32.97	MAL	18	0.4893	29.36	**0.1522**	0.827
EN	127	0.5496	32.97	HS	16	0.8303	49.82	**0.5536**	0.000
EN	127	0.5496	32.97	HG	10	0.6232	37.39	0.1890	0.858
EN	127	0.5496	32.97	LAL	138	0.4753	28.52	**0.1571**	0.069
EN	127	0.5496	32.97	OK	190	0.4939	29.63	**0.1438**	0.078
EN	127	0.5496	32.97	Ch	14	0.6068	36.41	**0.2975**	0.177
EN	127	0.5496	32.97	TC	24	0.5343	32.06	0.1237	0.897

K–S test results
"Set 1" = control sample; "x hours" = average cost distance between the class of assemblage/location and the resource "aguadas" expressed in hours; distance "x min" = the same value expressed in minutes; "Set 2" = sample that is compared with the control; "x hours" and "x min" = same as above referring to the different classes; "D" = maximum difference between the cumulative data distributions; "P" = critical value for an alpha = 0.05. Statistically significant differences are highlighted in bold font

(HSs) are 50 min away. These stands are spatially segregated from activities related to the extraction and processing of raw materials. However, during hunting trips, raw materials found by chance could have been collected and had their cortex removed in situ, perhaps for knapping during downtime (Binford, 1979; Nelson, 1991), as part of strategies embedded in others. The Ch class occurs at an average distance of 36 min from lithic resources, repeating a similar pattern to that of water resources. Future works should analyze the locations of the Ch and places where minerals of hematite or ochre pigments are available, as their extraction may be related to ritual activities. Ochre has been mentioned in chronicles and archaeological research as being associated with mortuary customs (Aguerre, 2000; Deodat, 1946; Martinic, 1995; Miotti, 2006; Vignati, 1934 and many others). As for workshops (W), they constitute a class with an n too low to apply the Kolmogorov–Smirnov test. However, it is interesting to note that existing workshops are located far from known quarries, outcrops, and secondary deposits. This suggests that raw materials were transported to workshop locations. However, this idea requires more evidence.

6.2.4 *Sheltered Sectors of the Landscape*

As current climate studies indicate, few places on the globe are as dominated by a single meteorological element as southern South America, where strong winds are persistent (Ariztegui et al., 2008). Winds have blown constantly from the west in this region since humans first occupied it, although records from the post-glacial period show an increase in their intensity (Haberzettl et al., 2009). This justifies analyzing sheltered and exposed sectors as another variable (Borrero, 1994–1995). Given the west-to-east direction of the winds, the most protected sectors are those facing east.

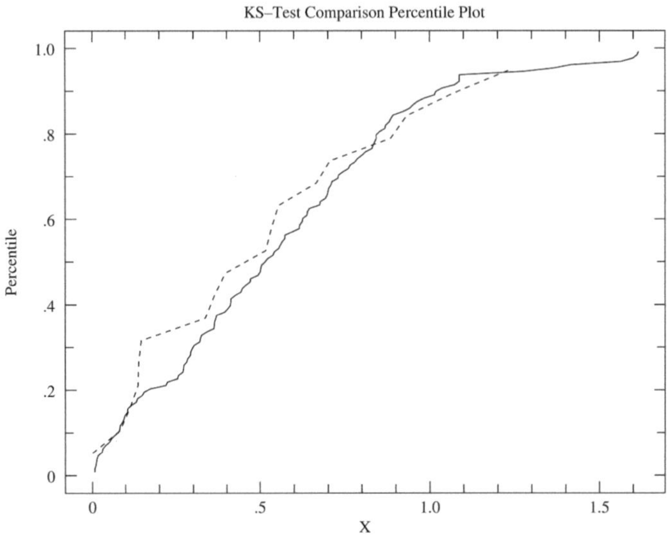

(1)

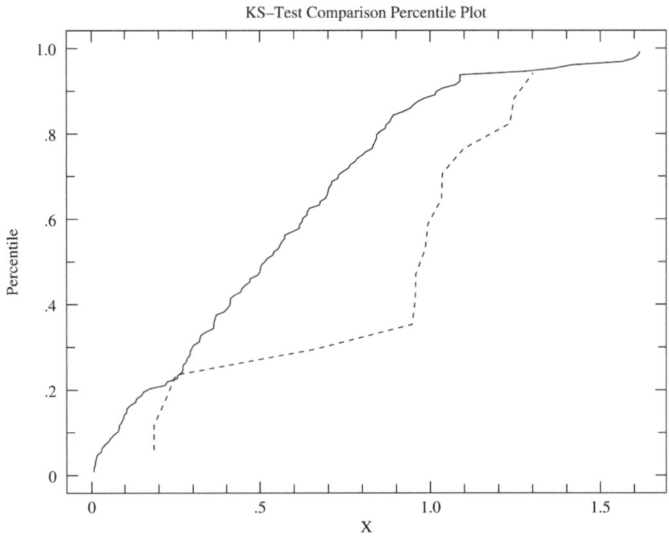

(2)

Fig. 6.28 Cumulative frequency diagrams showing the variable distance to quarries. The continuous line shows the control class: NE; the dashed line shows the analyzed class: 1) MAL: multiple activities locus; 2) HS: hunting stations; 3) HG: hunting grounds; 4) LAL: limited activities loci; 5) Ok: occasional knapping; 6) Ch: chenques; 7) TC: temporary camps

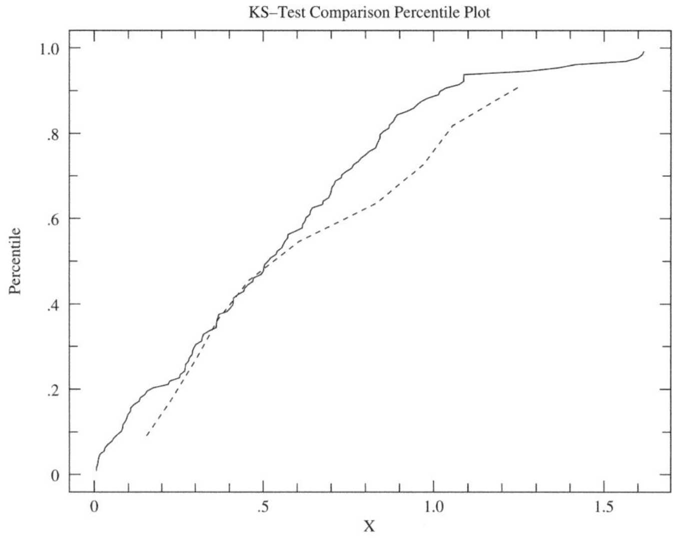

(3)

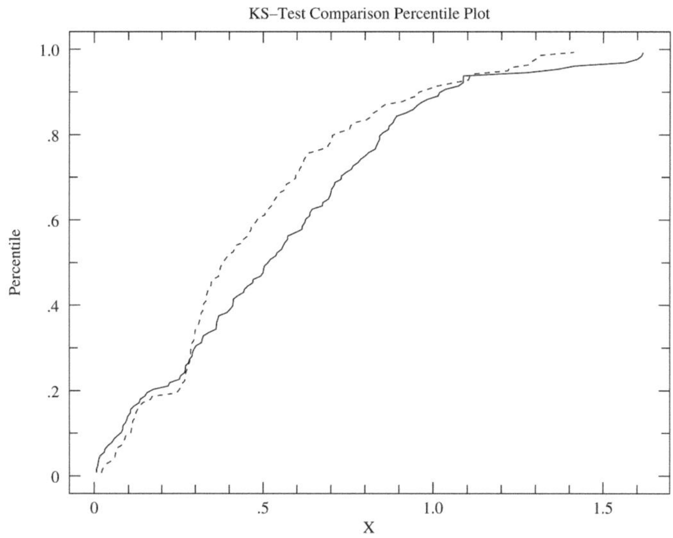

(4)

Fig. 6.28 (continued)

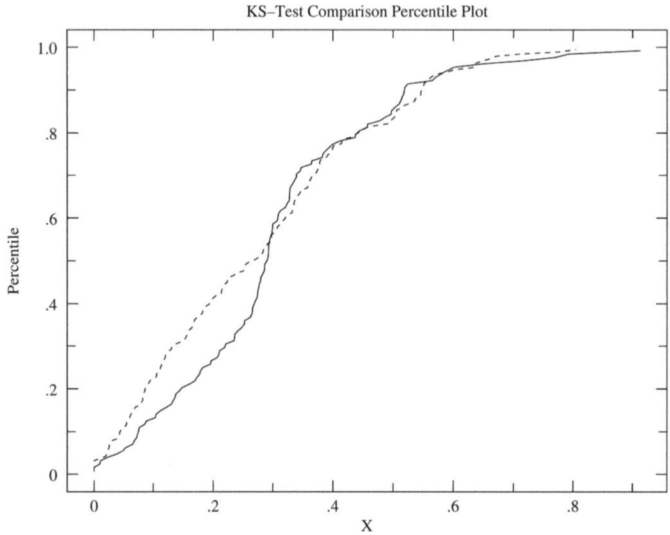

(5)

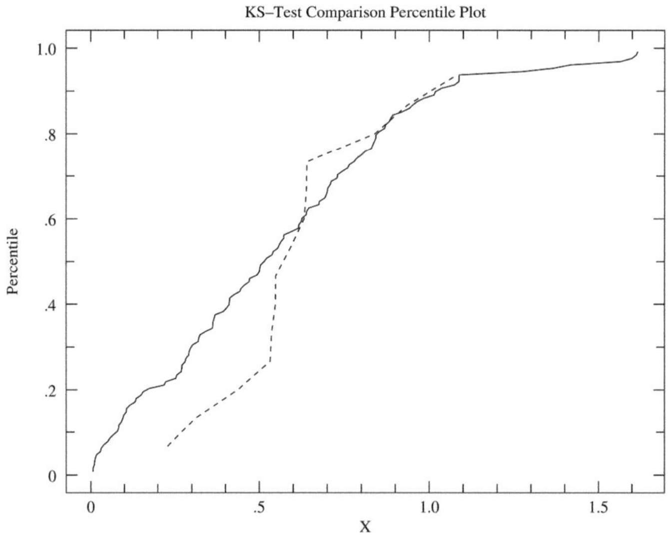

(6)

Fig. 6.28 (continued)

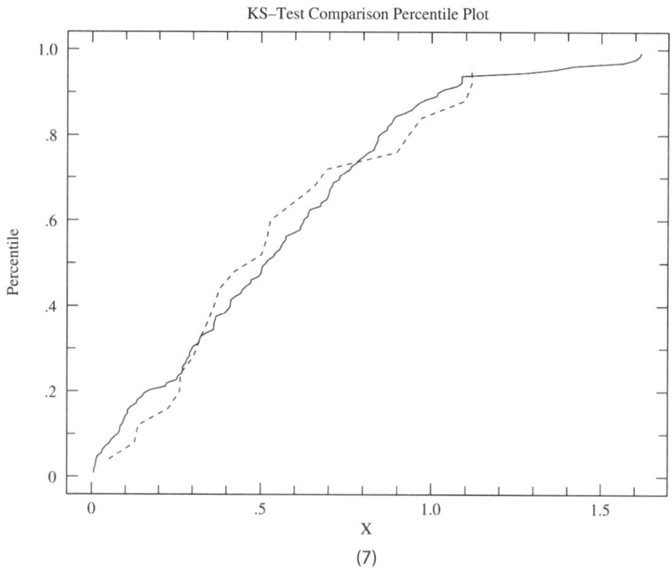

Fig. 6.28 (continued)

6.2.4.1 Sheltered Sectors and GIS

To model the shelter conditions of the landscape, a GIS digital elevation model (DEM) was used in conjunction with field information. A terrain aspect map, derived from the DEM, presents the orientation angle value for each cell. Although the shelter of a landscape is a difficult concept to bring to a map (Kvamme, 1988), sheltered and exposed sectors in this landscape were defined based on regional wind characteristics. The 30-m resolution ASTER GDEM (METI, Japan, and NASA, USA, see Table 5.1) was used to generate a terrain orientation map. On this map, the E, NE, and SE orientation cells were assigned to sheltered sectors, while the remaining values were assigned to exposed sectors. To verify the accuracy of the map, 458 GPS records were used to identify sectors that were relatively sheltered or exposed to the wind. In the field, shelter conditions were recorded as follows: "1-bushes" when shelter was provided by vegetation but not topography, "2-shelter" when topography was the main protection factor against wind, "3-shelter and bushes" when both were present, and "4-no shelter" when completely exposed. A chi-square test was used to evaluate the goodness of fit between the GIS terrain aspect values and the recorded shelter conditions (Tables 6.15 and 6.16). The X^2 test yielded observed and critical values of 57.22 and 32.67, respectively, indicating significant dependence between the rows and columns (for an alpha of 0.05 and 21 degrees of freedom). Analyzing x^2 per cell shows that the sectors with the greatest shelter are those oriented north and southeast. In sectors with north orientations, topography and vegetation act together as the main protection against winds. SE-facing sectors are mainly protected by the terrain, without bush vegetation present. Sectors with E

Table 6.15 Terrain orientation values derived from an aspect map (files) and wind shelter conditions recorded in the field (columns)

Aspect	1-bush	2-topography shelter	3-topography shelter and bushes	4-no shelter
E	14	11	10	22
N	57	15	29	28
NE	28	11	13	19
NW	6	5	5	19
S	10	5	3	25
SE	20	17	8	27
SW	4	3	0	16
W	6	3	6	13
Total = 458	145	70	74	169

and NE orientations also exhibit trends that classify them as sheltered. Conversely, the least sheltered aspects are the S, SW, and NW. Based on these results, the north (N), northeast (NE), east (E), and southeast (SE) orientation cells were classified as sheltered (including values between 337.5–360 °N and 0–157.5 °N), while the south (S), southwest (SW), west (W), and northwest (NW) orientation cells were classified as exposed (including values between 157.6 and 337.5 °N) (Fig. 6.29). To test for dependence between types of archaeological evidence and shelter from the wind, a chi-square goodness-of-fit test was performed. Expected frequencies were calculated, taking into account the total area of sheltered (44.28%) and exposed (55.71%) surfaces, to rule out the possibility that their different proportions, as they occur naturally in the landscape, might introduce a bias in the results (Connolly & Lake, 2006; Shennan, 1992) (Tables 6.17 and 6.18; Fig. 6.30).

6.2.4.2 Location with Respect to Sheltered/Wind-Exposed Sectors

With an α of 0.05 and 10 degrees of freedom (DOF), the observed value of x^2 is 2.314. Its critical value is 18.307, indicating that the variables are not significantly dependent. The different classes are located in sectors that are both exposed to and sheltered from the wind. However, the individual contribution to x^2 reveals that the Ch are located in sheltered places in lower proportions than the control sample. Regarding residential base camps (MAL), it was expected that they would be located in sheltered areas (Martinić, 1995; Miotti, 1989; Vignati, 1934, among others), but the analysis reveals a variety of locations in both sheltered and windy areas. Those MALs that present topographic protection from the wind are located in the La Primavera canyon and the Paso site (ID_60) in the Las Mercedes locality. Others are the El Pájaro Muerto sites (ID_25) and site D9 (ID_78) in La Lotita. Locations exposed to the wind may indicate the use of awnings in residential settlements where natural topographic shelter is lacking.

Table 6.16 Chi-square test per cell

Aspect	1-bush	2-topography shelter	3-topography shelter and bushes	4-no shelter
E	(−) NS	(+) NS	(+) NS	(+) NS
N	(+) ***	(−) NS	(+) **	(−) ***
NE	(+) NS	(+) NS	(+) NS	(−) *
NW	(−) *	(−) NS	(−) NS	(+) **
S	(−) NS	(−) NS	(−) NS	(+) ***
SE	(−) NS	(+) **	(−) NS	(+) NS
SW	(−) NS	(−) NS	(−) *	(+) ***
W	(−) NS	(−) NS	(+) NS	(+) NS

Test results and contributions to x^2. Shelter condition and cardinal orientation in the local relief. The following are the acronyms for the results of the statistical tests applied through the XLSTAT software: (+): observed frequency higher than the theoretical frequency; (−): observed frequency lower than the theoretical frequency; NS: Chi-square test per cell not significant at threshold alpha = 0.100; *: Chi-square test per significant cell at threshold alpha = 0.100; **: Chi-square test per significant cell at threshold alpha = 0.050; ***: Chi-square test per significant cell at threshold alpha = 0.010; DOF: degrees of freedom

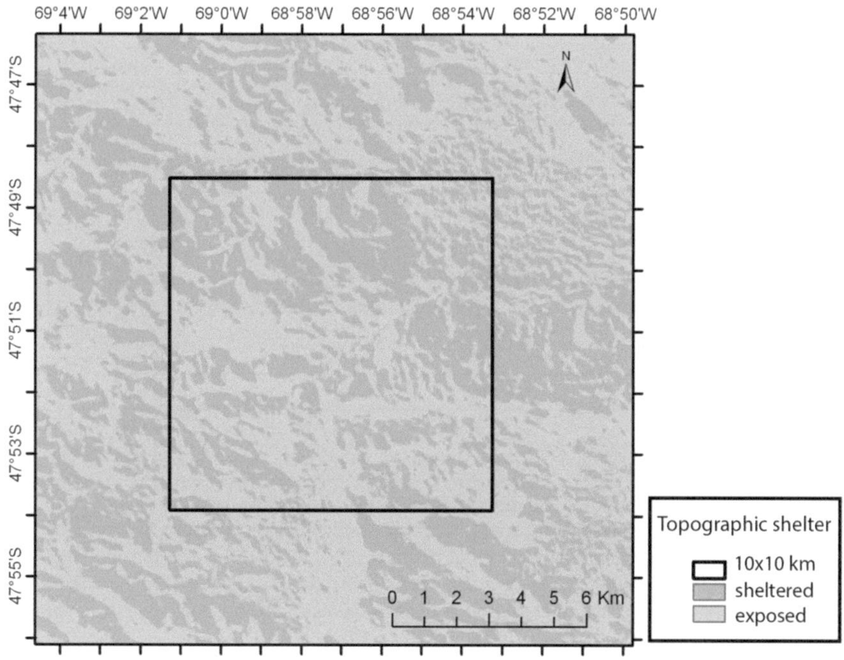

Fig. 6.29 Map of sectors repaired/exposed from the wind

Table 6.17 Observed (Of) and expected (Ef) frequencies in sheltered and exposed sectors of the landscape

Classes	(Of) sheltered	(Of) exposed	Total	(Ef) sheltered	(Ef) exposed
A	4	5	9	0.666	0.922
HS	9	7	16	1.499	1.291
Q	7	5	12	1.332	0.922
HG	4	6	10	0.666	1.107
Ch	**1**	**13**	**14**	**0.167**	**2.398**
TC	12	12	24	1.998	2.213
LAL	71	67	138	11.824	12.357
MAL	7	11	18	1.166	2.029
NE	54	73	127	8.993	13.464
W	3	3	6	0.500	0.553
OK	91	99	190	15.155	18.259
Total	263	301	564	44%	55.5%

Statistically significant differences are highlighted in bold font

Table 6.18 Chi-square test per cell

Class	Sheltered	Exposed	Sheltered	Exposed
A	(−) NS	(+) NS	0.002	0.001
HS	(+) NS	(−) NS	0.057	0.045
Q	(+) NS	(−) NS	0.113	0.090
HG	(−) NS	(+) NS	0.018	0.014
Ch	(−) **NS**	**(+) NS**	**0.825**	**0.653**
TC	(+) NS	(−) NS	0.010	0.008
LAL	(+) NS	(−) NS	0.121	0.096
MAL	(−) NS	(+) NS	0.043	0.034
EN	(−) NS	(+) NS	0.087	0.069
W	(+) NS	(−) NS	0.003	0.002
OK	(+) NS	(−) NS	0.010	0.008

Test results and contributions to x^2. Shelter conditions and archaeological evidence class. Acronyms for the results of statistical test: (+): observed frequency higher than the theoretical frequency; (−): observed frequency lower than the theoretical frequency; NS: Chi-square test for non-significant cell at threshold alpha = 0.100; *: Chi-square test per significant cell at threshold alpha = 0.100; **: Chi-square test per significant cell at threshold alpha = 0.050; ***: Chi-square test per significant cell at threshold alpha = 0.010; DOF: degrees of freedom

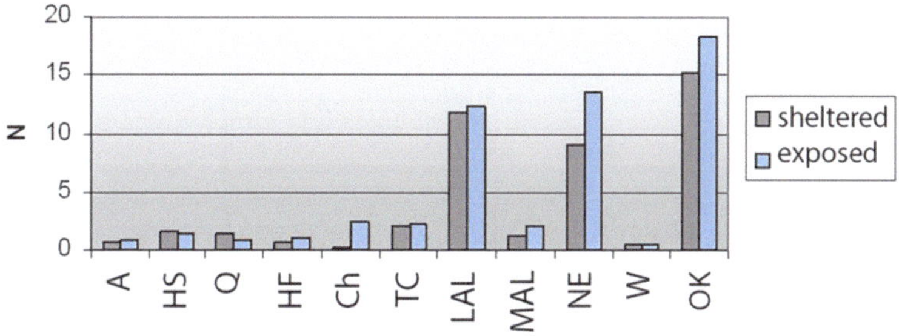

Fig. 6.30 Distribution of archaeological evidence classes in sheltered and windy sectors

6.2.5 Topography Units

Topography and terrain slope are important factors in the choice of location for certain classes of settlements (Dalla Bona & Larcombe, 1996; Kvamme, 1988). Flat surfaces are expected to be selected over rugged ones for residential settlements. These sites are also expected to be located in lower sectors of the landscape, rather than in higher ones that are more exposed to meteorological factors.

6.2.5.1 Topography and GIS

Terrain analysis made it possible to differentiate between flat and high sectors on the plateau and slightly inclined sectors in the lower parts of the landscape. Different classes of archaeological evidence may present different visibility and shelter conditions that could be relevant when searching for patterns between types of assemblages/locations. To map this variable, we combined two maps. The first is the terrain slope class map generated from the DEM. We then reclassified the values to discriminate the landscape into slopes of less than 10° (flat terrain and gentle to moderate slopes) and greater than 10° (moderate to strong slopes and cliffs), as defined by Butzer (1980). Finally, this map was transformed into vector format. The second map is based on basic geomorphological cartography in vector format at a scale of 1:50,000 and was also reclassified. The high geomorphological units (basaltic plateaus) were merged and designated as "high sectors." The remaining units (hills, rocky outcrops, alluvial plains, aggradation plains 1 and 2, and mass-removal areas) were merged into the "lower sectors" unit.

Finally, the two maps were combined to create a topography vector map with three categories resulting from the combination of high and low sectors, as well as gentle and steep slopes. These categories are steep slopes, gentle slopes in high sectors, and gentle slopes in low sectors. Thus, the aspects of topography relevant to the study are represented on a map. See Figs. 6.31 and 6.32.

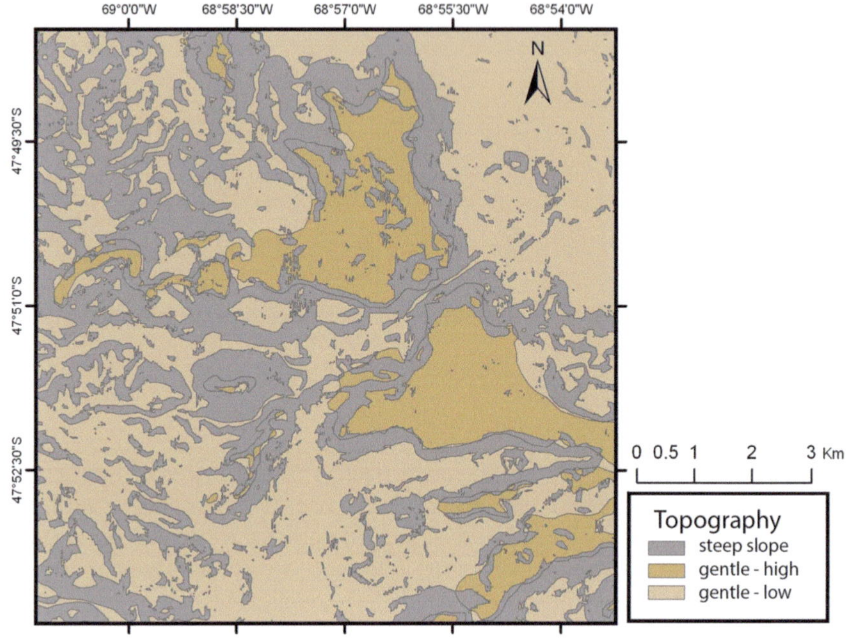

Fig. 6.31 Map of topography

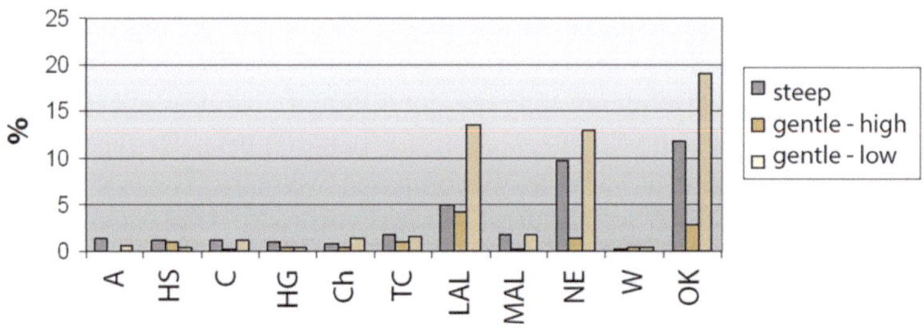

Fig. 6.32 Distribution of archaeological evidence classes in the defined topographic units

6.2.5.2 Location on Different Topographic Units

We tested the distribution of assemblage/location classes in relation to topographic units. Table 6.19 shows the observed and expected frequencies based on the natural occurrence of the classes in this landscape (steep slopes = 35.45%, gentle slopes in high sectors = 11.67%, and gentle slopes in low sectors = 52.88%). The goodness-of-fit test yielded an observed value of $x^2 = 9.965$, while the critical value is 31.410 (for an α of 0.05 and 20 degrees of freedom [DOF]). Therefore, there is no clear pattern, and the hypothesis of independence between the location of different types

Table 6.19 Observed (Of) and expected (Ef) values of occurrences in the defined topographic units

Classes	(Of) steep	(Of) gentle_ high	(Of) gentle_ low	Total	(Ef) steep	(Ef) gentle_ high	(Ef) gentle_ low
A	6		3	9	1.418	0	0.52529801
HS	5	9	2	16	1.18167	0.9908491	0.35019868
Q	5	1	6	12	1.18167	0.1100943	1.05059603
HG	4	4	2	10	0.94533	0.4403774	0.35019868
Ch	3	3	8	14	0.709	0.330283	1.4007947
TC	7	8	9	24	1.65433	0.8807547	1.57589404
LAL	21	38	77	138	4.963	4.2936792	13.482649
MAL	7	1	10	18	1.65433	0.1100943	1.75099338
EN	41	12	74	127	9.68967	1.3211321	12.957351
W	1	3	2	6	0.23633	0.330283	0.35019868
OK	50	26	109	190	11.8167	2.8624528	19.0858278
Total	150	105	302	564	35.45%	11.67%	52.88%

of evidence and the defined topographic units is accepted. However, analyzing individual contributions to x^2 (Table 6.20) suggests some trends, though they are not statistically significant. Rock art sites (A) appear more frequently on steep slopes due to the characteristics of caves and rock shelters. Conversely, HS tend to be located more frequently than expected in high sectors with gentle slopes, and LAL tend to be located in both high and low sectors with gentle slopes. The lack of patterns may partly be due to the fact that geomorphological definitions depend on the scale of analysis, making it particularly complex to define thresholds between units (Fisher et al., 2004; Kvamme, 1988). Though weak, these patterns may reflect the use of high, flat sectors for hunting activities, specifically stalking prey (HS). Other limited activities (LALs) occur in this unit as well, though it is unclear if they are related to hunting. Conversely, loci with art are located in terrain with steep slopes. We believe this reflects the natural structure of the landscape, as caves form in areas with vertical stone walls exposed to wind and water erosion.

6.2.6 Distance to the Paths Network

Past mobile societies that occupied the study area carried out periodic residential movements and daily logistical excursions (Binford, 1982). In consequence, studying the accessibility of the landscape is important when analyzing the past settlement system. Over time, mobility paths are consolidated by use and can be considered factors that attract recurrent use of space (Scheinsohn, 2001). In this study, we identify the paths with the highest topographical accessibility between the most stable or

Table 6.20 Chi-square test per cell

Class	steep	gentle_high	gentle_low	steep	gentle_high	gentle_low
A	**(+) NS**	(−) NS	(−) NS	**0.772**	0.227	0.246
HS	(+) NS	**(+) NS**	(−) NS	0.092	**1.648**	0.726
Q	(+) NS	(−) NS	(−) NS	0.149	0.098	0.029
HG	(+) NS	(+) NS	(−) NS	0.177	0.279	0.351
Ch	(−) NS	(+) NS	(+) NS	0.028	0.007	0.009
TC	(+) NS	(+) NS	(−) NS	0.027	0.335	0.164
LAL	(−) NS	**(+) NS**	(+) NS	1.191	**1.014**	0.177
MAL	(+) NS	(−) NS	(−) NS	0.134	0.220	0.006
EN	(+) NS	(−) NS	(+) NS	0.167	0.779	0.006
W	(−) NS	(+) NS	(−) NS	0.024	0.466	0.037
OK	(−) NS	(−) NS	(+) NS	0.002	0.295	0.085

Test results and contributions to x^2. Topographic units and archaeological evidence class. Acronyms for the results of statistical test: (+): observed frequency higher than the theoretical frequency; (−): observed frequency lower than the theoretical frequency; NS: Chi-square test for non-significant cell at threshold alpha = 0.100; *: Chi-square test per significant cell at threshold alpha = 0.100; **: Chi-square test per significant cell at threshold alpha = 0.050; ***: Chi-square test per significant cell at threshold alpha = 0.010; DOF: degrees of freedom

frequently used campsites (MAL). Distance to the network and the hierarchy of the closest path are other variables to include in the locational analysis performed here. The relationship between functional evidence classes and the path network helps characterize settlement strategies.

This modeling is consistent with recent approaches to resource use among hunter-gatherers. Reticular space use models have been proposed instead of concentric zone models (i.e., site catchment areas, according to Higgs & Vita Finzi, 1972, see Roper, 1979, for a discussion). Albert and Le Tourneau (2007) propose this mobility network model among the Yanomami, who currently live in the Brazilian jungle. These networks consist of frequently used paths and places where these paths intersect. These paths are part of mobility circuits of various lengths that are used daily for gathering, fishing, and occasional hunting, as well as for routes on everyday hunting excursions and long-distance collective hunting and gathering expeditions. The model also distinguishes *empty zones* that are not used for resource extraction (Albert & Le Tourneau, 2007: 589), indicating an inhomogeneous landscape use that is worth considering in the archaeological distributions of our case study. Interestingly, although these spaces may have an ecological explanation as reservoirs of natural resources, ethnographic evidence suggests that these areas were avoided due to the presence of forest spirits that hunt humans (Albert & Le Tourneau, 2007). This model of spatial and resource use has also been observed in Inuit groups (Sauvet, 2019) and in Vanuatu, Melanesia (Mondragón, 2011). It is also included in a model of territoriality and possession (*tenure*) that has been proposed for all hunter-gatherers (Ingold, 1986: 130).

6.2.6.1 Mobility Pathways and GIS

GIS has long been used in archaeology for accessibility modeling. In this sector, particularly where the local topography is uneven, these techniques have the potential to identify the most frequently used mobility paths between places of interest. For this analysis, we modeled mobility pathways between MAL, which likely indicate where residential base camps were commonly established. The variable analyzed is the access roads that were most commonly used as passage corridors between the main settlements. We generated an optimal displacement accumulation model (Modelo de Acumulación de Desplazamiento Óptimo, MADO) (Fábrega & Parcero, 2007). The generation of this model is explained below, and a brief description of the concepts and techniques used is pertinent at this point. Considering movement's dynamic nature, it has two components: orientation and direction. The analysis of movement cost depends on both. Modeling movement to define the least cost path (optimal path) between two points in the landscape has received much attention. There are numerous approaches to human movement on foot throughout the literature (Conolly & Lake, 2006). All of these approaches proceed in two stages. First, a map of accumulated costs is generated from one point. Then, the least cost path back to the origin is defined from another point on the landscape (Conolly & Lake, 2006; Van Leusen, 2002; Wheatley & Gillings, 2002). The accumulated cost surfaces always consider the slope of the terrain as the main factor, to which other friction factors can be added. In this study, we use a friction map obtained by applying Tobler's (1993) formula to calculate the energy cost of crossing the terrain on foot. Following Fábrega and Parcero's (2007) proposal, we apply an isotropic cost model since the direction of movement is unimportant. This model seeks to identify the most frequently used corridors in any direction. In this analysis, cultural elements, MAL, are used as nodes from which optimal paths are estimated. Then, hierarchies are defined in these paths: paths that overlap the most are called main paths, while those that overlap the least are called secondary paths. This procedure enables us to examine the spatial relationships between this assemblage and others. Evidence most associated with secondary paths can relate to tasks with specific objectives rather than activities incorporated into rounds involving various tasks (Binford, 1982; Nelson, 1991). Other assemblages located away from main paths and closer to secondary ones are likely to indicate specific resource extraction or processing activities. In this case, it is possible to propose that evidence associated with secondary paths is related to specific tasks, rather than to activities involving various tasks (Binford, 1979; Nelson, 1991).

The primary and secondary road networks were generated by calculating 18 accumulated cost surfaces, with each MAL serving as the origin (one at a time). The optimal paths were determined by considering all MALs as the origin and each MAL as the destination. This produced a network of paths that originated from all MALs and ended at each origin MAL. This procedure was repeated for each MAL until 18 path networks were obtained in vector format. These were then transformed into raster format to produce 18 maps. The cells on these maps were reclassified to give a value of 1 to the cells coinciding with the paths and a value of 0 to the rest. The 18 maps were then added together in a cell-by-cell procedure. The result is a map presenting

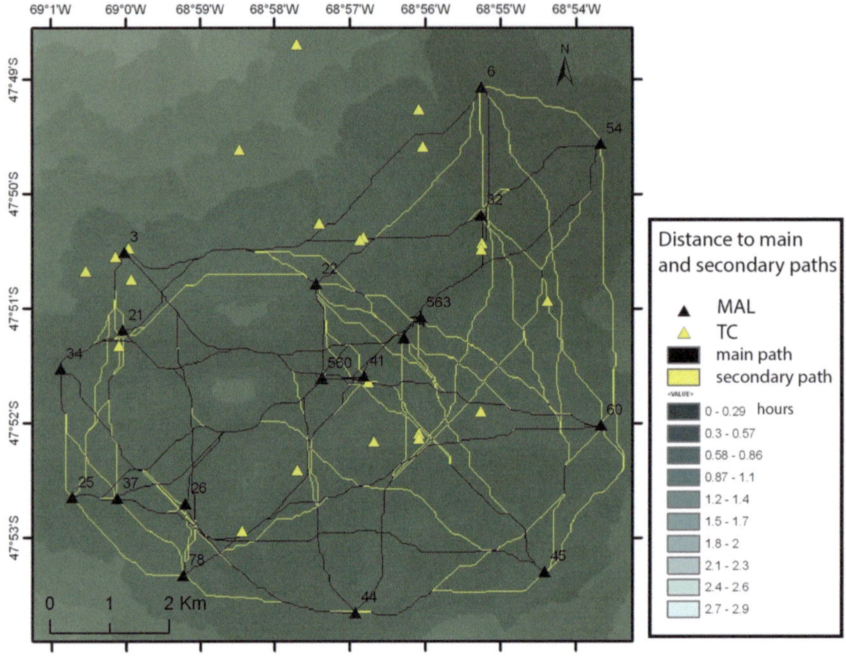

Fig. 6.33 Distance to main and secondary paths between MAL

higher values in the cells where the greatest number of optimal paths from the total calculated networks coincide. Fábrega and Parcero (2007) refer to this product as a lower-cost accumulated flow map. To define the main and secondary paths, the values of this map were reclassified using a classification by natural breaks (ranging from 0 to 18). Cells with values between 1 and 3 were classified as secondary paths, and cells with values between 4 and 18 were classified as main paths. Finally, cumulative cost distances toward the two defined path classes were calculated to measure the distance between any point of archaeological evidence and these mobility routes (Fig. 6.33). Figure 6.34 shows similar patterns within each class regarding distances to main and secondary roads. This indicates that the hierarchy of access paths is not related to a particular assemblage/location. However, the classes differ from one another. Both observations were further investigated by applying the K–S test.

6.2.6.2 Observed Pattern of Distance to Main and Secondary Paths

As shown in Table 6.21 and Fig. 6.35, the TC, OK, LAL, and Q classes are closer to the main access paths than the NE control sample. Conversely, the Ch are located at greater average distances. This pattern suggests that the Ch shows a placement that differs from that of other types of evidence. In this case, the trend indicates that the Ch are located far from the paths between the campsites where domestic activities

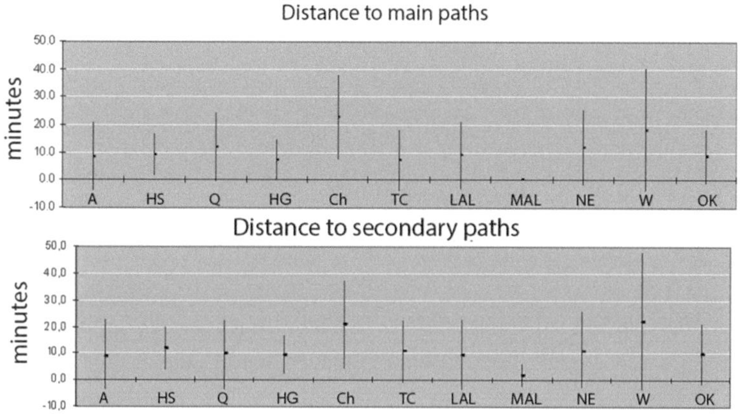

Fig. 6.34 Average distance values and standard deviations to the main and secondary paths

are carried out (MAL). TC, LAL, and OK are closer to the main roads than NE, suggesting that these activities were associated with residential locations and the movement paths between them.

Table 6.22 and the graphs in Fig. 6.36 show four significant patterns regarding secondary paths. The MAL and the hunting grounds (HG) are located near the secondary paths. Conversely, the hunting stations (HS) and chenques (Ch) are far from them. Since all the paths were drawn between the MAL, this correspondence is expected. It is interesting, however, that HGs and HSs exhibit significant patterns in relation to secondary paths. These patterns are opposite: HGs are close to secondary

Table 6.21 Distance to main accesses

Set 1	*n*	*x* hours	*x* min	Set 2	*n*	*x* hours	*x* min	*D*	*P*
NE	127	0.1949	11.69	MAL	18	–		–	–
NE	127	0.1949	11.69	HS	16	0.1547	09.28	0.1654	0.795
NE	127	0.1949	11.69	HG	10	0.1204	07.22	0.2567	0.509
NE	127	0.1949	11.69	LAL	138	0.1483	**08.90**	**0.1913**	0.013
NE	127	0.1949	11.69	OK	190	0.1435	**08.61**	**0.1269**	0.160
NE	127	0.1949	11.69	Ch	14	0.3788	**22.73**	**0.5253**	0.001
NE	127	0.1949	11.69	TC	24	0.1183	**07.10**	**0.3123**	0.030
NE	127	0.1949	11.69	Q	12	0.1958	**11.75**	**0.1609**	0.870

K–S test results

"Set 1" = control sample; "x hours" = average cost distance between the class of assemblage/location and the resource "aguadas" expressed in hours; distance "x min" = the same value expressed in minutes; "Set 2" = sample that is compared with the control; "x hours" and "x min" = same as above referring to the different classes; "D" = maximum difference between the cumulative data distributions; "P" = critical value for an alpha = 0.05. Statistically significant differences are highlighted in bold font

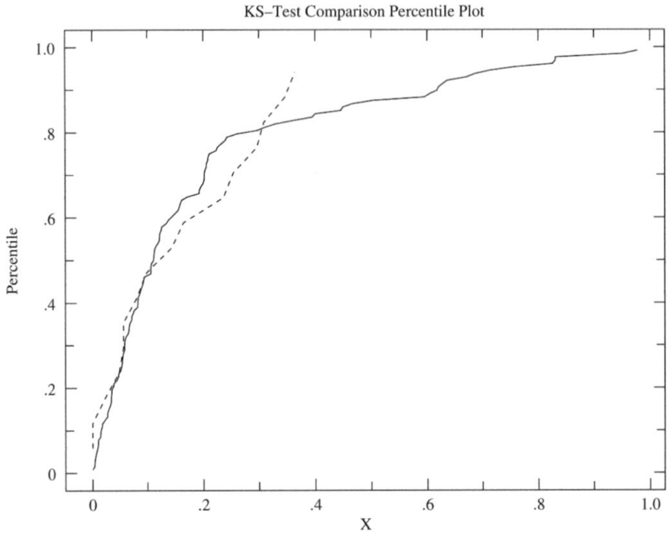

(1)

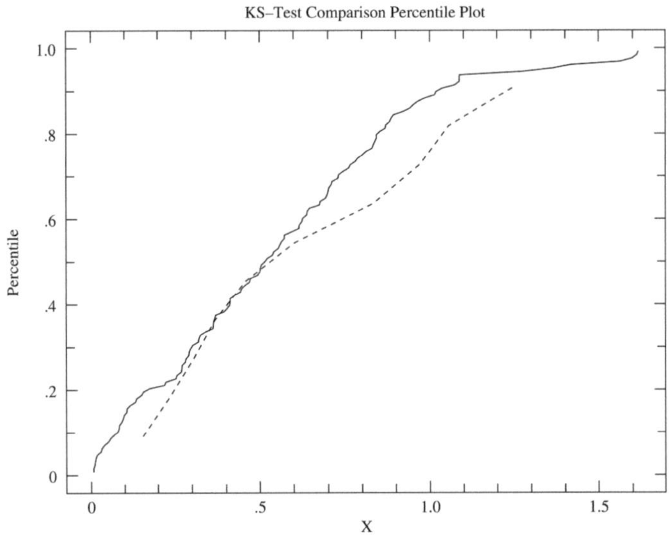

(2)

Fig. 6.35 Cumulative frequency diagrams showing the variable distance to the main paths network. The continuous line shows the control class: NE; the dashed line shows the analyzed class: 1) HS: hunting stations; 2) HG: hunting grounds; 3) LAL: limited activities loci; 4) Ok: occasional knapping; 5) Ch: chenques; 6) TC: temporary camps; 7) Q: quarries

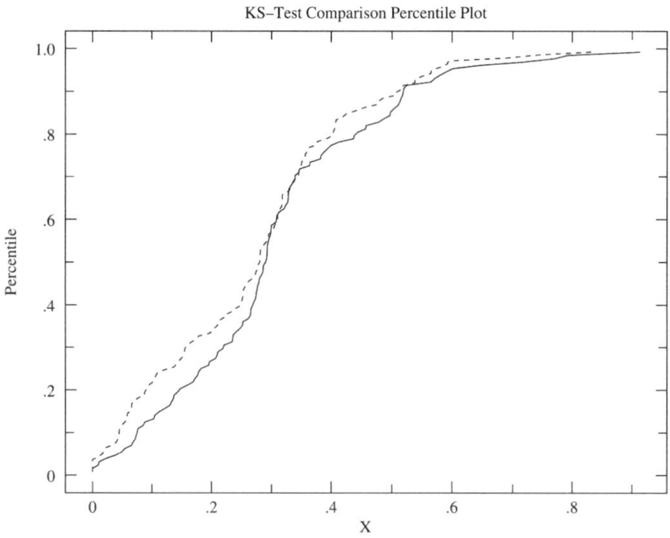

(3)

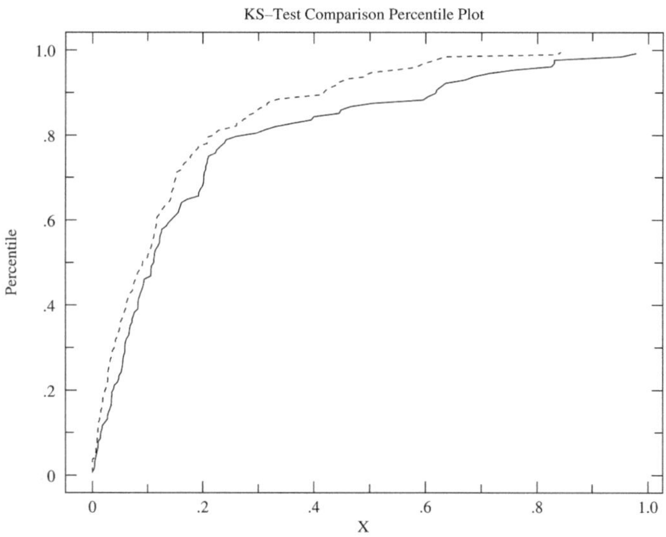

(4)

Fig. 6.35 (continued)

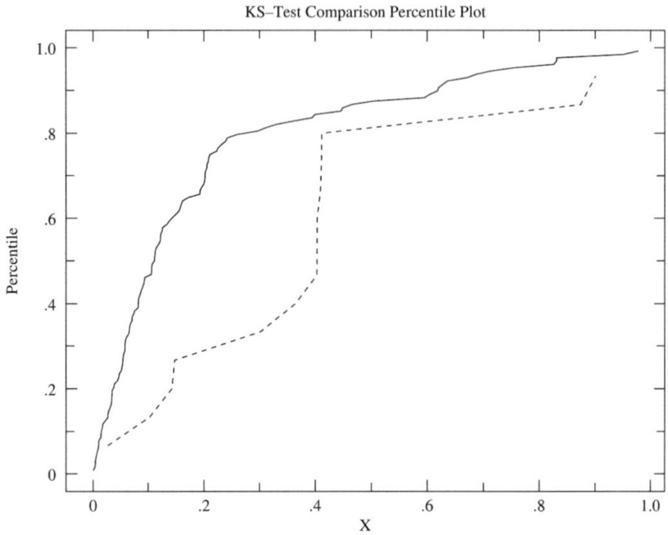

(5)

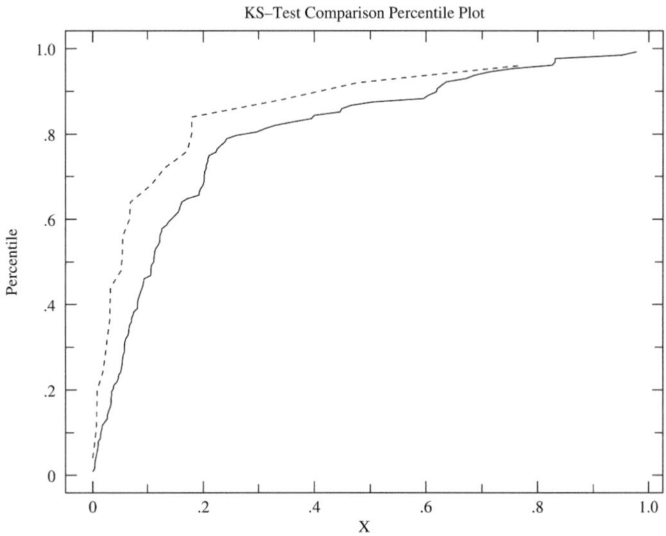

(6)

Fig. 6.35 (continued)

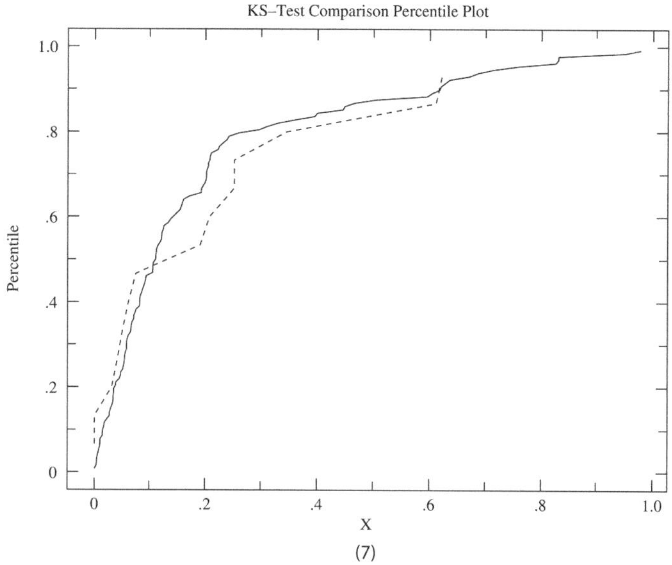

Fig. 6.35 (continued)

paths, and HSs are far from them. Since both classes are related to hunting activities, their locations may indicate the structuring of spaces away from residential areas. In this regard, HSs seem to be intentionally located far from paths, but the outcome of the hunting strategy occurs closer to them. Another important fact is that fifteen of these HSs (total $n = 24$) are associated with type 1 parapets (small stone structures/ hunting blinds) that are associated with hunting activities in regional archaeology (Gradín, 1959–1960). This pattern is further explored in Chap. 7.

6.2.7 Visibility of the Environment and Other Cultural Features

In archaeology, the way people relate to the landscape has been addressed through GIS visibility analysis, in addition to terrain accessibility. The visual properties of a location—or the sensory accessibility of the landscape—can influence settlement decisions (Chapa et al., 2009; García Sanjuán, 2005; Llobera, 1996; Van Leusen, 2002; Wheatley, 1995). A good view of the environment can influence the choice of a location. Although the reasons may be functional, for example, due to their importance for the location of surveillance and control buildings (Fábrega & Parcero, 2007), aesthetic reasons or others may be important (for example Masai, in a context of pastoral society, Kvamme, 1988). In the specific case of the location of burial sites and monuments, the visual dimension of the landscape allows approaching the

Table 6.22 Distance to secondary accesses

Set 1	*n*	*x* hours	*x* min	Set 2	*n*	*x* hours	*x* min	*D*	*P*
NE	127	0.1828	10.97	MAL	18	3.05E−02	1.83	**0.6369**	0.000
NE	127	0.1828	10.97	HS	16	0.1946	11.68	**0.3322**	0.067
NE	127	0.1828	10.97	HG	10	0.1549	09.29	**0.3063**	0.292
NE	127	0.1828	10.97	LAL	138	0.1492	8.95	0.1252	0.235
NE	127	0.1828	10.97	OK	190	0.1647	9.88	0.1206	0.204
NE	127	0.1828	10.97	Ch	14	0.3482	20.89	**0.5017**	0.002
NE	127	0.1828	10.97	TC	24	0.1764	10.58	0.1955	0.383
NE	127	0.1828	10.97	Q	12	0.1564	9.38	0.1777	0.780

K–S test results
"Set 1" = control sample; "x hours" = average cost distance between the class of assemblage/location and the resource "aguadas" expressed in hours; distance "x min" = the same value expressed in minutes; "Set 2" = sample that is compared with the control; "x hours" and "x min" = same as above referring to the different classes; "D" = maximum difference between the cumulative data distributions; "P" = critical value for an alpha = 0.05. Statistically significant differences are highlighted in bold font

archaeological record from the spheres of perception and symbolism (García Sanjuán et al., 2006; Renfrew & Bahn, 1993; Tilley, 1994; Wheatley, 1995; Woodman, 2000).

6.2.7.1 Visibility and GIS

At this point, it is worth delving into the topic outlined in Chap. 5: the debates on the application of GIS technologies for studying human perception of the environment (García Sanjuán, 2006). These debates sparked discussions among archaeologists and others regarding the techniques used and the epistemological possibility of recreating human perception through these systems. Regarding the technology used, visibility analysis is available to archaeologists using standard computer GIS tools. The advantage lies in the ease of accessing these tools and applying them to various archaeological topics. However, this is not without problems. Applying the technique without knowing the procedures and transformations through which the data passes can lead to erroneous conclusions. Furthermore, these popular, simple techniques are being replaced by innovative analytical tools that consider not only two-dimensional, static space, but also the structure around a mobile subject (Llobera, 2003), and even enable reconstructions of individual actors through virtual reality (Winterbottom & Long, 2006). Beyond the evolution of these analytical tools and their increasing availability in archaeology, important theoretical questions remain: To what extent do visibility studies say something about perception? (Mlekuz, 2004; Winterbottom & Long, 2006). First, perception of space does not occur only through vision; other forms of perception, such as sound, can also be modeled (Mlekuz, 2004).

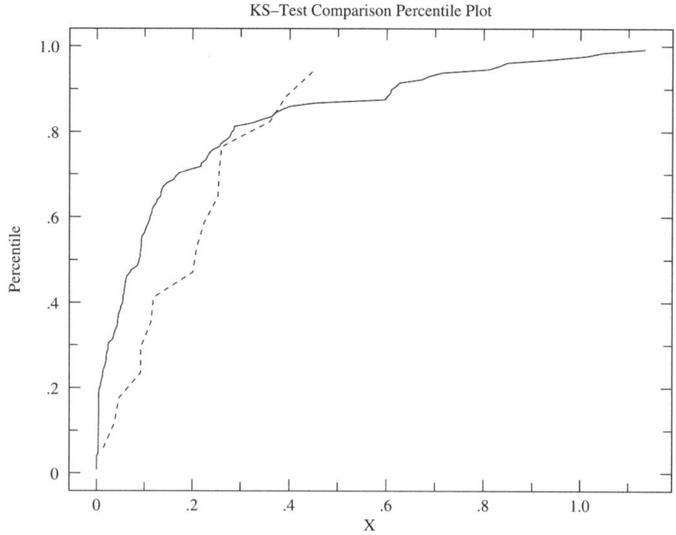

(1)

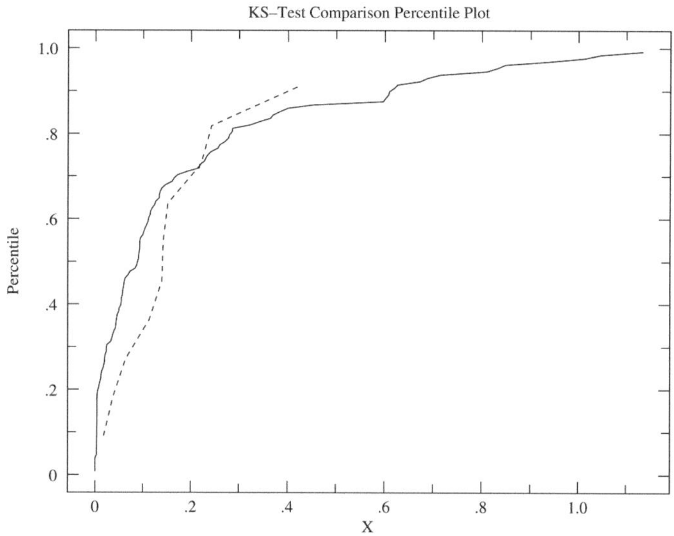

(2)

Fig. 6.36 Cumulative frequency diagrams showing the variable distance to secondary paths network. The continuous line shows the control class: NE; the dashed line shows the analyzed class: 1) HS: hunting stations; 2) HG: hunting grounds; 3) LAL: limited activities loci; 4) Ok: occasional knapping; 5) Ch: chenques; 6) TC: temporary camps; 7) Q: quarries; 8) MAL: multiple activities loci

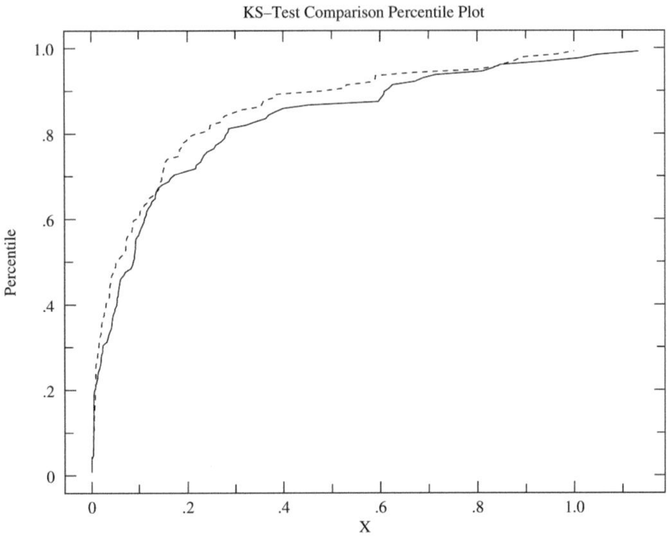

(3)

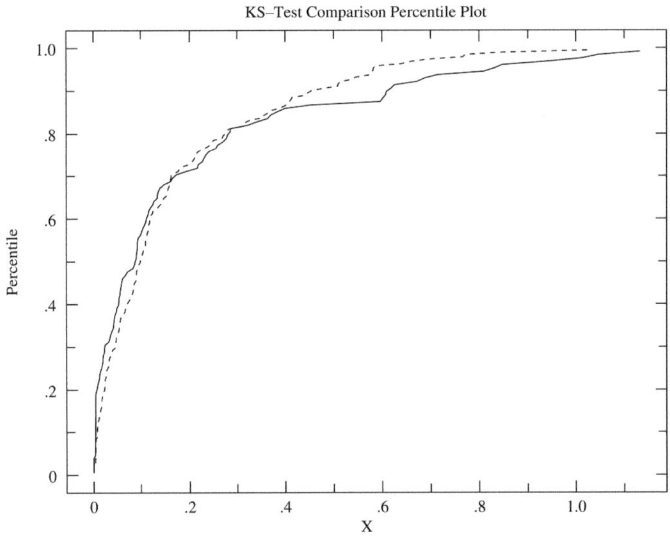

(4)

Fig. 6.36 (continued)

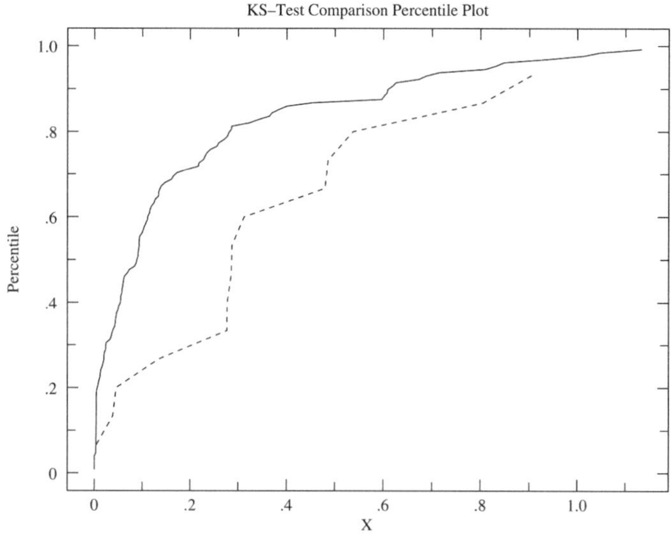

(5)

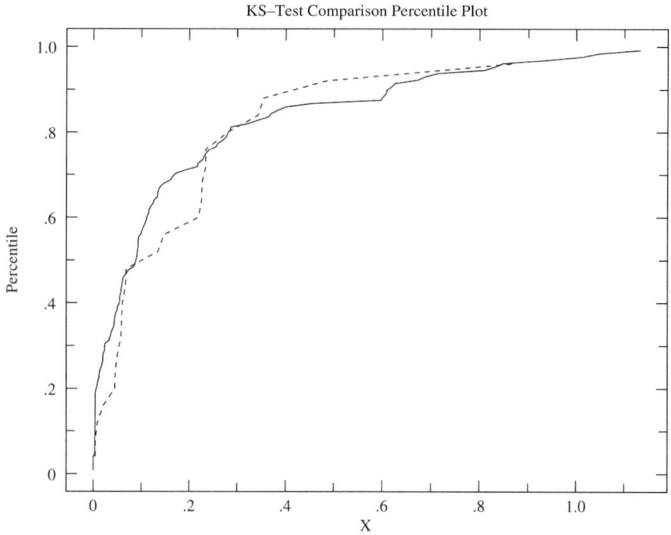

(6)

Fig. 6.36 (continued)

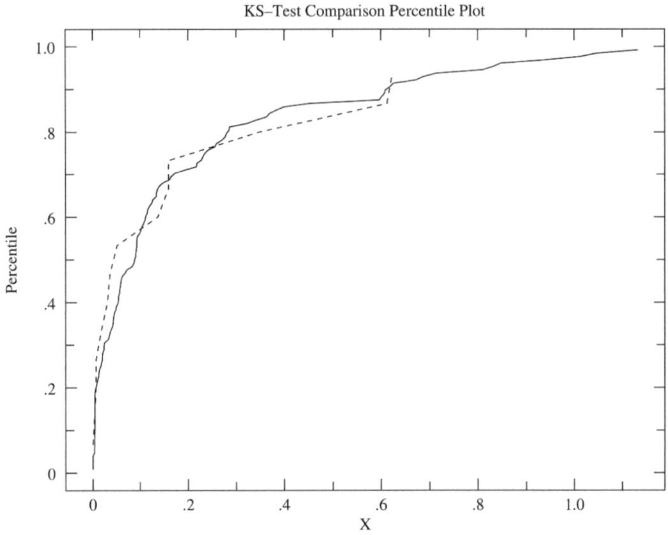

(7)

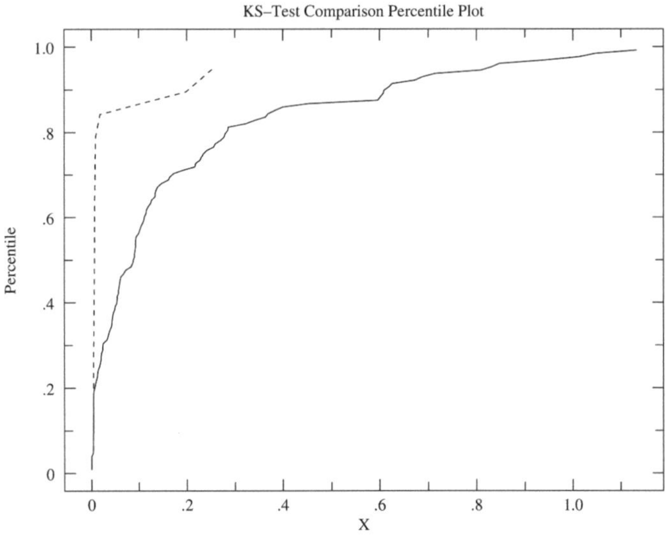

(8)

Fig. 6.36 (continued)

According to Lake and Woodman (2003), perception is the reception of information. However, it can also refer to the perception of space in a subject-environment relationship. Perception involves physical factors affecting the subject's visual field, such as mechanical body restrictions, as well as cognitive factors referring to cultural preconceptions directing attention to certain environmental stimuli (Lake & Woodman, 2003). In other words, perception of space is culturally dictated

(Wheatley, 1993) and an individual act (Mlekuz, 2004). Thus, we conclude that visibility studies of this type allow us to define environmental constraints on views. These studies state the limits of possible perception without addressing the subjectivity of individual actors. The main benefit of applying these GIS analyses to visibility-related problems is the ability to provide a formal framework for these studies. This formalizes the analysis, enabling us to measure the visual characteristics of space by applying hypothesis testing to address perception (Lake & Woodman, 2003; Wheatley, 1993). It also allows us to use quantitative analyses with statistical methods (Fisher et al., 1997; Lake et al., 1998; Llobera, 2003). The possibility of approaching symbolism from the archaeological record depends on the theoretical paradigm within which the research is developed (Lake & Woodman, 2003; Magnin, 2009a). Whether it is possible to interpret or explain what is visible from certain locations in terms of symbolism or functionality depends on the type of questions formulated from one theoretical framework or another. According to Lake and Woodman (2003), visibility analyses have been conducted in archaeology using both processual and processual-cognitive frameworks. In our study, we calculate viewsheds for all defined evidence classes, characterizing each according to the methodological scheme. We describe other visual analysis techniques later in Chap. 7, when we analyze visibility as a locational factor in two spheres of activity: structuring the hunting space and structuring the funerary space. In the first case, we combine the calculation of viewsheds with the analysis of optimal paths and lines of sight. In the second case, we analyze accumulated viewsheds and lines of sight. A viewshed analysis uses a digital elevation model in raster format and a vector file containing observation points as input data. The algorithm considers the shape of the terrain to define the viewshed, determining which cells are visible and which are invisible from that observation point. The result is a binary map in which cells with a value of 1 are visible, and cells with a value of 0 are invisible. The total number of visible cells constitutes the viewshed (Wheatley & Gillings, 2002). Wide viewsheds offer panoramic views from the observation point. Conversely, narrow viewsheds of a few cells represent a smaller visible area. For the analysis, the following parameters were established: (a) The observer's height was set at 1.76 m. This parameter should correlate with the population's average height. It was established based on Imbelloni (1949), as cited by Gradín and Aguerre (1992), who studied southern Tehuelches; (b) A maximum distance radius was established for each visibility calculation from the observation point. This eliminates problems related to the edge effect (Wheatley & Gillings, 2002). The radius was set at 10 km around each observation point. It was defined based on the extension of the study area. Since the base DEM exceeds the limits of the 10 km radius circles, none of the calculated basins are incomplete. This allows the viewsheds to be compared since they have the same potential extent. Due to computing requirements, this study used a maximum of $n = 14$ randomly selected assemblages/locations of each class to calculate viewsheds and their average value. The results are shown in Fig. 6.37, which displays the calculated average values and standard deviations, as well as in Table 27 (Supplementary Material). The chenque class (Ch) stands out for having broader viewsheds than the other classes. Between

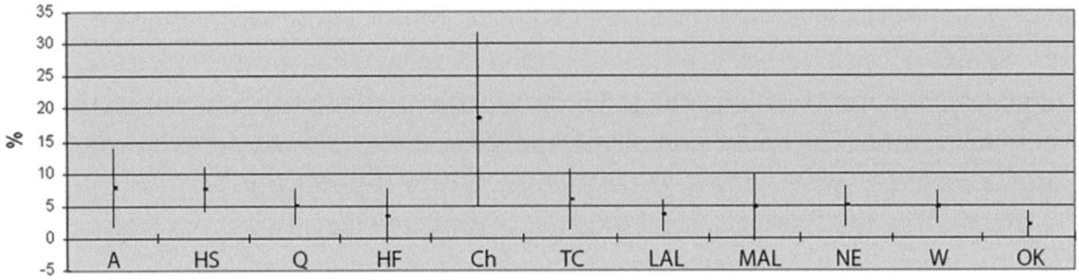

Fig. 6.37 Average values and standard deviations of the visible percentage of the total area (within a 10 km radius)

15 and 20% of the landscape within a 10 km radius is visible from each chenque location.

6.2.7.2 Observed Pattern of Viewshed Width

A K–S test was applied to explore these observed trends. The results are presented in Table 6.23 and Fig. 6.38. These results indicate that chenques (Ch), hunting stations (HS), and rock art sites (A) have significantly broader viewsheds than the control sample. Chenques have visual access to 18% of the surrounding landscape, while HS and A sites see only 8%. In contrast, the viewshed pattern detected in the occasional knapping (OK) and hunting ground (HG) classes indicates very reduced viewshed areas, with only 2% and 3% of the surrounding area visible, respectively. These results coincide with initial expectations. Wide visibility seems to be an important factor in the location of chenque sites of probable funerary function and hunting stations. In this case, wide viewsheds probably serve a functional purpose, as a clear view of potential prey is essential for the placement of these sites. The symbolic and economic spheres coincide in this and other variables explored. In the case of the HG class, once again, there is a pattern that is opposite and perhaps complementary to that of the HS (this pattern is analyzed further in Chap. 7).

Table 6.23 Width of the visual basin

Set 1	*n*	*x* (m^2)	Set 2	*n*	*x* (m^2)	*D*	*P*
NE	28	15,766,200	MAL	14	14,607,900	0.2500	0.540
NE	28	15,766,200	HS	14	23,870,700	**0.5714**	0.002
NE	28	15,766,200	HG	10	2,478,268	**0.4786**	0.045
NE	28	15,766,200	Ch	14	57,660,300	**0.7857**	0.000
NE	28	15,766,200	TC	14	18,432,000	0.2857	0.369
NE	28	15,766,200	LAL	14	10,911,600	0.1786	0.897
NE	28	15,766,200	OK	14	6,139,800	**0.6071**	0.001

K–S test results
Statistically significant differences are highlighted in bold font

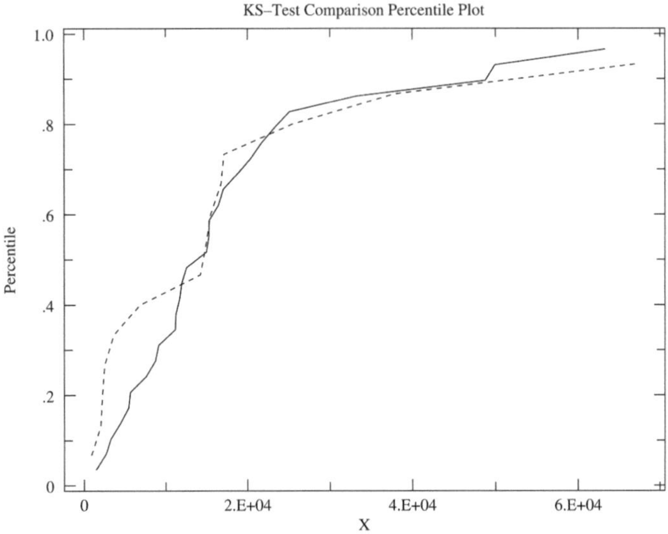

(1)

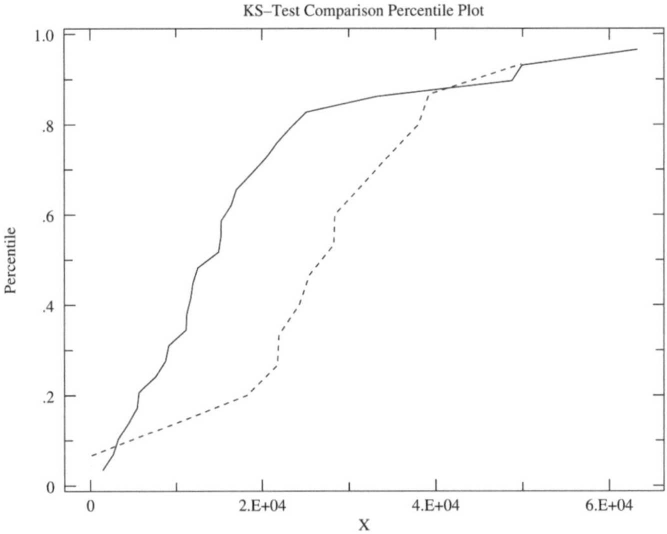

(2)

Fig. 6.38 Cumulative frequency diagrams showing the variable width of the visual basins. The continuous line shows the control class: NE; the dashed line shows the analyzed class: 1) MAL: multiple activities locus; 2) HS: hunting stations; 3) HG: hunting grounds; 4) LAL: limited activities loci; 5) OK: occasional knapping; 6) Ch: chenques; 7) TC: temporary camps

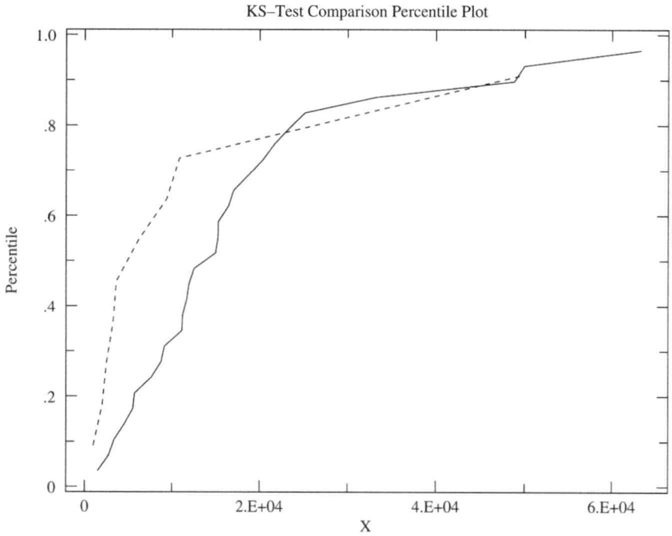

(3)

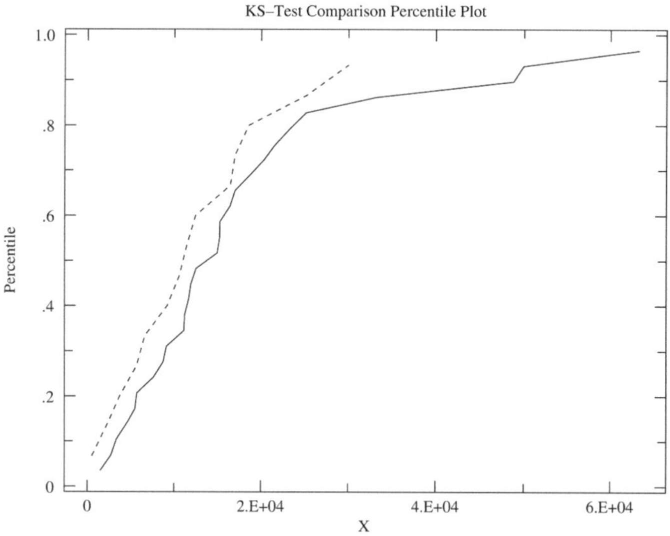

(4)

Fig. 6.38 (continued)

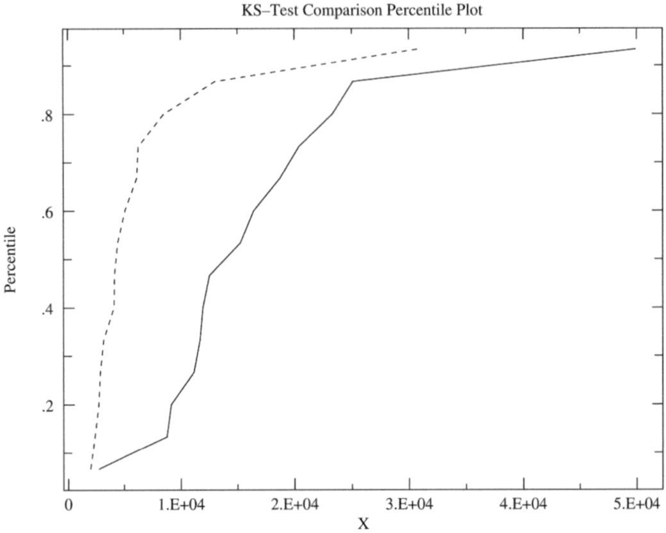

(5)

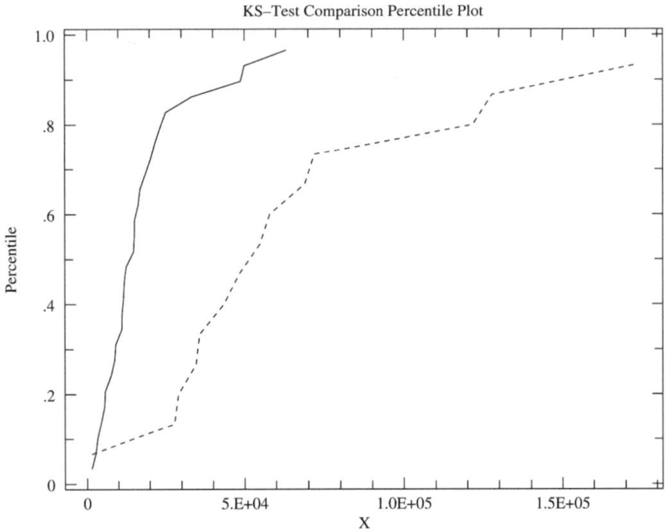

(6)

Fig. 6.38 (continued)

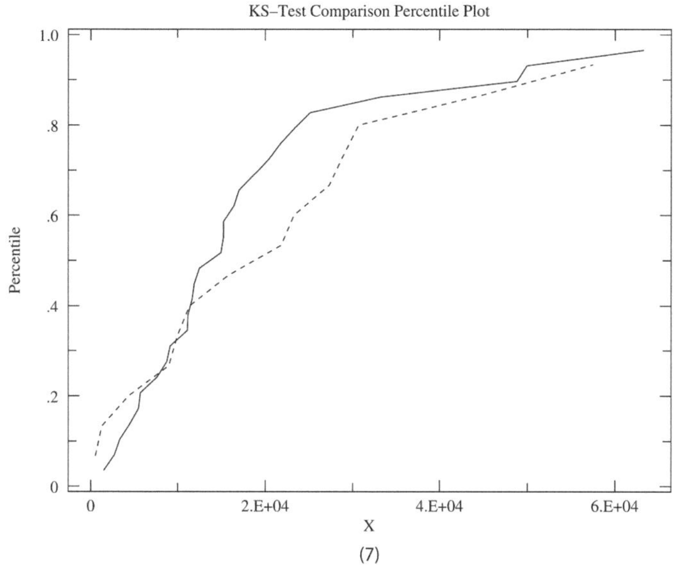

Fig. 6.38 (continued)

6.3 Conclusion

This chapter defines eleven classes of archaeological evidence based on artifact density, assemblage diversity, artifact types, and features. These classes were established without considering locational variables. Then, we modeled aspects of the environmental context that are important for interpreting past landscape use (see Chap. 4): distance to water sources, distance to lithic raw materials, distance to mobility paths, shelter condition, and viewshed amplitude. Exploring this data yielded statistically significant patterns and tendencies, summarized in Table 6.24.

Rock art sites (A) are not numerous enough for the statistical significance tests applied to the other classes; however, they are characterized by their measures of central tendency. They are located in areas with steep slopes, close to water sources, and with wide visual basins. This is due to the fact that the paintings were made in caves and rock shelters. These formations are primarily the result of hydro-aeolic action in the ignimbrites and tuffs of the Chon Aike Formation (Panza, 2001). The location of rock art is strongly influenced by the natural occurrence of cave and rock shelter walls. The proximity of rock art to water sources demonstrates an association with natural conditions conducive to the formation of caves and rock shelters. Are wide visual basins also a natural factor associated with caves and rock shelters in this area? The next chapter will conduct an exploratory analysis to generate new information that will allow us to further explore possible locations for this type of evidence (see Chap. 7).

Table 6.24 This is a summary of the relationships between types of evidence and environmental variables that were explored

1-Topography	**A***	**HS**	Q	HG	Ch	TC	**LAL**	MAL	W*	OK	NE
	Steep	**High**	Steep	High steep	Steep	High steep	**High low**	Steep low	High	Low	Steep low
2-Shelter (%)	Q	HS	LAL	W*	TC	OK	A*	NE	HG	MAL	**Ch**
	59	54	49	47	47	45	44	40	38	36	**6**
3-Dist. springs	A*	MAL	Q	NE	OK	HG	**LAL**	**TC**	W*	**Ch**	**HS**
	9	12	13	*13*	16	17	**17**	**18**	22	**23**	**23**
4-Dist. temporary water	**A***	MAL	**LAL**	Q	NE	**OK**	HG	**TC**	**W***	**HS**	**Ch**
	4	7	**9**	9	*10*	**11**	11	**12**	**16**	**19**	**20**
5-Dist. quarries	Q	A*	**LAL**	**MAL**	TC	**OK**	NE	**Ch**	W*	HG	**HS**
	0	25	**29**	**29**	29	**30**	*33*	**36**	37	37	**50**
6-Dist. main paths	MAL	HG	**TC**	A*	HS	**LAL**	**OK**	**Q**	NE	W*	**Ch**
	0	7	**7**	8	9	**9**	**9**	**12**	*12*	18	**23**
7-Dist. secondary paths	**MAL**	A*	Q	**HG**	LAL	OK	TC	NE	**HS**	W*	**Ch**
	2	9	9	**9**	9	10	11	*11*	**12**	22	**21**
8-Viewshed (%)	**OK**	**HG**	LAL	NE	MAL	W*	Q	TC	**A***	**HS**	**Ch**
	2	**3**	3	*5*	5	5	5	6	**8**	**8**	**18**

Bold typography indicates significant results in the X^2 test (variables 1 and 2) and the K–S test (variables 3–8). The values of variables 3–7 are the mean cost distances in minutes. The values of variable 8 are the mean percentage of visible terrain (*) Rock art values (A) and workshops (W) present low *n* and are shown here for visualizing tendencies in the data

The supply quarries (Q) of lithic raw materials should be in a similar situation, where the natural distribution of rock outcrops determines the location of extraction activities. However, the area is abundant in lithic raw materials, and the quarries are close to the main paths between the MAL, suggesting that the quarries are associated with residential movement vectors in the landscape. Conversely, we can investigate whether Q influenced the location of MAL, as the analysis of distance to raw material sources suggests. To analyze this and learn about site provisioning strategies, we will study raw material extraction activities and analyze the circulation of raw material varieties at different scales.

The MAL is located in various topographies, including low, gently sloping areas. This trend is expected for the establishment of a residential base camp in the lower, warmer sectors of the landscape. On the other hand, the distance to water sources indicates a clear tendency to minimize distances to this critical resource without establishing the settlement next to it. The average distance from temporary and permanent water sources is 7–12 min on foot. Since the presence of water implies the concentration of other resources, such as plants and fauna, which were not modeled here, it is proposed that the MAL's settlement pattern may indicate an interest in those resources.

The location of the MAL is central to explaining the distribution of other satellite classes. Because MALs are considered the structuring centers of mobility and displacement, their possible influence on the location of other activities is analyzed through the layout of the main and secondary path networks. The distances of each class to the main and secondary path networks are compared to determine which classes are more or less associated with domestic activities in the MAL. The classes that show a pattern of closeness are the temporary camps (TC), limited activity loci (LAL), and occasional knapping loci (OK). Among the classes closest to secondary paths are the hunting grounds (HG). These are places where prey was hunted, suggesting that this activity occurred during domestic residential moves. An opposite pattern was observed in the location of the chenques (Ch), which are interpreted as burial sites. In this case, they are far from both the main and secondary access paths. The class seems to separate itself from the others in this and the rest of the locational variables. They are distant from resources and from the communication network paths between MAL. They are located where there is no shelter from meteorological conditions and frequently in steep topography. Studies of viewshed basins and intervisibility, detailed in the following chapter, reveal an additional distinctive pattern for chenques.

Chapter 7
Exploratory Locational Analysis

Abstract Throughout this chapter, we deepen our study of some of the trends detected in Chap. 6 through new exploratory analyses and a review of specific data. The locations of rock art sites, chenques, quarries, and hunting stands are characterized and discussed in relation to the existing regional knowledge corpus. Integrating this information with that from other environmental, ethnographic, archaeological, and ethnoarchaeological studies, as well as the patterns obtained, allows us to characterize the locational decisions of the hunter-gatherers who inhabited the La Primavera locality in the past.

Keywords Chenques location · GIS analysis · Locational decisions · Past human activities · Mobility network · Quarries · Residential sites location · Resources procurement strategies · Rock art location

7.1 Residential Activities

An analogy can be made with the Aónikenk, a group of hunter-gatherers who inhabited the province between the Santa Cruz River and the Strait of Magellan. Various ethnographic records define the characteristics of nomadic life and the places chosen as campsites.

> Si la provisión de alimentos había impuesto e imponía la vida nómade, otros requerimientos complementarios también condicionaban en idéntico sentido. Tal el aprovisionamiento de determinadas materias primas…distribuidas naturalmente en distintos lugares, que obligaban a la marcha periódica en procura de su obtención. También la búsqueda de sectores más reparados que permitieran superar de mejor manera las contingencias climáticas en un territorio señalado por la variabilidad, con el añadido de la casi permanente vigencia eólica que acentuaba el rigor de elementos como las precipitaciones y las bajas temperaturas (Martiníc, 1995:60).
>
> If the provision of food imposed the nomadic life, other complementary requirements were also conditions in the same sense. Such was the supply of certain raw materials… naturally distributed in different places, which required periodic rounds of searching to obtain them. Also, the search for more repaired sectors to better overcome climatic contingencies in a territory marked by topographic variability, with the addition of the almost permanent presence

L. A. Magnin, *The Landscape and Mobility of Hunter-Gatherer Societies*, The Latin American Studies Book Series, https://doi.org/10.1007/978-3-032-08127-8_7

of wind that accentuated the rigor of elements such as precipitation and low temperatures (Martiníc, 1995:60, author's translation).

Este sistema de asentamientos originaba una trama ocupacional conformada por distintos sitios, que iban desde los más simples, paraderos de alojamiento en las rutas de tránsito entre áreas de importancia, hasta los campamentos-base, en parajes elegidos por su disponibilidad de recursos (agua, leña combustible, caza, materia prima, etc.) y de reparo, en los que se erigiría la toldería y permanecía la mayor parte del tiempo el grueso del grupo, en particular las mujeres, niños y ancianos, en tanto que los cazadores recorrían los alrededores y parte de aquéllos realizaban actividades de extracción o colecta y otras complementarias… (Martiníc, 1995:62).

This settlement system originated an occupational network made up of different sites, which ranged from the simplest accommodation stops on transit pathways between important areas to base camps in places chosen for their availability of resources (water, fuel wood, hunting, raw materials, etc.) and shelter, in which the awning would be erected. There, the bulk of the group would remain most of the time, particularly women, children, and the elderly, while the hunters toured the surroundings and while some of them carried out resource extraction or collection and other complementary activities… (Martiníc, 1995:62, author's translation).

Figure 7.1 shows the distribution of residential sites in the La Primavera area. The following pages explore the main locational patterns and tendencies.

7.1.1 Multiple Activity Loci (MAL)

7.1.1.1 Multiple Activity Loci (MAL) Close to Water Resources

As shown in the previous chapter, multiple activity loci (MALs) tend to be located near permanent and temporary water sources. However, there are also cases in which MALs are located away from water sources. This variability creates interesting opportunities for studying landscape use and group mobility in relation to long-term paleoenvironmental information and seasonal or short-term climatic variations.

Under arid environmental conditions, the proximity of residential sites to permanent springs could have been of greater importance. Paleoenvironmental studies indicate that aridification advanced throughout the Holocene (Salemme & Miotti, 2008). MAL locations associated with permanent water sources may date back to the Late Holocene. Another possibility is that the locality was used on a seasonal basis during the dry summer and fall months.

The location of MAL near currently dry and temporary shallow lakes suggests two possible scenarios: shorter versus more prolonged stays at the sites. In the first scenario, occupations occurred on a seasonal basis, with visits at the end of winter and in spring. The second scenario is that the locality was occupied during periods when the availability of water was higher than it is today.

Throughout the Holocene, the trend toward constant aridification may have been interrupted by periods of higher temperatures and humidity. These occurred during the middle Holocene, when the *Optimum climaticum* period occurred (from 7000 BP and lasting approximately 1000 years), as well as 500 years ago during the Little Ice

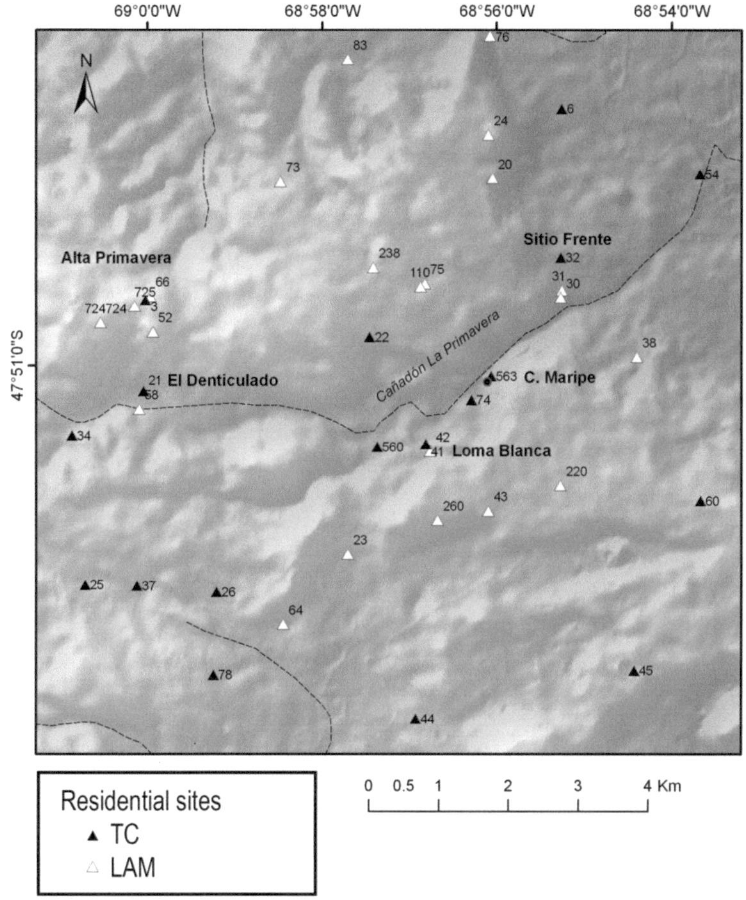

Fig. 7.1 This is a residential site map of the micro-study area. The locations of the MAL and TC are indicated with ID numbers. The names of the MALs are listed next: 3—Alta Primavera; 6—Camino Clandestino; 18—Cueva Mora; 21—El Denticulado; 22—El Descanso; 25—El Pájaro Muerto; 26—El Piche Lento; 32—Sitio Frente; 34—GB; 37—La Herradura; 41—Loma Blanca Bajo; 44—Mirador; 45—Oasis; 54—Pico Blanco; 60—Sitio del Paso; 74—La Quinta; 78—D9; 560—La Playita; 563—Cueva Maripe

Age. The Little Ice Age was wetter but also colder than today and lasted 400 years (Morales et al., 2009; Salemme & Miotti, 2008, see Chap. 2).

Three sites in La Primavera Canyon and one on the plateau illustrate the variability of MAL locations in relation to permanent and temporary water sources.

The open-air MAL La Playita (Magnin & Blanco, 2006) and Loma Blanca Bajo are protected by 30-m-high ignimbrite walls that provide excellent shelter. La Playita is close to temporary shallow lakes, and La Quinta is close to a temporary tributary.

The middle section of the canyon, where the La Playita and Loma Blanca Bajo sites (MAL ID_560 and MAL ID_41) are located, is currently filled with sediments but probably housed shallow lakes in the past (personal communication with Quaternary geologist Dr. Jorge Rabassa). Those sites were probably occupied during periods of high humidity. Though there are currently active springs at the bottom of the canyon, the arrangement of the archaeological sites around the limits of dry shallow lakes suggests that occupations occurred during periods when the shallow lakes contained water. These periods may have been of short duration after rainfall or snowmelt in dry climates, or of longer duration in more humid climates when more water was available than today.

In the case of La Quinta (MAL ID_74), the site is closely associated with four active springs that open into a wet meadow and a temporary tributary that runs through the canyon. During fieldwork in 2008, a 2 by 2 m grid was excavated at the La Quinta site. The excavation sequence revealed lithic and bone archaeological materials, as well as charcoal spicules. Radiocarbon dating (see Chap. 8) revealed that occupation began around 1000 BP. This period coincides with the Medieval Climate Anomaly, which had a predominantly hot and dry climate. Therefore, it is likely that hunter-gatherers used the springs as their water source, rather than the temporary tributary.

In the case of Site Oasis (MAL ID_45), it is associated with a temporary shallow lake in a high-elevation area of the landscape. However, there are no nearby springs (Figs. 7.1 and 7.2). The site has partial topographical shelter due to uneven terrain. However, it is in a topographically elevated position on a plateau 125 m above the base level of the Las Mercedes valley. The dependence on water that is only temporarily available in the shallow lake, coupled with the absence of permanent springs nearby, gives this site a unique character. We also found evidence that this shallow lake was used for hunting (isolated projectile points in different sectors of the depression). The MAL and hunting sectors coincide spatially, though they likely weren't used for these different purposes simultaneously. Another sector of the plateau, located 1 km away from the Oasis site, has hunting blinds. This suggests that the high plateau was used for hunting activities (see point 7.4 below). Based on this information, we propose that the record at the Oasis site was generated by successive occupations of temporary hunting camps rather than by the placement of MAL for prolonged use.

A series of data classified as limited activity loci (LALs), are also located next to temporary sources of water and away from springs. This, indicates the use of certain spaces during periods of high water availability. They suggest that specific tasks were carried out from base camps in another part of the landscape (for example, the MAL sites mentioned). Examples of LALs include Laguna Cuarzosa (LAL ID_141), which is located on the plateau north of Cueva Maripe, and El Pedimento (LAL ID_56), located next to a temporary stream.

To better understand the locational strategies of MAL in the La Primavera area, it is necessary to excavate these sites to determine their chronology, duration, and indicators of seasonality. A detailed study of the chronology of these sites is necessary. If the MAL residential sites were associated with temporary water sources and their occupations were dated to between 6000 and 7000 years BP, they could explain the archaeological silence proposed by Salemme and Miotti (2008), which is based mainly on cave records from this period in the Central Plateau. Additionally, a deeper study of the environment is needed to understand the cycles of water availability in the landscape and their impact on hunter-gatherer's mobility.

7.1.1.2 A Case of MAL Far from Water

The Alta Primavera site (MAL ID_3) is relatively far from temporary water sources and even farther from springs (see Figs. 7.1 and 7.2). It is 227 m above the valley floor on a basaltic plateau covering the rocky outcrops of the Bajo Grande Formation (Panza, 2001). Bajo Grande (Panza, 2001) is at the headwaters of the canyon. The nearest water source is a small, temporary shallow lake formed on the basalt, located 1.2 km from the site. There is also a permanent water spring at the bottom of the valley, 1.5 km away with a slope of more than 200 m. This pattern differs from those mentioned above, suggesting that other factors influenced the location of this LAM. The Alta Primavera sector (hereinafter "AP") is located on an elongated plateau measuring 1.5 km by 0.2 km that extends west into the Alta Primavera Oeste sector ("APW"), another plateau measuring 1 km by 0.2 km. Stone structures were found

Fig. 7.2 Archaeological record in upper sectors of the study micro-area, 1) Oasis Site (MAL ID_45); 2) hunting blind (ID_262), both in Las Mercedes Location; 3) residential structure (ID_724) in the Alta Primavera sector. The width of its stone wall is 1.5 m; 4) hunting blind in the same sector; 5) view from the hunting blind (ID_96) in the Alta Primavera sector; 6) hunting blind (ID_49) in the Alta Primavera Oeste sector. Other stone structure (ID_48) is visible at the top of the hill

2

3

Fig. 7.2 (continued)

4

5

Fig. 7.2 (continued)

6

Fig. 7.2 (continued)

in both sectors. In Sector APW, five Type 1 parapets (smaller structures interpreted as hunting blinds) and two Type 2 parapets (larger residential parapets; see Chap. 6) were located. Four Type 1 parapets were found in APW. Based on the lithic artifats found on the surface (Table 6.11), the Type 1 parapets were interpreted as possible hunting stands. The Type 2 parapets did not present any archaeological materials on the surface. However, their size and shape coincide with the residential parapets described for other sectors. Therefore, we propose that they could have served a similar function. Given the location of MAL, it is reasonable to assume that water was transported from nearby shallow lakes or springs using leather containers or animal bladders (Martinić, 1995). Based on the current state of knowledge, it is impossible to infer the humidity or aridity conditions during the occupation period. In any case, water must have been brought to the sites, either from the temporary shallow lake (the closest source) or the spring (the farthest source).

Regarding the registration of MAL and HS in the AP and APW sectors, these high-terrain occupations may indicate summer use of these areas due to their low winter temperatures compared to spaces at the bottom of canyons. Summer use of these sectors coincides with the behavior of guanacos, which tend to occupy the highest parts of the landscape in spring and summer (Gradín & Aguerre, 1992). This location is ideal for hunting guanacos because ethological studies indicate that, when threatened, these animals use steep slopes as escape routes.

> …the guanaco prefers flat or open terrain., but is associated with steep reliefs that allow it to climb quickly and stay out of the reach of its pursuer (Tavarone, 2004:37).

The archaeological evidence found suggests a diverse setting of residential sites. Its interpretation requires further paleoenvironmental and geoarchaeological studies and indicates a diverse occupation panorama, where the central tendencies and exceptions do not respond to one simple factor.

7.1.1.3 MAL and Their Location in Relation to the Quarries

Due to the abundance of lithic raw materials in the region, it is not expected for MAL to be located near quarries (see Chap. 2). However, although the MAL are located an average of 29 min away, they are located at shorter distances than the control group and other evidence classes. Were the Q attractors for MAL locations? To answer this question, we need to analyze the archaeological artifacts included in the assemblages/locations classified as MAL and compare them with the rocks present in nearby quarries. The exploratory analysis performed later aims to characterize the structure of lithic resources at La Primavera and discuss MAL locations and sourcing strategies.

7.1.2 Transitory Campsites (TC)

7.1.2.1 TC and Redundancy in the Occupation of Spaces

It is striking that, according to the results obtained, temporary camps (TCs) are located relatively far from permanent water sources (see Chap. 6). Similar locational patterns than those for MAL were expected, especially regarding the critical resource, given that similar tasks were probably performed in both, the TC as in the MAL. As previously mentioned, this can be explained by considering the high residential mobility of these groups. It is likely that the TCs were located in places previously occupied by the MAL. Therefore, material evidence from these shorter occupations may have overlapped with other MAL occupations, resulting in low archaeological resolution. Due to the redundancy of occupation of the same spaces, only TCs that deviate from the most typical MAL pattern can be recognized in the registry. Some of these assemblages are located very close to the MAL. Examples include the TC Alta Primavera (ID_3), Sitio Frente (ID_32), El Denticulado (ID_21), and Sitio Loma Blanca (ID_42) (Fig. 7.3). Other TCs, however, exhibit a different pattern; they are not near the MAL. The main differences are their greater distance from water and their location in higher sectors of the landscape.

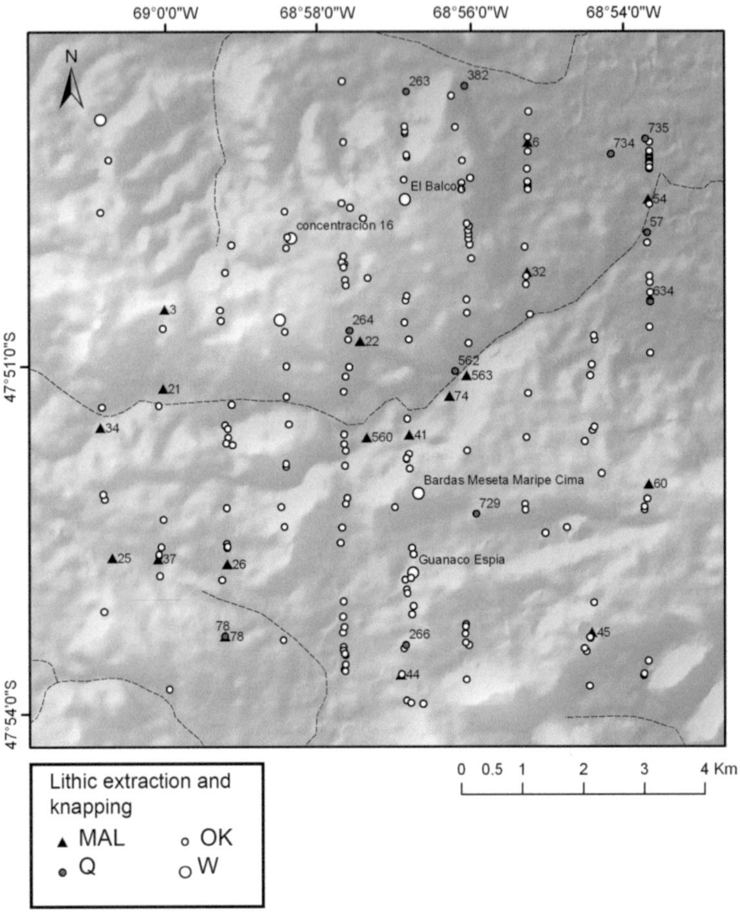

Fig. 7.3 Distribution of quarries, workshops, occasional knapping loci and MAL. The names of the MALs are listed next: 3—Alta Primavera; 6—Camino Clandestino; 18—Cueva Mora; 21—El Denticulado; 22—El Descanso; 25—El Ppájaro Mmuerto; 26—El Piche Lento; 32—Sitio Frente; 34—GB; 37—La Herradura; 41—Loma Blanca Bajo; 44—Mirador; 45—Oasis; 54—Pico Blanco; 60—Sitio del Paso; 74—La Quinta; 78—D9; 560—La Playita; 563—Cueva Maripe

7.1.2.2 TCs and Their Location in High Sectors

Unlike the MAL, the CT are located in areas with higher topography. Searching for locations for temporary camps suggests that these sporadic and less frequent occupations occurred during the summer, when temperatures permitted camping in higher, colder areas of the terrain and when winds were strong. The longest occupations, which produced the densest and most diverse records, occurred in lower sectors of the terrain. Settling in the highest locations probably required using awnings or light windbreaks, such as the leather screens supported by a framework of branches described by Gusinde (1982) for the Selk'nam. These windbreaks were proposed as possible items of the southern Tehuelche groups' material culture (Casamiquela, 2000).

7.1.3 Paths Networks Between MAL

Despite the differences in location between MAL and the TCs, we can identify the TCs as belonging to the same sphere of activity as MAL. This is evident in their proximity to the main access roads between MAL, as shown in Chap. 6. Base camps, which are indicated by archaeological assemblages classified as MAL, constitute the structuring element of the pattern of hunter-gatherer place location. Models of spatial organization based on archaeological and ethnoarchaeological evidence propose this (Bettinger, 1980; Binford, 1980, 1982, 1988; Gamble, 1990). This concept is equivalent to the concept of home base in Foley's regional archaeological structure model (1981). If we consider MAL to be the center of hunter-gatherer mobility organization, then the interweaving of the access networks modeled between them, based on topographic properties, can suggest possible relationships between different kinds of evidence. TCs' proximity to main paths is consistent with their function as short-stay camps along mobility routes between more frequently occupied places. Locations suitable for campsites were surely periodically used for residential purposes, and the paths between them must have been an important organizing axis for activities. The nomadic lifestyle was based on territorial knowledge, which led to the establishment of a differentiated spatial occupation system consisting of simple stops on transit routes between important areas and base camps in places chosen for their abundance of resources (Martinić, 1995; Miotti, 1989). This pattern coincides with Silvana Chapalala's account of favorite locations for temporary camps:

> ...esos toldería de gente que viene de viaje no es para establecerse, es para pasar la noche o dos o tres días que a lo mejor hay agua o pasto para la caballada...Por eso los toldos se hacían a la par de la meseta o al borde de las lagunas al reparo de los matorrales...que tiene una loma o un cerro...siempre del lado que no pega el viento (Aguerre, 2000:17).
>
> ...those *tolderías* of people who come on a trip are not to settle, they are to spend the night or two or three days where maybe there is water or grass for the horses... That's why the awnings were made next to the plateau or on the edge of the shallow lakes, sheltered by the bushes... that have a hill... always on the side that the wind doesn't hit (Aguerre, 2000:17, author's translation).

7.2 Quarrying Activities (Q) and Use of Lithic Resources

7.2.1 Quarries, MAL, Occasional Knapping Places and Workshops

Lithic raw material quarries (Q) are a type of evidence tied to the locations of natural outcrops and deposits of useful rocks. For this reason, the behavior of other types of evidence in relation to them is analyzed here. First, the MAL are significantly closer to the Q than the other classes of evidence. This suggests that when choosing the location of the residential base camp, the critical resource (water) and the raw material for knapping were both considered. Conversely, the occasional knapping loci (OK) and limited activity loci (LAL) are also close to the Q. Regarding the workshops (W), there are too few to establish statistically significant patterns ($n = 6$). The assemblages classified as W include evidence of lithic manufacturing activities, but not of resource extraction or processing activities. There is not yet enough evidence to hypothesize how this activity could be connected with others in the systemic context (Fig. 7.3). To advance in this sense, the next section analyzes the structure of the lithic resources in the area.

7.2.2 The Structure of Lithic Resources in the Studied Sector

The Central Plateau landscape was described as rich in high-quality raw materials for manufacturing lithic artifacts, with homogeneous distribution on the macro-regional scale (Cattaneo et al., 2006). However, fieldwork revealed that this characterization depends on the scale of analysis. At the micro-regional scale, the heterogeneity in location, variety, and quality of knapping materials is evident.

7.2.2.1 Raw Materials Present

The analyses performed here are based on a field survey design that enabled the collection of field data at regular intervals, covering the study microarea (see Chap. 5). The natural distribution of raw materials suitable for artifact manufacturing was geolocated using GPS. This information, together with what is known from previous research, constitutes the environmental context against which the distributions of knapped artifacts will be compared. The doctoral thesis work of Hermo (2008) is the main background to the study of lithic technology in this locality. His research was conducted at various scales of analysis. His work focused on the Middle Holocene component of Cueva Maripe (CM) and its relationship with the quarries Rocky (CR), Cantera del Verde (CDV), Cantera del Rojo (CDR), and the secondary deposit Pediment 1 (LP-P1). This work enabled Hermo to examine lithic raw material sourcing at the regional level and develop a circulation model of lithic raw materials for Cañadón

La Primavera and its surroundings. The studies performed include an analysis of thin sections of the rocks from the four sites and the Cueva Maripe site using a petrographic microscope. The analysis resulted in a classification of eight groups, which includes different varieties of chalcedony, opals, silicified ignimbrites, and siliceous rocks (see Chap. 2). This study was useful for comparing the different raw materials using microscopic methods after macroscopic recognition. However, since petrographic determinations cannot be observed with the naked eye on rock outcrops or lithic artifacts, we could not use the rock varieties defined by Hermo (2008). For this reason, the classification applied here includes the most specific categories that could be approximated by the naked eye in hand samples in the laboratory and during fieldwork. The raw materials were classified as follows: siliceous rocks (including opals and silicified rocks); rhyolite; chalcedony (translucent, homogeneous siliceous rocks); andesite; gabbro/diabase; obsidian; basalt; petrified wood; and other undetermined rocks. The general characteristics that allowed us to recognize the different raw materials were taken from Tarbuck and Lutgens (2007). Several volcanic rocks, such as highly silicified tuffs, ignimbrites, and breccias, were included in the siliceous rock category. These rocks have conchoidal fractures and knapping qualities ranging from regular to good (Aragón & Franco, 1997). This group can also include rhyolites, except for those with a porphyritic texture and visible quartz crystals. In this case, the sample is classified as rhyolite. Conversely, the category of chalcedony does not align with what can be defined as such from a mineralogical perspective. It is impossible to identify this mineral in a hand sample. In this case, the term is used descriptively to refer to a very homogeneous variety of siliceous rocks with a conchoidal fracture and a vitreous luster that are translucent.

Regarding the color of the rocks, as observed in the field (Hermo, 2008), it varies considerably within the same outcrop. To evaluate the association between rock classes and colors, the colors and tones within the collection were reduced to seven categories: red, yellow, green, brown, white, black, and indeterminate. Applying an X^2 test rejected the null hypothesis of independence between the raw material and color variables ($\alpha = 0.050$, $X^2 = 3864.601$, critical value $= 65.17$, Cramer's V $=$ 0.327) (Tables 7.1, 7.2, and 7.3). In other words, raw materials tend to appear in specific colors. Most andesites are green; basalts are black; chalcedonies are light colors (tabulated as white); gabbros and diabases are always green; obsidian is black; rhyolite is red; most siliceous rocks are yellow, brown, or green; and petrified wood is mainly white, black, or brown. Unidentified raw materials are generally black. Conversely, the analysis of contributions to X^2 revealed the most significant associations: black obsidian, red rhyolite, white chalcedony, black basalt, green gabbro/diabase, and brown siliceous rocks. Petrified wood, however, occurs with equal frequency in white, black, and brown.

Figure 7.4 illustrates the observed trends in the variety of raw materials and colors represented in the archaeological sample.

Table 7.1 Frequency distribution of artifacts raw materials and colors

	Yellow	White	Black	Brown	Red	Green	Other	Total
Andesite	1	1	1	0	0	3	0	6
Basalt	0	0	33	3	0	1	0	37
Chalcedony	22	133	9	82	7	2	0	255
Gabbro/diabase	0	0	0	0	0	3	0	3
Obsidian	0	0	153	1	0	0	0	154
Rhyolite	38	60	1	94	993	2	0	1188
Siliceous rock	324	404	462	1608	1068	157	2	4025
Petrified wood	22	60	79	119	0	0	0	280
Other/indeterminate	10	3	68	1	0	1	1	84
Total	417	661	806	1904	2072	169	3	6032

Table 7.2 Significance of the Chi-square test per cell for artifacts raw materials and colors

	Yellow	White	Black	Brown	Red	Green	Other
Andesite	(+) NS	(+) NS	(+) NS	(−) NS	(−) NS	**(+) *****	(−) ***
Basalt	(−) NS	(−) *	**(+) *****	(−) ***	(−) ***	(−) NS	(−) ***
Chalcedony	(+) NS	**(+) *****	(−) ***	(+) NS	(−) ***	(−) *	(−) NS
Gabbro/diabase	(−) NS	(−) NS	(−) NS	(−) NS	(−) NS	**(+) *****	(−) ***
Obsidian	(−) ***	(−) ***	**(+) *****	(−) ***	(−) ***	(−) *	(−) NS
Rhyolite	(−) ***	(−) ***	(−) ***	(−) ***	**(+) *****	(−) ***	(−) NS
Siliceous rock	**(+) *****	(−) ***	(−) ***	**(+) *****	(−) ***	**(+) *****	(−) NS
Petrified wood	(+) NS	**(+) *****	**(+) *****	**(+) *****	(−) ***	(−) ***	(−) NS
Other/indeterminate	(+) NS	(−) **	(+) ***	(−) ***	(−) ***	(−) NS	(+) **

Test results and contributions to x^2. Raw materials used and their colors. Associations that occur more frequently than expected are highlighted in bold typography. Acronyms for the results of statistical test: (+): observed frequency higher than the theoretical frequency; (−): observed frequency lower than the theoretical frequency; NS: Chi-square test for non-significant cell at threshold alpha = 0.100; *: Chi-square test per significant cell at threshold alpha = 0.100; **: Chi-square test per significant cell at threshold alpha = 0.050
***: Chi-square test per significant cell at threshold alpha = 0.010; DOF: degrees of freedom

7.2.2.2 Forms of Availability of Raw Materials

In La Primavera, lithic raw materials are naturally found as rock outcrops, secondary deposits, and scattered nodules. The quarries identified in the study area are described below. Information on lithic raw material sources in the macro-study area is presented in Table 2.3 (Chap. 2).

Rock outcrops

Table 7.3 Values of the Chi-square test per cell for artifacts raw materials and colors

	Yellow	White	Black	Brown	Red	Green	Other
Andesite	0.826	0.178	0.049	1.894	2.061	**47.707**	0.003
Basalt	2.558	4.055	**159.212**	6.450	12.710	0.001	0.018
Chalcedony	1.084	**394.9**	18.451	0.028	74.152	3.704	0.127
Gabbro/ diabase	0.207	0.329	0.401	0.947	1.031	**101.16**	0.001
Obsidian	10.646	16.876	**852.175**	46.631	52.899	4.315	0.077
Rhyolite	**23.710**	37.837	156.748	210.550	**838.395**	29.405	0.591
Siliceous rock	7.521	**3.115**	10.690	**89.660**	71.582	**17.348**	0.000
Petrified wood	0.361	**28.012**	**46.224**	**8.017**	88.347	7.845	0.139
Other/ indeterminate	3.028	4.183	287.193	24.552	28.854	0.778	21.97
Total	49.941	489.556	1531.14	388.734	1170.03	212.264	22.93

Test results and contributions to x^2. Colors of different rocks that were used. The values that correspond to the associations that occur more frequently than expected are highlighted in bold typography

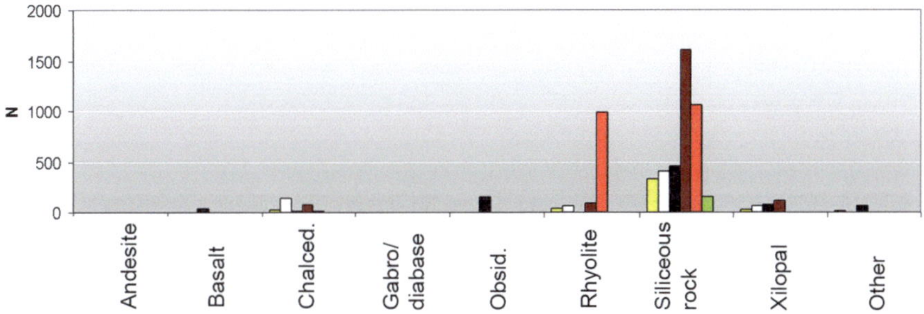

Fig. 7.4 Frequency of rocks used as raw materials and their colors. The values represent the total collection of lithic artifacts in the study area

The Cantera del Gris (CDG) quarry is located in the Las Mercedes area. This vein-shaped, light greenish-gray siliceous rock outcrop measures approximately 260 by 20 m, and flaking negatives were recorded on it. Some cores of the same rock were also found next to the seam. Because this is an extensive outcrop, the quality for knapping varies. In some areas, internal fracture planes that facilitate the detachment of clasts due to weathering were observed. Based on these observations, the quality of the rock can be classified as poor to regular, with some areas of the vein where the quality is good (Fig. 7.5).

The Platense Quarry (CP) is also located in Las Mercedes. It is a siliceous vein that extends approximately 50 by 25 m, and the rock is reddish-brown with small black and white parts. This seam has flake negatives, indicating that it was used as a source of raw materials. The waste density in its surroundings is lower than in

CDG (Fig. 7.5). Quarry CF4 is located in the La Primavera locality. It shows clear signs of exploitation, including large cores with isolated flakes, primary flakes of significant size, and high debitage density. In this case, the raw material is siliceous rock that emerges in the form of blocks beneath a lava flow. Its color varies from brown to yellow. The rock appears as large, subangular nodules, which indicates an absence of transport (Fig. 7.5). The D9 site at the La Lotita locality presents a different record. It is located at the foot of a rhyolite ridge where nodular flakes with cortex, cores, and a high density of unformatted artifacts of red rhyolite were densely distributed. Formatted artifacts made from other siliceous rocks are also part of the surface assemblage (Fig. 7.5). Finally, the Rocky quarry (CR) is located near Cueva Maripe in the center of the studied area. It contains brown rhyolitic ignimbrite (Fig. 7.5) (Hermo, 2008). Its location in the landscape is similar to that of CF4. It is associated with the Las Mercedes Basalt in a sub-cropping position. Other quarries in areas adjacent to the prospected sector complement the raw material landscape. The Rojo quarry (CDR), located 10.5 km south, contains only red silicified rhyolite. The CDV quarry, located 1.5 km north, contains high-quality green silicified breccia (Hermo, 2008; Fig. 7.5).

1

Fig. 7.5 Raw materials from rock outcrops study area. 1) Sample of light greenish-gray siliceous rock with small veins that break the rock from the gray quarry (CDG); 2) Details of the flakes on the brown siliceous rocks outcrop from the Platense quarry (CP), light brown and pink colors were also recorded; 3) Primary flake in the CF4 quarry. In addition to yellow, other colors predominate such as brown, light brown, and pink; 4) Outcrop of rhyolite and scattered artifacts in D9. The predominant color is red, the cortex is pink; 5) Red-brown siliceous rocks sample from Rocky (CR) quarry; 6) Sample of rhyolitic ignimbrite from Cantera del Rojo (CDR). In this sample, the yellow color predominates, but the most common is red, although brown and violet colors also appear; 7) Green siliceous rocks sample from the CDV quarry

2

3

Fig. 7.5 (continued)

4

5

Fig. 7.5 (continued)

6

7

Fig. 7.5 (continued)

Table 7.4 Raw materials registered in secondary deposits

Transect section	Secondary deposit	Gravel (2–6 cm)	Siliceous rock nodules (6–25 cm)	Petrified Wood nodules (6–25 cm)	Basalt nodules (6–25 cm)	Ignimbrite blocks (>25 cm)
G0-G1	Alluvial plains		x			
J2-J1	Alluvial cones	x	x			
K2-K1		x	x	x	x	
K3-K2	Pediment	x	x	x	x	
K4-K3		x	x	x	x	x

Secondary deposits

Raw materials are also present in secondary deposits in the locality. These deposits cover pediment levels (recorded in transects K4-K3 and K3-K2) and alluvial cone deposits (J2-J1, K2-K1, and the Cantera Los Bloques CB quarry). These deposits are characterized by sub-angular to rounded cobbles of various raw materials and knapping qualities. The diversity of recorded lithologies and sizes is shown in Table 7.4. Large boulders (≥10 cm) were registered in all the surveyed transects, particularly in those that crossed alluvial cone deposits. However, large sections of gravel too small for knapping were also present.[1]

The archaeological evidence in these sectors consists of extensive dispersions of archaeological materials, primarily isolated flaked cores and diagnostic artifacts indicating early stages in the production chain of lithic artifacts (characterized by large size and high proportions of cortex; see Chap. 6). The predominant visibility values are good (0) and very good (1 and 2), and the artifact densities vary (Table 7.5).

Based on this evidence, it is interpreted that the analyzed sectors were used as supply quarries. These sectors are characterized by extensive distributions of low- and medium-density artifact assemblages, although there are also sectors with higher density and marked continuity in the record. The lithic dispersions were so widespread that their limits could not be defined, so this prompted a change in the field survey strategy. In these cases, the observation unit was reduced. Instead of using 20-m-diameter circular survey units, two-meter-diameter circles were arranged along the transect line, separated by 20-m intervals (see Chap. 5).

Figure 7.6 shows quarries in primary outcrops and secondary deposits on the geomorphological map (see Chap. 6). Table 7.6 presents geological information (Panza, 2001) and the types of rock found in each quarry.

Disperse nodules

Another way that raw materials suitable for knapping appear in the landscape is in the form of siliceous rock nodules dispersed throughout different parts of the landscape.

[1] The knapping of these small cobbles may be possible by the use of a bipolar technique, but this region has not been recorded.

Table 7.5 Indicators of quarrying activities on secondary deposits

ID	Transect section	Density	Visibility	First-stage indicator (%)
672	K2-K1	High	0	50
634	K4-K3	High	0	50
708	K4-K3	High	0	100
57	K3-K2	Low	1	100
210	K3-K2	Low	1	50
211	K3-K2	Low	0	100
667	K2-K1	Low	2	33
154	G0-G1	Medium	0	50
263	G0-G1	Low	1	100

ID: assemblage/location identifier, density values, archaeological visibility values, first stage (%): percentage of artifacts with indicators of raw materials procurement (see definition in Chap. 6)

These nodules often show evidence of extraction, such as cores with isolated flake blanks (sensu Aschero, 1975, 1983). The distribution of nodules (knapped or not), based on 176 survey forms, was analyzed in relation to the various geomorphic features in which they were located (Fig. 7.6). An x^2 goodness-of-fit test was carried out to analyze their distribution (Shennan, 1992). The analysis results indicate the presence of isolated nodules in all geoforms ($\alpha = 0.01$, critical value $= 15.08$, $X^2 = 363.92$). However, there is a significant difference between the observed and expected distributions. This difference mainly occurs in the higher than expected frequencies of nodules observed on basaltic plateaus and associated mass-removal deposits (Table 7.7).

On the other hand, the frequency of nodules in the hills was lower than expected. On the plateaus, siliceous nodules are highly visible due to the low vegetation cover and their contrasting color with the basalt rocks. The nodules in this sector are generally ochre to yellow, making them obtrusive against the dark basalt. Therefore, nodule counts are probably overrepresented on the plateaus. However, this trend allows us to revise our previous ideas about the basaltic plateaus as sectors without resources for knapping. The nodules observed in the hills are along the channels of temporary tributaries. Based on the available information, we think that the isolated nodules were not an important source of raw materials, but were used occasionally.

7.2.3 Raw Materials Sourcing

Studies of the distribution of lithic raw materials are one way to approach mobility ranges and strategies for obtaining them, as well as the dynamics of technological behavior (Franco & Borrero, 1999; Nelson, 1991). Here, we are interested

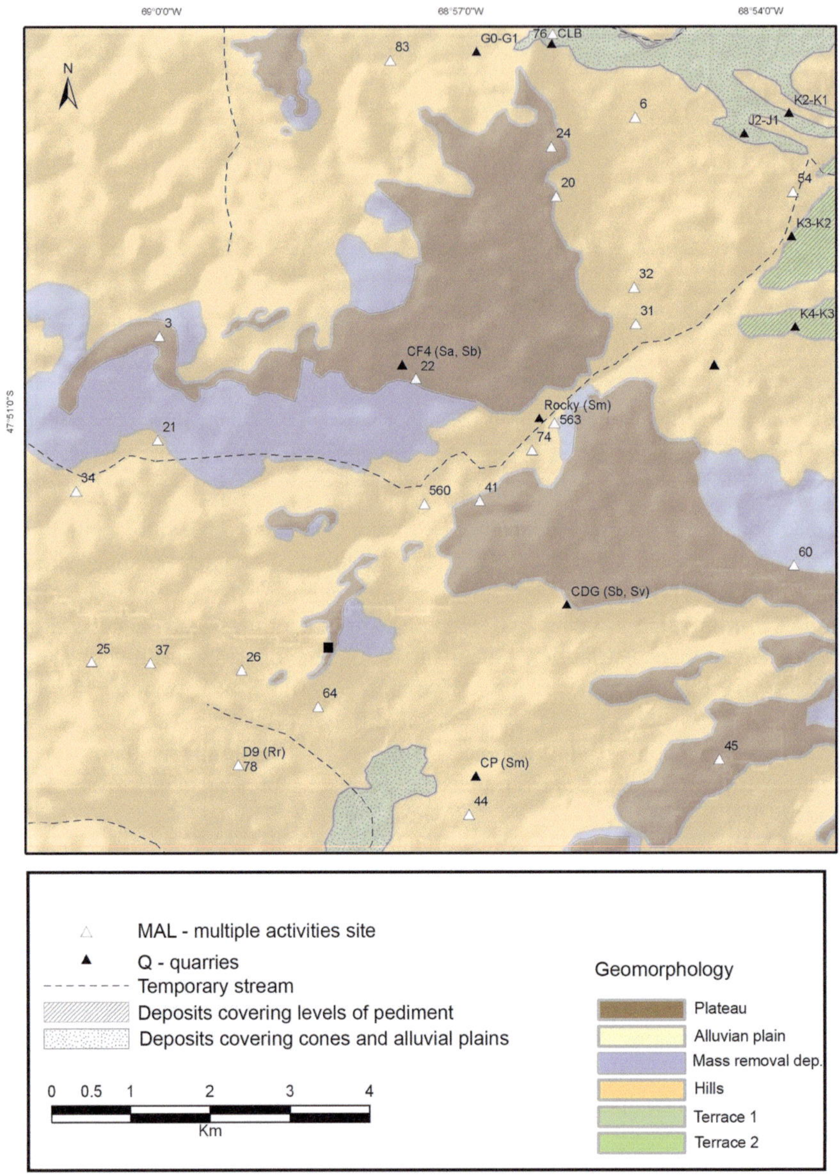

Fig. 7.6 Location of multiple activity loci (MAL) and quarries (Q) is shown on the geomorphological map. The closest quarry to the Pico Blanco site (MAL ID_54) is the secondary deposit K3-K2. Cueva Maripe (ID_563) is located near the Rocky quarry. El Descanso (ID_22) is located near the Cantera F4 (CF4) quarry. El Mirador (ID_44) is near the Cantera Platense (CP) quarry. At site D9, the functions of the MAL and the quarry coincide

Table 7.6 Rocks used to manufacture artifacts in the studied sector, the geological formations with which they are associated, and the known quarries in the study area (rock outcrops and secondary deposits)

Raw material	Geological FMs where these rocks/minerals are present	Quarries (rock outcrops)	Secondary deposits used as quarries
Andesite	Roca Blanca Bajo Pobre Chon Aike	–	Deposits that cover pediment levels K4-K3, K3-K2 Alluvial cone deposits J2-J1, K2-K1 Rockslide deposits CLB
Basalt	Las Mercedes	–	
Gabbro/diabase	Co. León Basalt Co. El Doce Basalt Alma Gaucha	–	
Chalcedony	Chon Aike Bajo Grande Baqueró	–	
Siliceous rocks		CR CDV CP CF4	
Rhyolite (Rhyolitic tuffs and ignimbrites)	Chon Aike	CDR D9	
Petrified wood	Baqueró	–	
Obsidian	–	–	–

Table 7.7 Frequency of siliceous rock nodules recorded in the landscape units

Geoform	Of	area (m^2)	Ef	X^2
Basalt plateau	**89**	10.19	17	304.94
Alluvial plain	1	0.05	1	0
Mass removal deposits	17	5.12	8	10.12
Hills	**58**	84.63	141	48.86
Total	165	100	167	**363.92**

Large pebble deposits were excluded from the analysis. $\alpha = 0.01$; critical value = 15.08

in studying their importance for locational decisions in past hunter-gatherer societies. This approach requires an analysis of the regional base of mineral resources (Berón, 2006). Due to the abundance of raw materials in this landscape, studying their sources presents difficulties (Hermo, 2008). To address these difficulties, we conducted an exhaustive search of published data and supplemented it with information from fieldwork. The gathered information presents different spatial scales ranging from regional geological studies (Belardi et al., 2006; Panza, 2001; Stern, 1999, 2004), to the local geomorphology map (Gómez & Magnin, 2008; Chap. 6), and to the intra-site studies of the CR, CDR, CV, and LP-P1 quarries (Hermo, 2008) and inter-site field information obtained in the surveys. For the provenance analysis, we determined the possible origin of the raw materials found in each collected artifact. While some raw materials have an extra-regional origin, others are local. In these cases, we expect to have more specific information about their origin. One problem we encountered, however, is the difficulty of distinguishing with certainty between rocks from quarries and secondary deposits. One way to determine the source of each raw material is to analyze the proportion of cortex in the artifacts collected throughout the sector.

7.2.3.1 Analysis of Cortex Proportion

Outcrops are large lithic masses, of which a small relative proportion is constituted by the weathered surface or cortex. The opposite is expected from raw materials in the form of smaller nodules and rounded or angular blocks. In this case, their greater surface areas are exposed to weathering and, therefore, to the formation of a cortex. Considering this, the percentage of cortex in lithic artifacts provides an idea of the source of the raw materials in a lithic assemblage (Andrefsky, 2005). Rock varieties significantly associated with values "0" (absent cortex) and "1" (less than 50%) are more likely to come from geological outcrops, while those associated with values "2" (less than 100%) and "3" (all cortex) probably come from nodules. This analysis aims to provide information about the form in which rock is found (in outcrops or secondary deposits) within the study area. Applying the Chi-squared test to a contingency table and analyzing the maximum differences allowed us to identify significant associations between rock type and cortex percentage (Tables 7.8 and 7.9). The results indicate that the variables are not independent ($\alpha = 0.050$, $X^2 = 529.96$, critical value $= 74.46$, Cramer's $V = 0.149$). This information is included in the summary of the lithic resource structure below.

The siliceous rocks in the studied area are found in geological outcrops and veins, with evidence of extraction negatives that confirm their use as quarries. They are immediately available (According to Bayón and Flegenheimer, 2004), within a 10 km radius from any point in the studied sector. Of the known siliceous rock quarries in the surveyed area, two were used to extract brown and red siliceous rocks (CR and CP), one produced brown and yellow siliceous rocks (CF4), and one produced greenish-gray siliceous rocks (CDG). To the north of the surveyed area, another source of green siliceous rocks (CDV) is known. Based on the results in Table 7.9, green siliceous rocks have low cortex percentages. This indicates that this raw material mainly occurs in primary outcropping quarries. CDV is located near the bottom of the basin, so transporting materials downslope would be quick, preventing cortex formation. Within the brown and black siliceous rocks, other trends are less clear. They appear with variable percentages of cortex, indicating that, in addition to being found in outcrops and veins, they are also found as cobbles in secondary deposits.

Conversely, red siliceous rocks (SRr) have very low cortex proportions, indicating that they originate from outcrops and suggesting the Rocky Quarry (CR) as their source. Finally, yellow and white siliceous rocks (SRy and SRw) have equally represented percentages of cortex without major differences. This suggests that the

Table 7.8 Frequency of artifacts according to their percentage of cortex (0 = absence; 1 = less than 50%; 2 = less than 100%; 3 = 100%; 4 = indeterminate)

Cortex % in dorsal surface	F(0)	F(1)	F(2)	F(3)	F(4)	Total
Raw material variety						
A, B, G	37	5	2	2	0	46
Cy, w	101	46	7	1	0	155
Cg, bl, br	71	22	7	0	0	100
O	99	36	14	3	1	153
Ry, w	86	7	2	3	0	98
Rg, bl, br	64	23	7	2	0	96
Rr	919	48	12	7	0	986
Sra	220	61	29	9	1	320
SRw	290	64	35	11	1	401
SRbr	1077	310	153	60	1	1601
SRbl	309	84	52	13	0	458
SRr	840	132	60	27	1	1060
SRg	73	54	23	5	0	155
PW	130	69	55	26	1	281
Other	64	11	4	3	1	83
Total	4380	972	462	172	7	5993

The acronyms used for raw materials are composed of uppercase letters indicating the lithology (A: andesite; B: basalt; C: chalcedony; O: obsidian: P: petrified wood; R: rhyolite; SR: silicified rock), and lowercase letters indicating the color (bl: lack; br: brown; g: green; r: red; y: yellow; w: white)

source of the yellow siliceous rocks is in outcrops, such as those in the CF4 quarry, as well as in secondary deposits. The strongest pattern in the analyzed data was found among the rhyolites, indicating that this raw material occurs without cortex more often than expected (see its contribution value to x^2 in Table 7.9). This can be explained by their exclusive origin from large outcrops near the bottoms of basins, meaning they were not transported. A rhyolite quarry that was previously studied by Hermo (2008) is the CDR quarry. Due to its distance from the study area (more than 10 km), its availability is considered local, not immediate (Bayón & Flegenheimer, 2004). Another immediately available source of rhyolite from any part of the studied sector is the D9 quarry. This site is in the early stages of research and has evidence of a large number of knapped artifacts.

In the case of petrified wood, we recorded outcrops of formations containing fossil wood; however, no quarry sites of this material were found. The high cortex percentages (Table 7.9) can be explained by the fact that the Baqueró Formation's petrified wood clasts in this area are small. This is in contrast to the petrified tree trunks in the Estación F quarry in the Jaramillo Petrified Forest National Park, which extend tens of meters long and exceed two meters in diameter (Hermo, 2005; Hermo & Vázquez, 1999; Panza, 2001). In this study area, the largest petrified wood recorded

Table 7.9 Raw materials and percentage of cortex.Test of x^2 per cell and contributions to x^2

0	1	2	3	4	% c	0	1	2	3	4
					Raw material					
(+) NS	(−) NS	(−) NS	(+) NS	(−) *	A, B, G	0.34	0.81	0.67	0.35	0.05
(−) **	(+) ***	(−) NS	(−) NS	(−) NS	Cy, w	1.33	17.31	2.05	2.67	0.18
(−) NS	(+) NS	(−) NS	(−) NS	(−) NS	Cg, bl, br	0.05	2.06	0.06	2.87	0.12
(−) **	(+) **	(+) NS	(−) NS	(+) NS	O	1.47	5.04	0.41	0.44	3.77
(+) NS	(−) NS	(−) NS	(+) NS	(+) NS	Other	0.18	0.45	0.89	0.16	8.41
(+) ***	(−) **	(−) *	(+) NS	(−) NS	Ry, w	2.88	4.98	4.08	0.01	0.11
(−) NS	(+) *	(−) NS	(−) NS	(−) NS	Rg, br, bl	0.54	3.55	0.02	0.21	0.11
(+) ***	(−) ***	(−) ***	(−) ***	(−) NS	Rr	54.61	78.33	53.90	16.03	1.15
(−) *	(+) NS	(+) NS	(−) NS	(+) NS	SRy	0.82	1.59	0.76	0.00	1.05
(−) NS	(−) NS	(+) NS	(−) NS	(+) NS	SRw	0.03	0.02	0.54	0.02	0.60
(−) ***	(+) ***	(+) ***	(+) **	(−) NS	SRbr	7.41	9.76	7.08	4.29	0.40
(−) ***	(+) NS	(+) ***	(−) NS	(−) NS	SRbl	1.98	1.27	7.89	0.00	0.53
(+) ***	(−) ***	(−) ***	(−) NS	(−) NS	SRr	5.50	9.27	5.77	0.38	0.05
(−) ***	(+) ***	(+) ***	(+) NS	(−) NS	SRg	14.32	33.13	10.22	0.07	0.18
(−) ***	(+) ***	(+) ***	(+) ***	(+) NS	P	27.66	12.04	51.30	39.88	1.37

Percentage of cortex (0 = absence; 1 = less than 50%; 2 = less than 100%; 3 = 100%; 4 = indeterminate). Acronyms for the results of statistical test: (+): observed frequency higher than the theoretical frequency; (−): observed frequency lower than the theoretical frequency; NS: Chi-square test for non-significant cell at threshold alpha = 0.100; *: Chi-square test per significant cell at threshold alpha = 0.100; **: Chi-square test per significant cell at threshold alpha = 0.050 ***: Chi-square test per significant cell at threshold alpha = 0.010; DOF: degrees of freedom

is trunks no longer than one meter and no greater than 30 cm in diameter that do not present conchoidal fractures. On the other hand, the highest quality silicified wood for knapping is fragmented into nodules smaller than 10 cm.

No similar quarries are known for the remaining raw materials used. For example, although there are numerous basalt outcrops, no evidence of extraction was found. This is probably because the area covered by the basalt flows is extensive. Additionally, the small number of basalt artifacts collected shows that few outcrops were identified in the field (Table 7.6). For this reason, the probability that they are immediately available cannot be ruled out. We are inclined to think that their origin is not immediate and could be local. In the case of gabbros and diabases, which are used in artifacts manufactured by chipping and abrasion, no extraction quarries were found. The geological map shows these rocks in formations outside the studied area, 1.8 km north. Thus, gabbros and diabase are locally available (between 10 and 60 km, Bayón & Flegenheimer, 2004). The rarity of the aforementioned raw materials makes it difficult to draw inferences from the percentages of cortex present. Unlike the other siliceous rocks, the yellow and white chalcedony (translucent siliceous rocks) were not found in outcrops or veins. Instead, they were found in deposits and within rock formation cavities. In the first case, small, translucent, gravel-sized nodules were found in sectors of the volcanic plateau and near temporary shallow lakes, forming accumulations. These nodules have very high amounts of cortex, which does not coincide with the small amount of cortex recorded in the sampled artifacts (less than 50% on their dorsal side). This suggests that the nodules formed in tuff cavities of the Chon Aike Formation are the most likely source of this raw material (personal communication from Drs. Eugenio Aragón and Raúl Fernández and own observations), as they have a greater volume of rock than the aforementioned cobbles. However, no evidence of extraction has yet been found in the local Chon Aike formations.

Finally, obsidian has the longest sourcing trajectory because no quarries were found in the area, and the closest known sources are in Pampa del Asador, Pampa de La Chispa, and Abanico del Cerro Bayo, which are more than 150 km away (Belardi et al., 2006). All of these trends are summarized in Table 7.10.

As previously mentioned, the secondary deposits contain raw materials of various lithologies, including siliceous rocks, petrified wood, basalt, and ignimbrite. They extend much farther than the outcropping rocks, covering 9.2 km^2 of cone deposits and alluvial plains and 2.3 km^2 of deposits on pediment levels. The rocks that form them originate from broader sectors of the landscape. Colors: g: green, b: black, b: brown, r: red, y: yellow.

7.2.4 *Spatial Distribution of the Raw Materials Used*

The next sections use all the information analyzed so far about the natural distribution of raw material sources in the study area to characterize the organization of settlements in relation to lithic resources. Two spatial analysis techniques will help

Table 7.10 Scheme resulting from the analysis of availability and sourcing of the raw materials used in the study area

Rock	Color	Availability	Origin: rock outcrop (RO)/secondary deposit (SD)	Associated quarry site
Siliceous rock	br	Immediate	RO and SD	CR*, CP
Siliceous rock	bl	Immediate	RO and SD	Small vein outcrops
Siliceous rock	g	Local	RO	CDV*
Siliceous rock	y, w	Immediate	–	CDG, CF4
Siliceous rock	r	Immediate	RO	CR*
Rhyolite	r	Immediate	RO	CDR*
Rhyolite	r	Unknown. immediate	RO	D9
Petrified wood	br, bl, w	Unknown. immediate	RO and SD	–
Andesite	g	Unknown. immediate	RO and SD	–
Basalt	bl	Unknown. immediate	–	–
Gabbro/ diabase	g	Unknown. immediate	–	–
Chalcedony	y, w	Unknown. immediate	–	–
Obsidian	bl	Regional	–	–

Provenance "RO" = rock outcrop, "SD" secondary deposit. Colors: "br": brown, "y": yellow, "w": white, "bl": black, "g": green, "r": red. *: indicates that the lithology is confirmed by microscopy studies

achieve this goal: (1) a map showing the distribution of lithic artifacts by transect, and (2) an analysis of rock catchment basins.

7.2.4.1 Raw Materials Distribution Map by Transect

The raw materials of the collected artifacts were registered in the database (see Supplementary Materia Chap. 5) using a GPS reference corresponding to their find place. Each piece is identified; in cases of concentrations and archaeological sites, several pieces share the same georeference. Using this data, thematic maps were created to provide a preliminary overview of the distribution of the collected evidence. To identify trends in their distribution, the data were grouped by the transect sections in which they were collected. Information on each section was graphed within 1 by 1 km areas for visual analysis of the distributions. Thematic maps were created for each type of raw material using proportional symbols. This analysis revealed the distribution of raw materials used in different transect sections. Each map also includes the known quarries of the corresponding rock type, as well as the geological formations from which the raw materials may have come (see Fig. 7.7 and Table 7.10). The map in Fig. 7.8 summarizes the total information and allows for a relative comparison of the dispersion of raw materials in the study area. A pie chart was used to show the percentages of all raw materials in the artifacts collected in each transect section. This map shows the proportions of raw materials present in the samples taken.

Several trends can be noted from these thematic maps. For example, red rhyolites are grouped in the southwestern sector of the study area. Brown and red siliceous rocks are widely dispersed, and green siliceous rocks are distributed more discreetly toward the north. Additionally, some landscape sectors lack archaeological evidence, such as transects B, C, D, and G (note that this observation excludes dark gray areas that were not prospected). These spaces may have been used less frequently, or for activities that do not leave durable archaeological evidence on the surface.

Based on the pattern found in Chap. 6, it would be interesting to determine whether the most immediately available raw materials were discarded at the MAL sites, which would confirm that the MAL sites were located near quarries. Another possibility is that rocks from more distant sources were used. The distance between the discarded raw materials and their sources could be used to analyze the circulation of different raw materials. This would allow us to quantify transport distances from sources to other places in the landscape (Hodder & Orton, 1992; Shennan, 1992). However, it is difficult to measure sourcing distances in this landscape because there are primary rock outcrops as well as secondary deposits of the same useful rocks.

7.2.4.2 Analysis of Rock Catchment Basins

Secondary deposits receive rocks from different formations that undergo weathering and are transported toward the base of basins by gravity and fluvial action (Panza, 2001). Hunter-gatherers may have extracted each variety of raw material from outcropping rocks in the study area or from deposits after natural transport to another point in the sector. This creates uncertainty about the location of a raw material's source, preventing reliable measurements of the distance between where an archaeological artifact was discarded and where the raw material was sourced. For this reason, the analysis of raw material circulation will have a maximum spatial resolution restricted to drainage basins, which behave like rock catchment basins.

Delimitation of basins using GIS

The basins in the sector were delineated using the TauDEM (Terrain Analysis Using Digital Elevation Models) module in ESRI ArcGIS (Taborton, 2002). This module performs a hydrological analysis of the terrain. It uses a digital elevation model (DEM) as base data and presents several specific functionalities. The one used here is the delineation of drainage basins. One parameter required to define a basin is the number of cells in the base DEM. Here, 111 cells (with a spatial resolution of 90 by 90 m) were used as a threshold value since this value was found to adequately define the basin areas and their divisions at the current scale of analysis (Fig. 7.9). As a result, three main drainage basins, each formed by several minor sub-basins, could be identified in the studied area. Basin 1 is located to the north and covers part of Estancias La Primavera, La Dorita, and Bajo Grande. Its surface area is 75.31 km^2. Basin 2 is located to the south and includes the fields of Estancias La Lotita and Las Mercedes. Its surface area is 21.50 km^2. Basin 3 is located in the extreme southeast

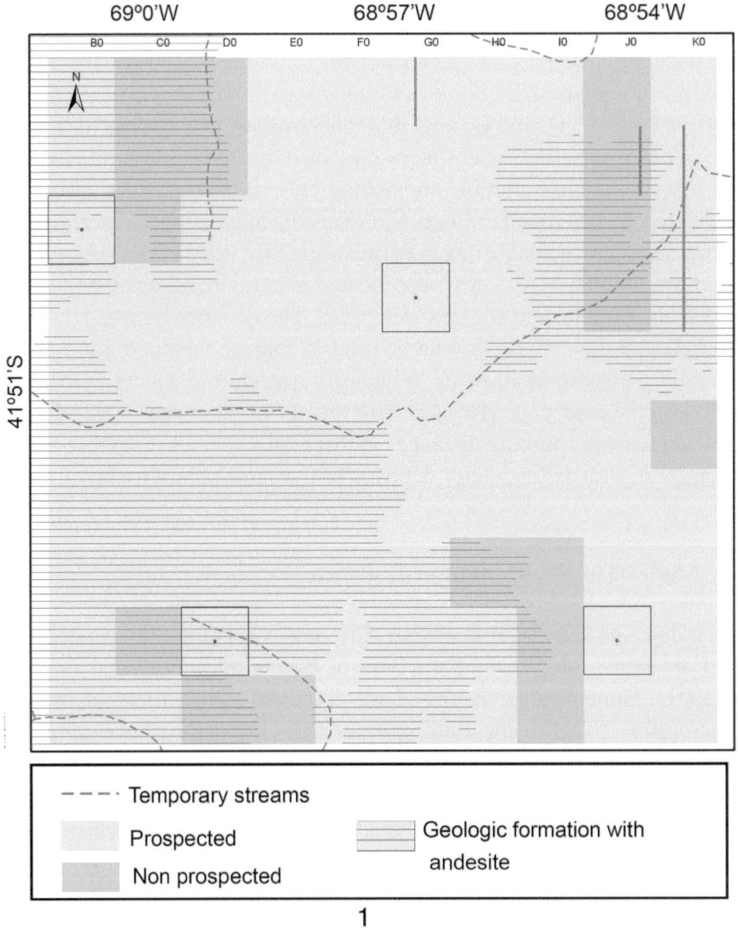

Fig. 7.7 Artifact distribution of the different raw materials in each transect section. 1, Andesite; 2, Basalt; 3, Gabbro/diabase; 4, Black obsidian; 5, Black chalcedony; 6, Red chalcedony; 7, Green chalcedony; 8, Yellow chalcedony; 9, White chalcedony; 10, Brown chalcedony; 11, Yellow rhyolite; 12, White rhyolite; 13, Red rhyolite; 14, Brown rhyolite; 15, Red siliceous rock; 16, Yellow siliceous rock; 17, Brown siliceous rock; 18, White siliceous rock; 19, Black siliceous rock; 20, Green siliceous rock; 21, Yellow petrified wood; 22, White petrified wood; 23, Brown petrified wood; 24, Black petrified wood. In all the maps in figure SM1, "CCG" corresponds to quarry F4; "CMB": Platense quarry; "CDG": Gris quarry and "C. Rocky": Rocky quarry

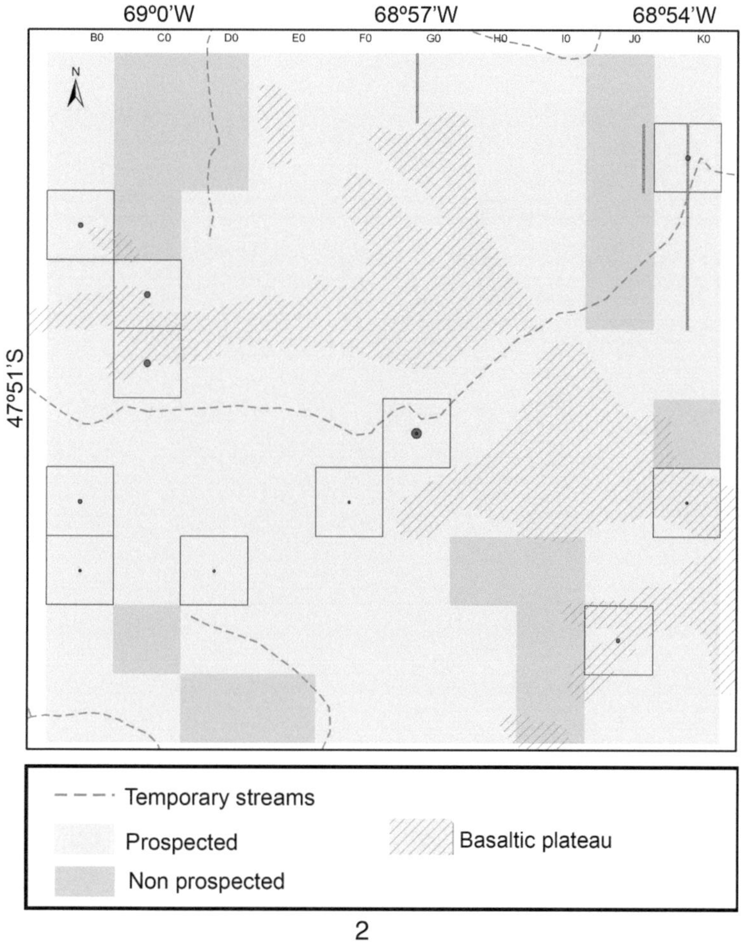

Fig. 7.7 (continued)

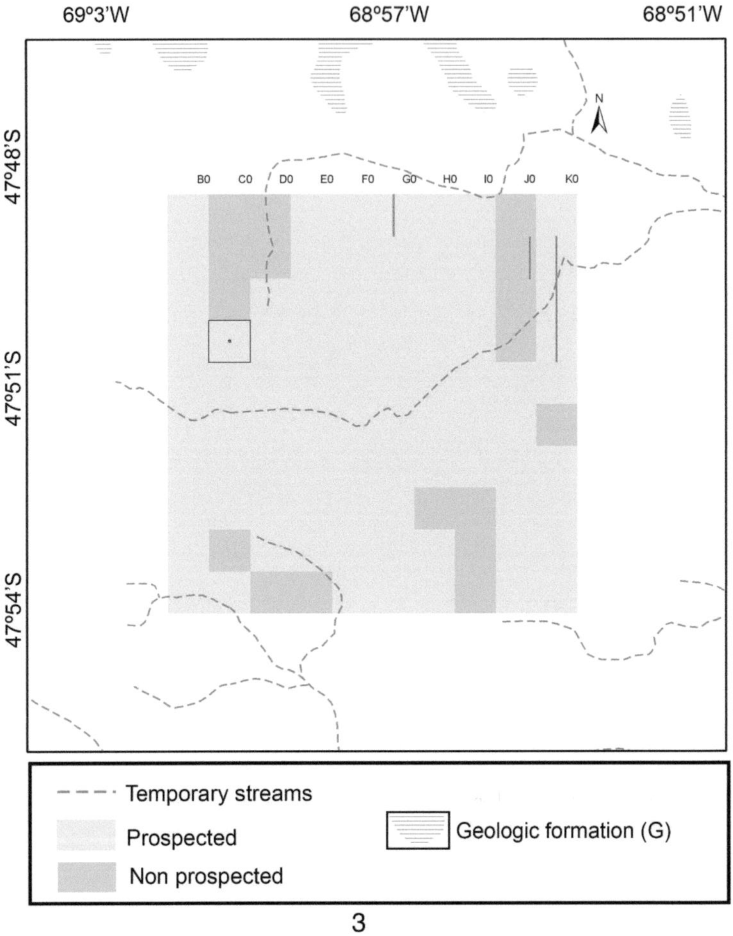

Fig. 7.7 (continued)

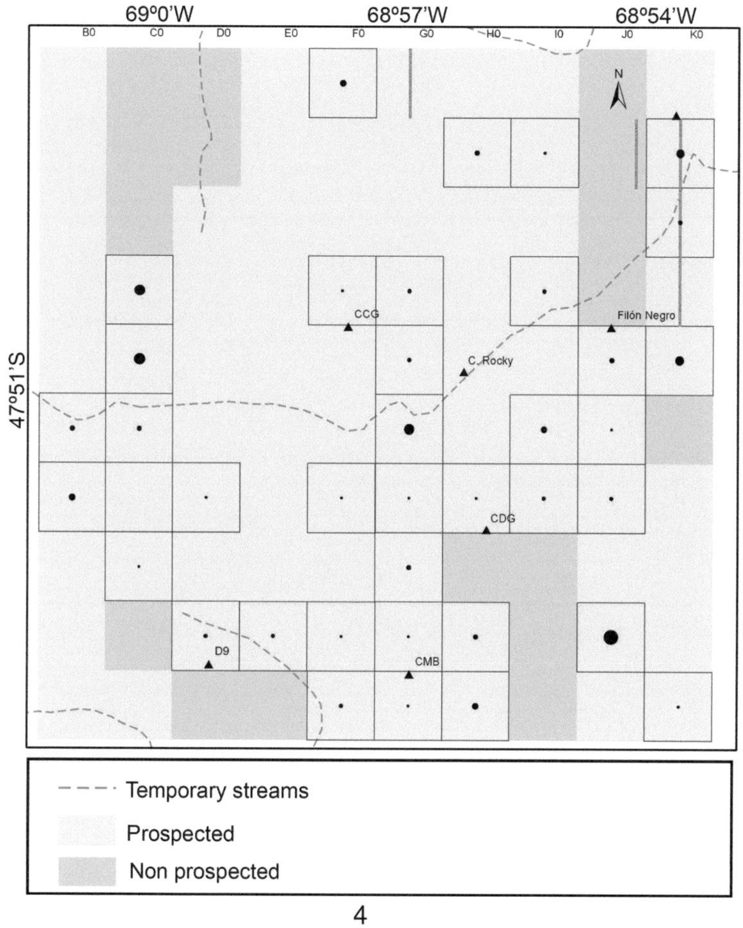

Fig. 7.7 (continued)

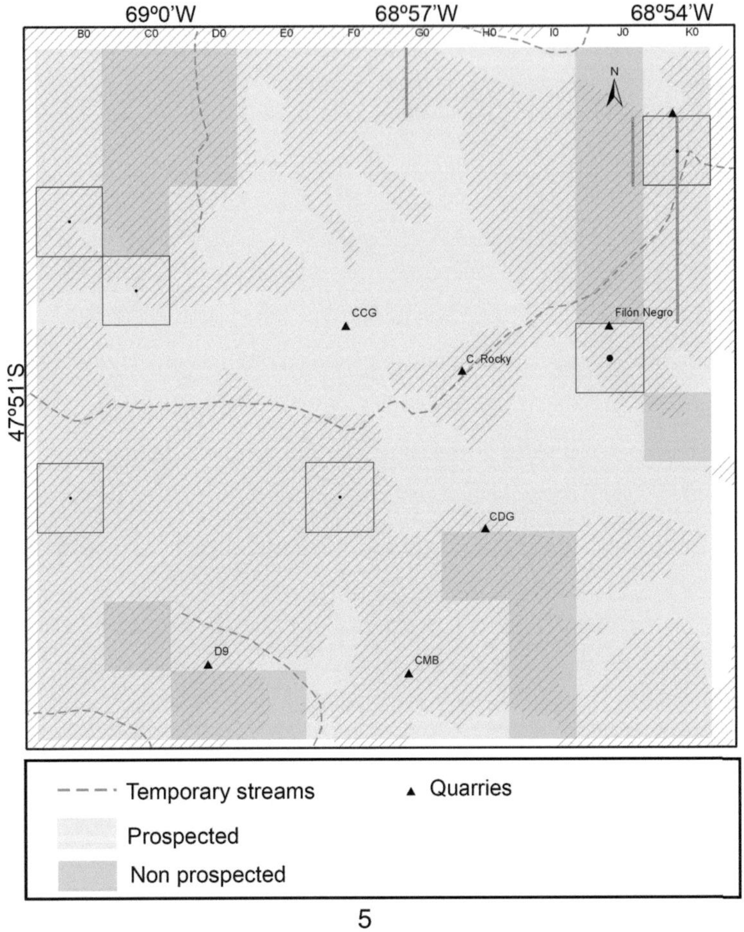

Fig. 7.7 (continued)

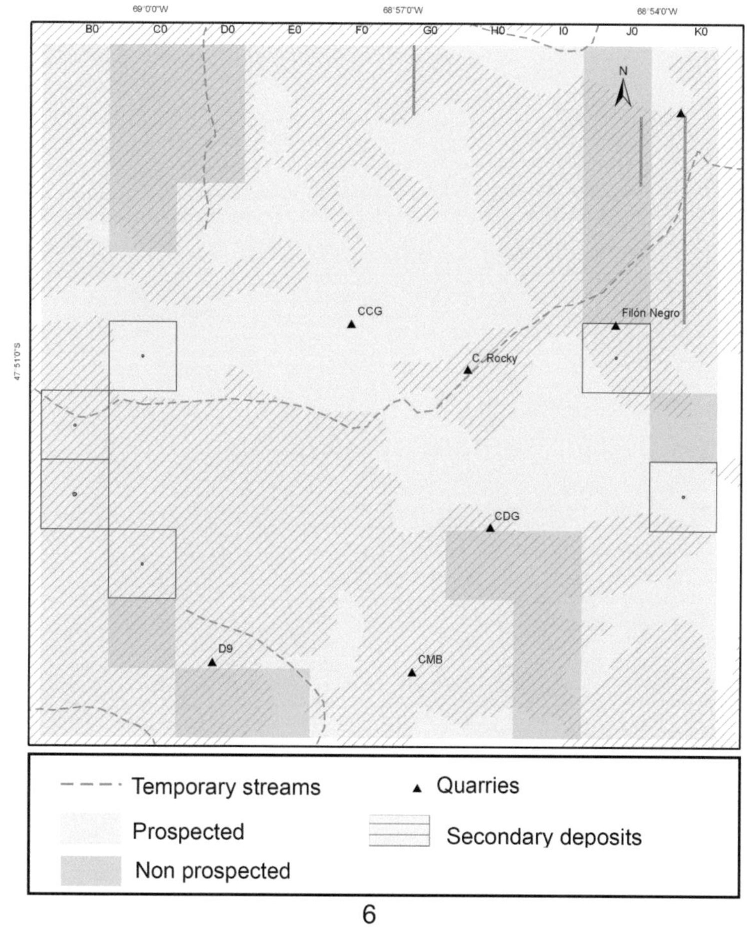

Fig. 7.7 (continued)

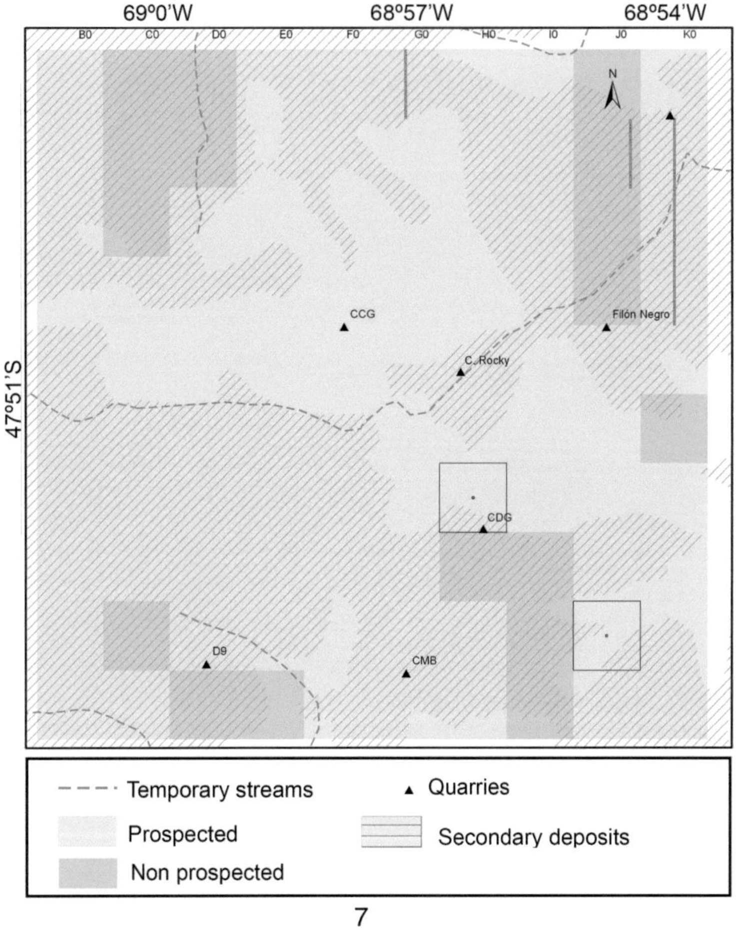

Fig. 7.7 (continued)

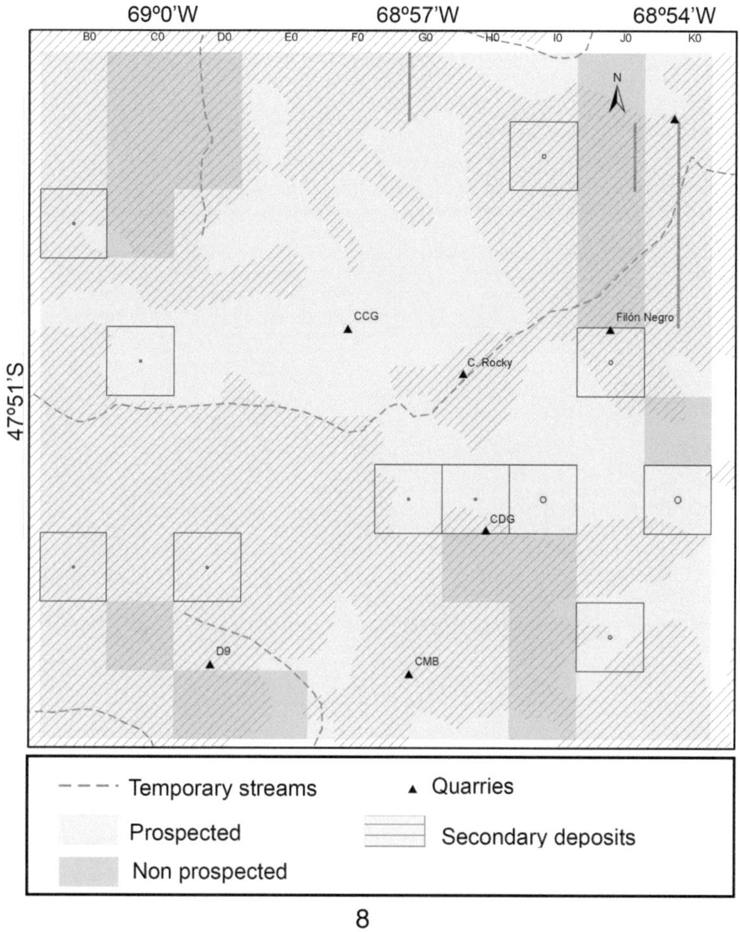

Fig. 7.7 (continued)

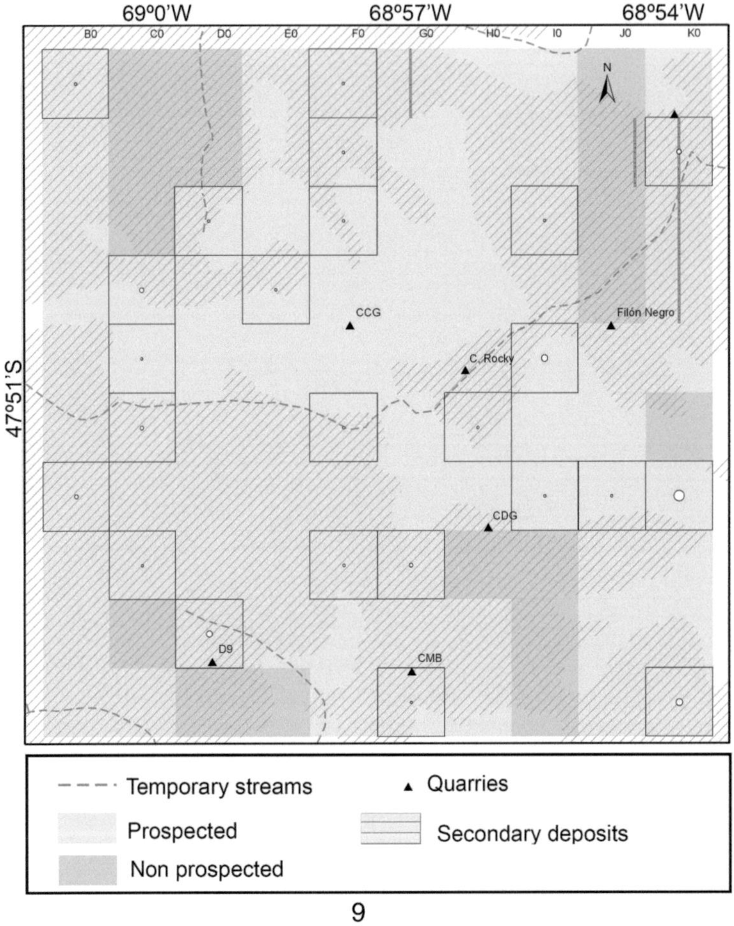

Fig. 7.7 (continued)

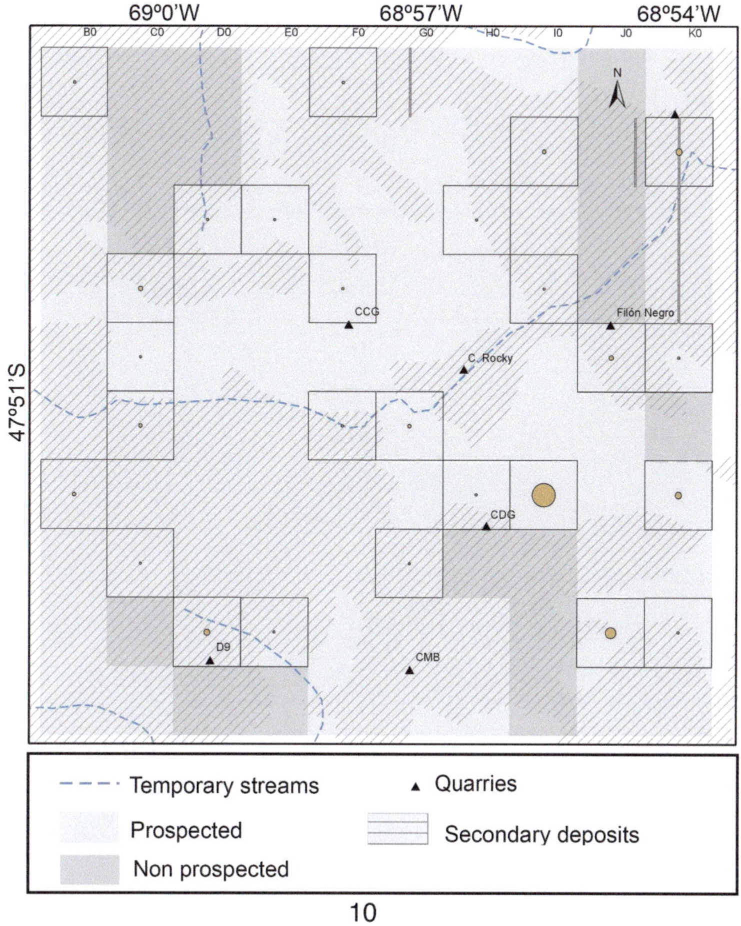

Fig. 7.7 (continued)

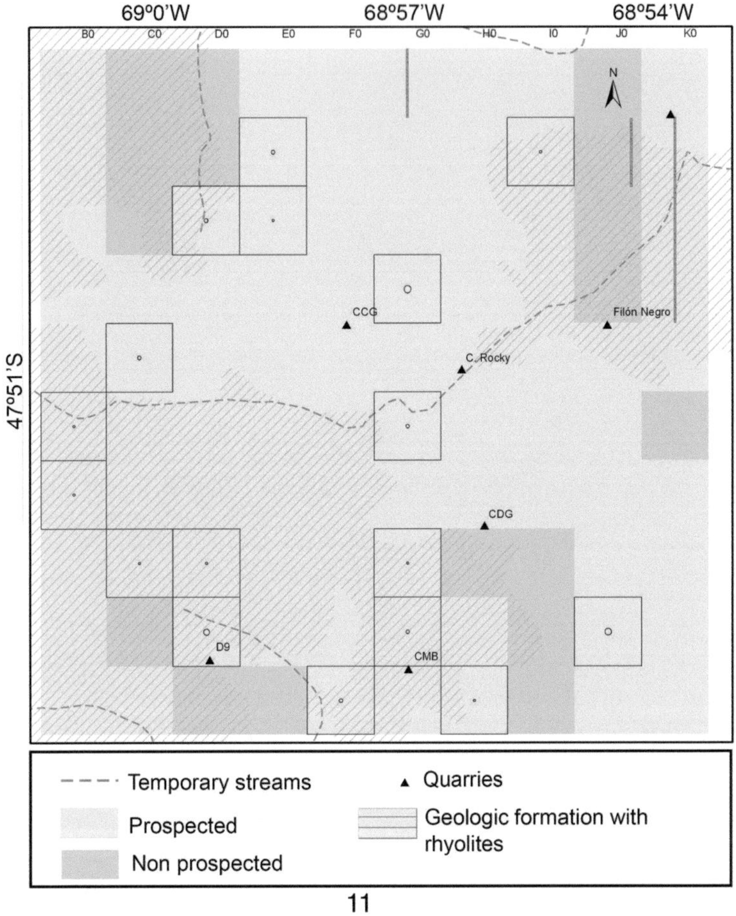

Fig. 7.7 (continued)

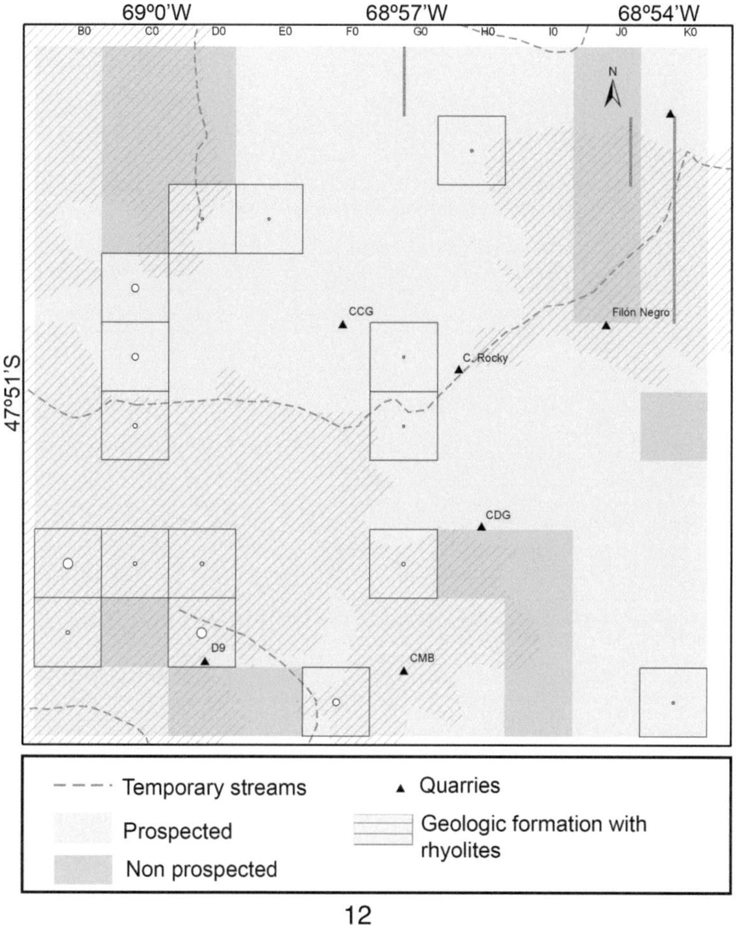

Fig. 7.7 (continued)

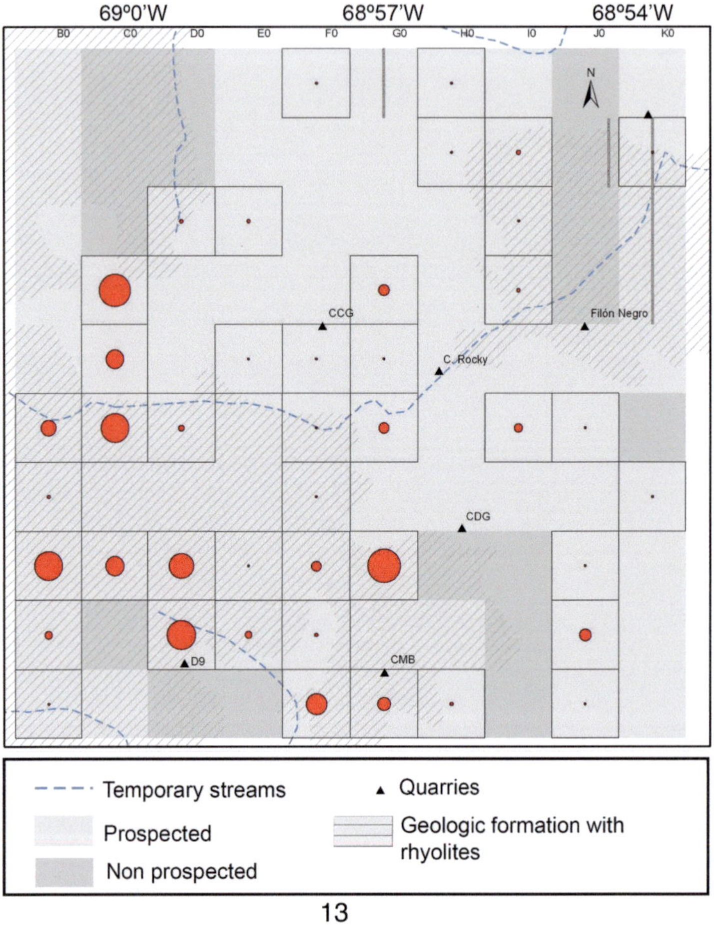

Fig. 7.7 (continued)

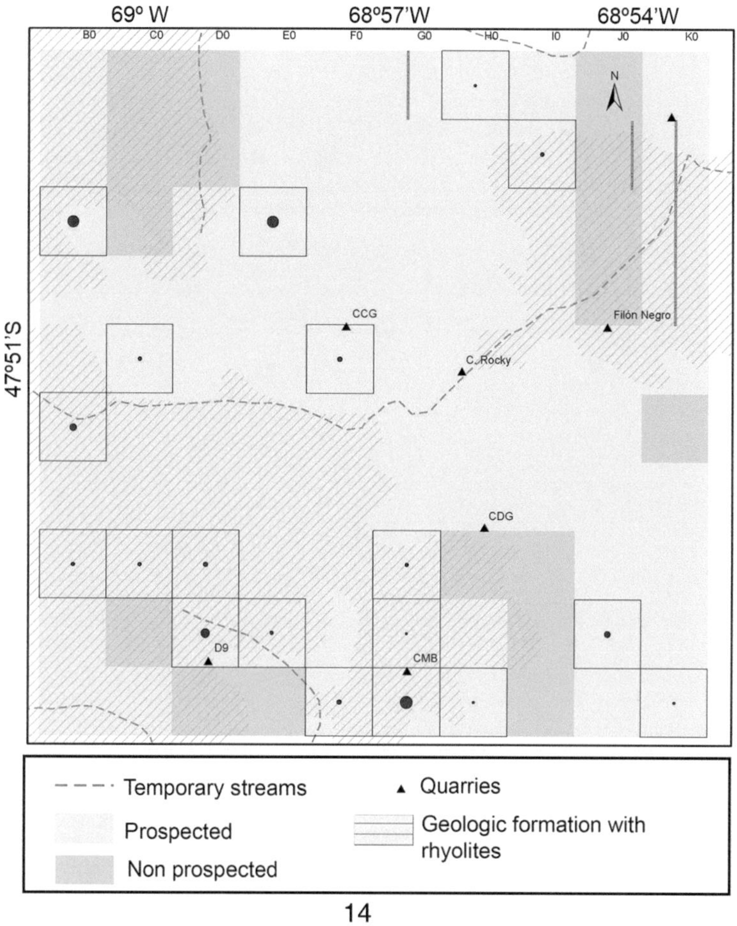

Fig. 7.7 (continued)

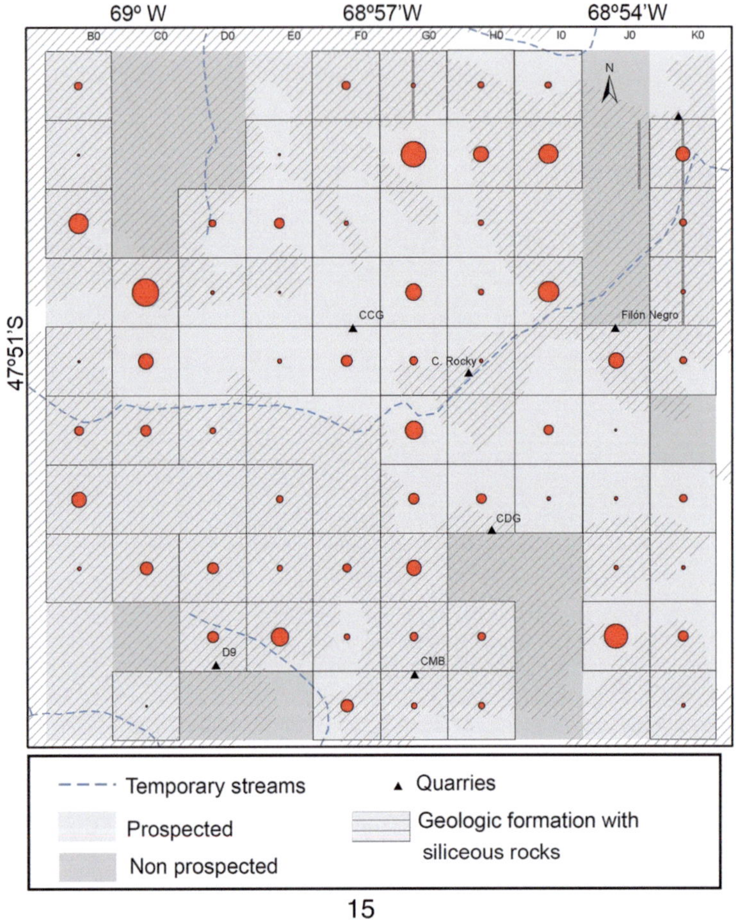

Fig. 7.7 (continued)

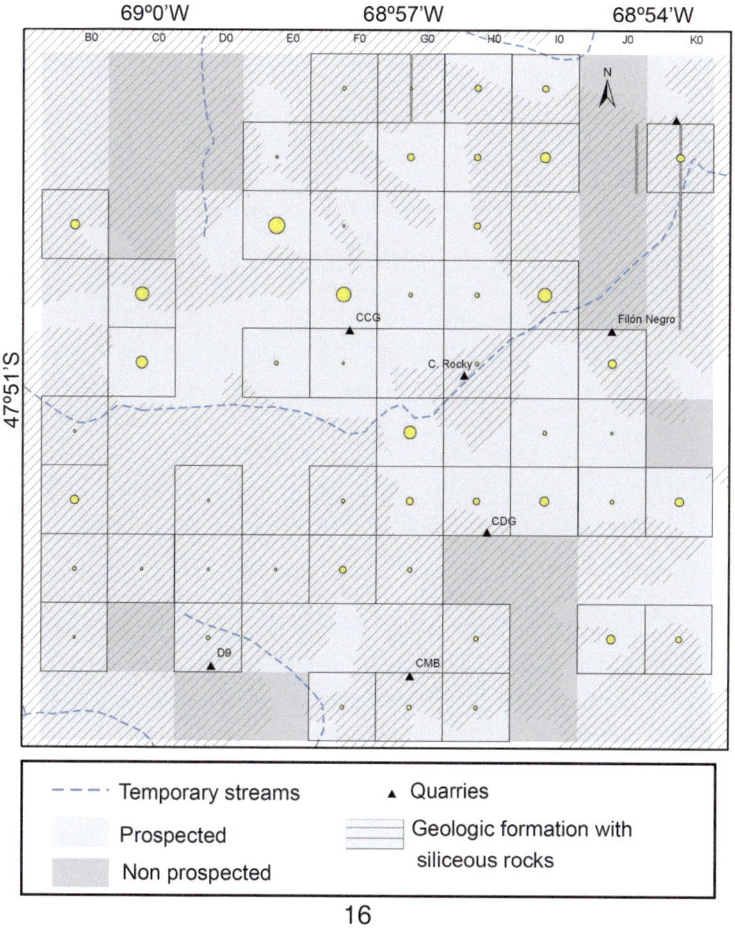

Fig. 7.7 (continued)

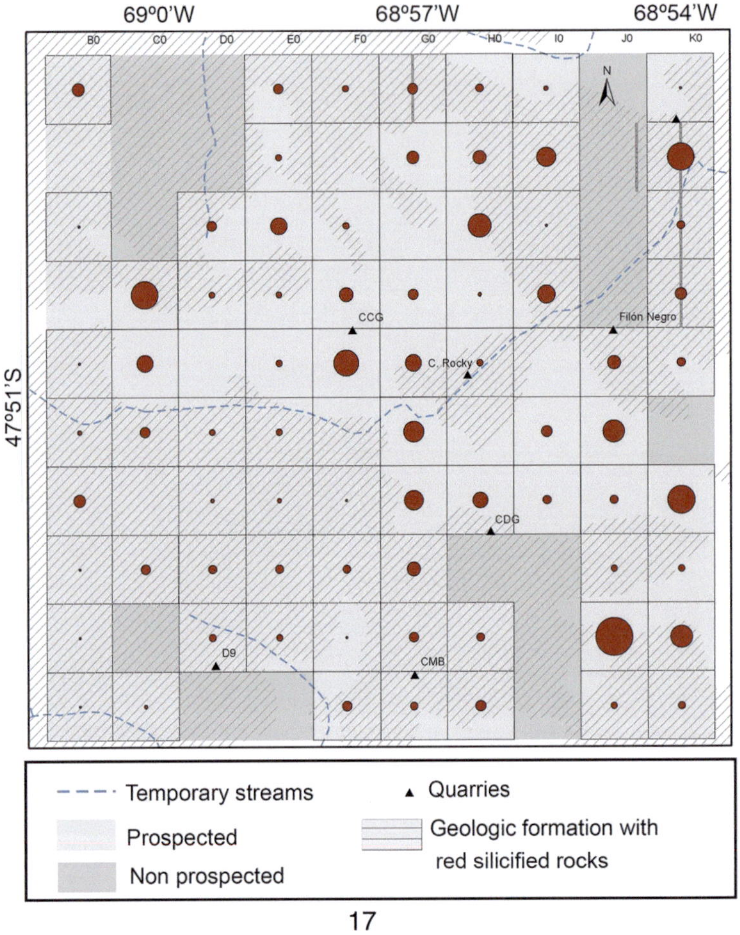

Fig. 7.7 (continued)

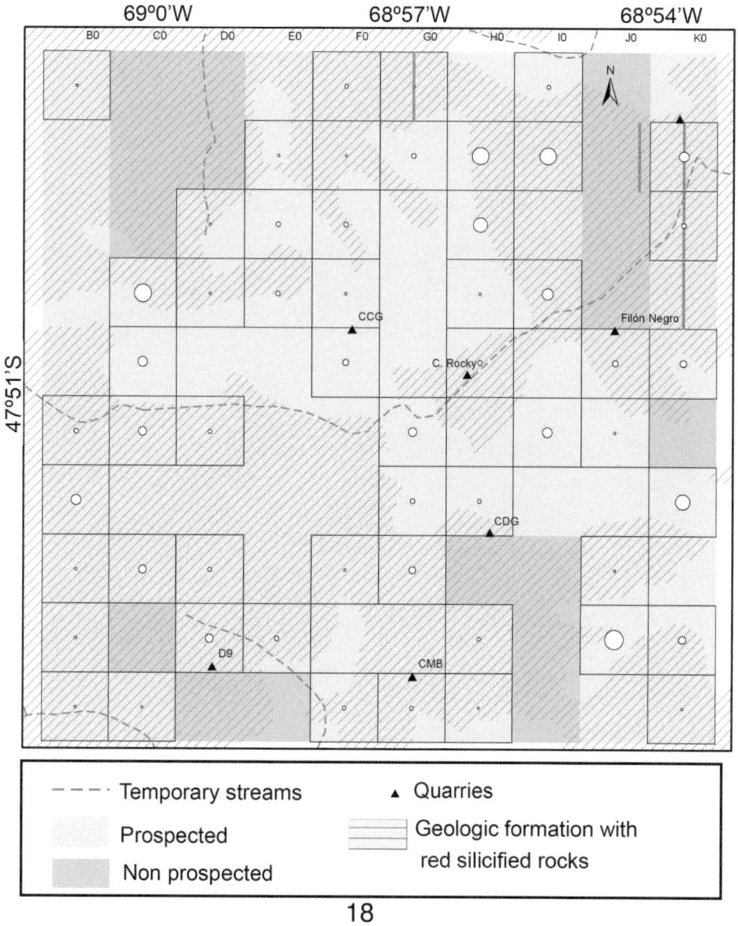

Fig. 7.7 (continued)

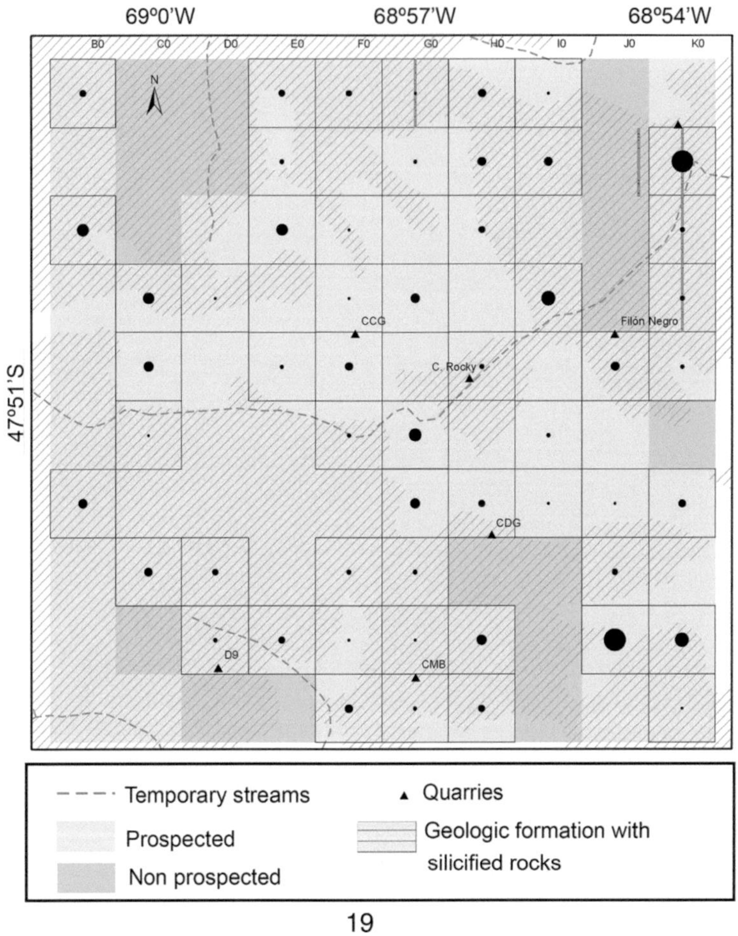

Fig. 7.7 (continued)

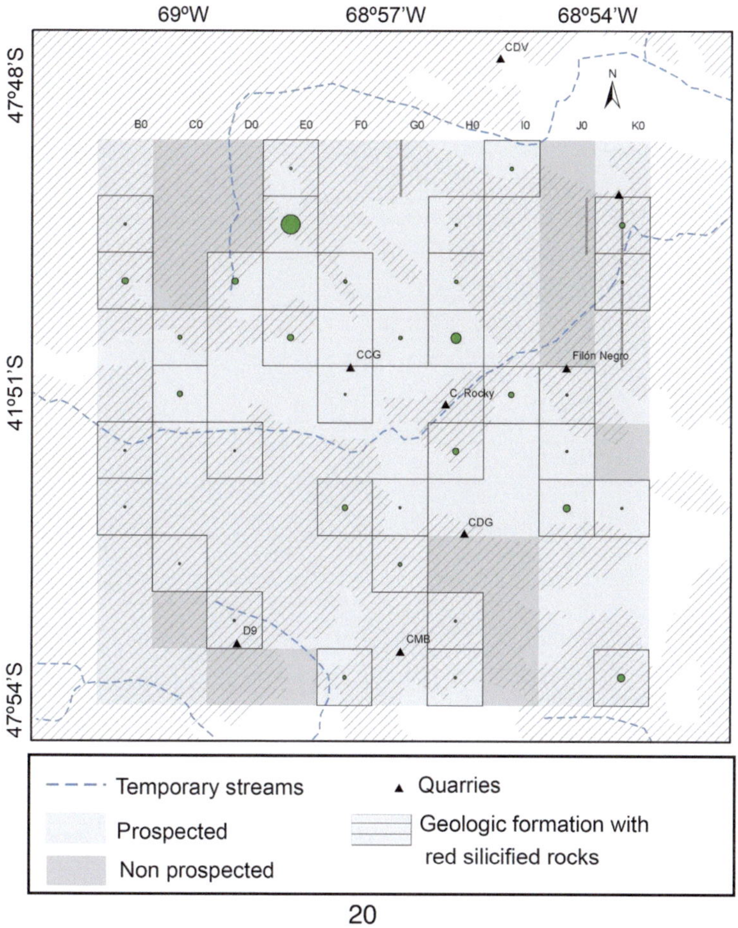

Fig. 7.7 (continued)

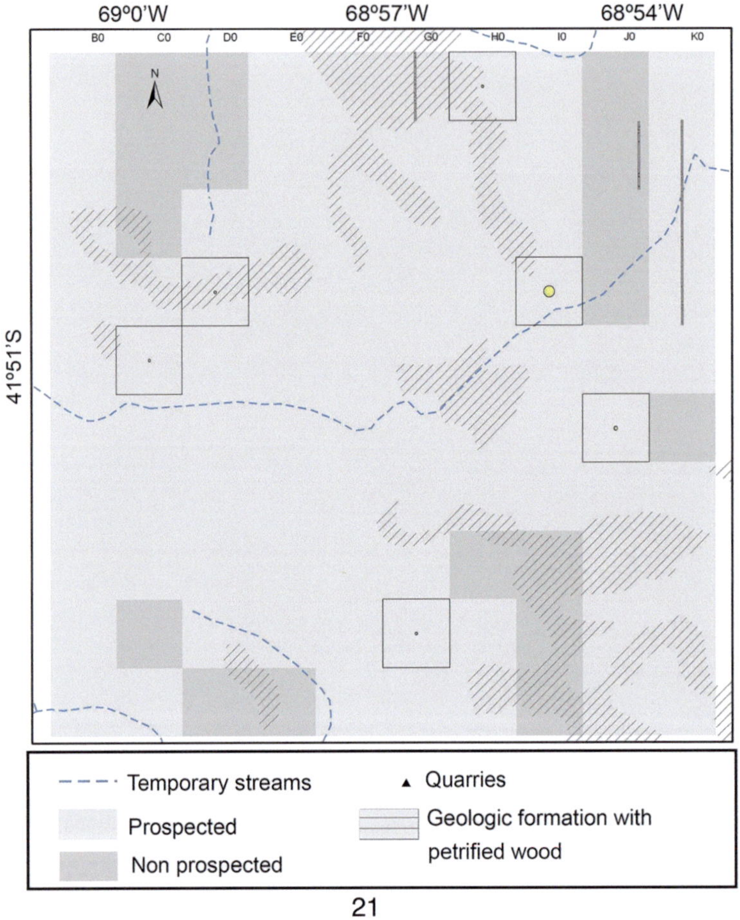

Fig. 7.7 (continued)

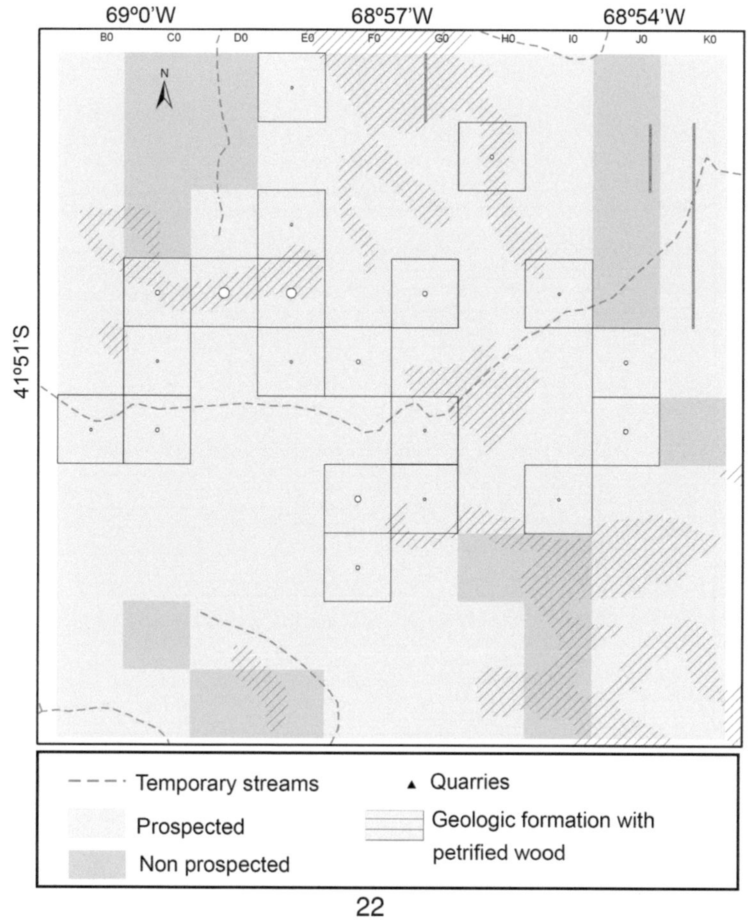

Fig. 7.7 (continued)

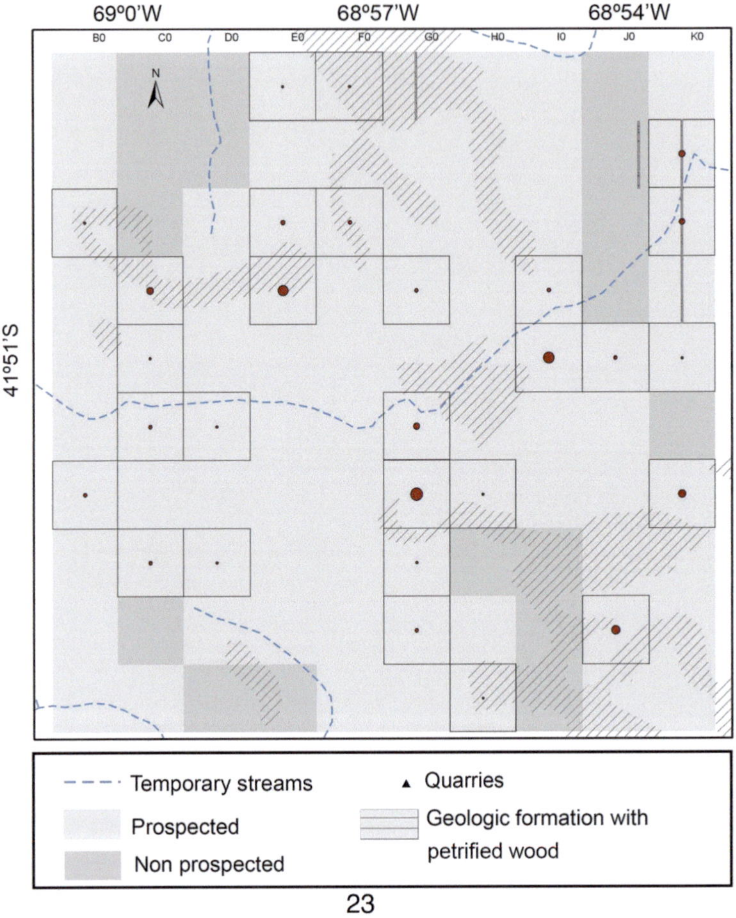

Fig. 7.7 (continued)

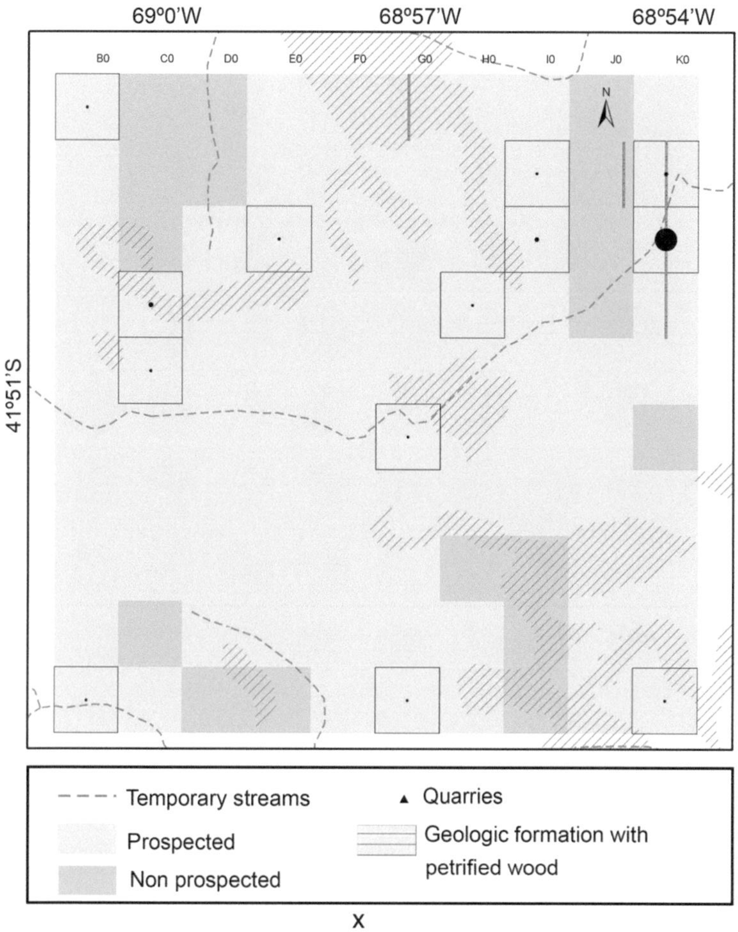

Fig. 7.7 (continued)

and covers part of Estancia Las Mercedes. Its surface area is 3.17 km^2. Basin 1 is crossed by the Cañadón de La Primavera and another canyon farther north. Both are tributaries of the Zanjón Blanco. Basin 2 is crossed by the Zanjón de Aquilino and its tributaries. Basin 3 is also a tributary of the Zanjón Blanco.

Watersheds are the limits of the possible mobility of rocks by natural agents, excluding their natural transport from distant regions. This allows us to study the circulation of raw materials since any rock not found naturally in the basin may have been transported there by humans. The distributions of raw materials are analyzed below according to the basin in which they were found. Then, the variability in the rocks used to manufacture artifacts is compared to the known rock sources in each basin. This allows us to trace evidence of their trajectories and the possible decisions made by the human groups that used them.

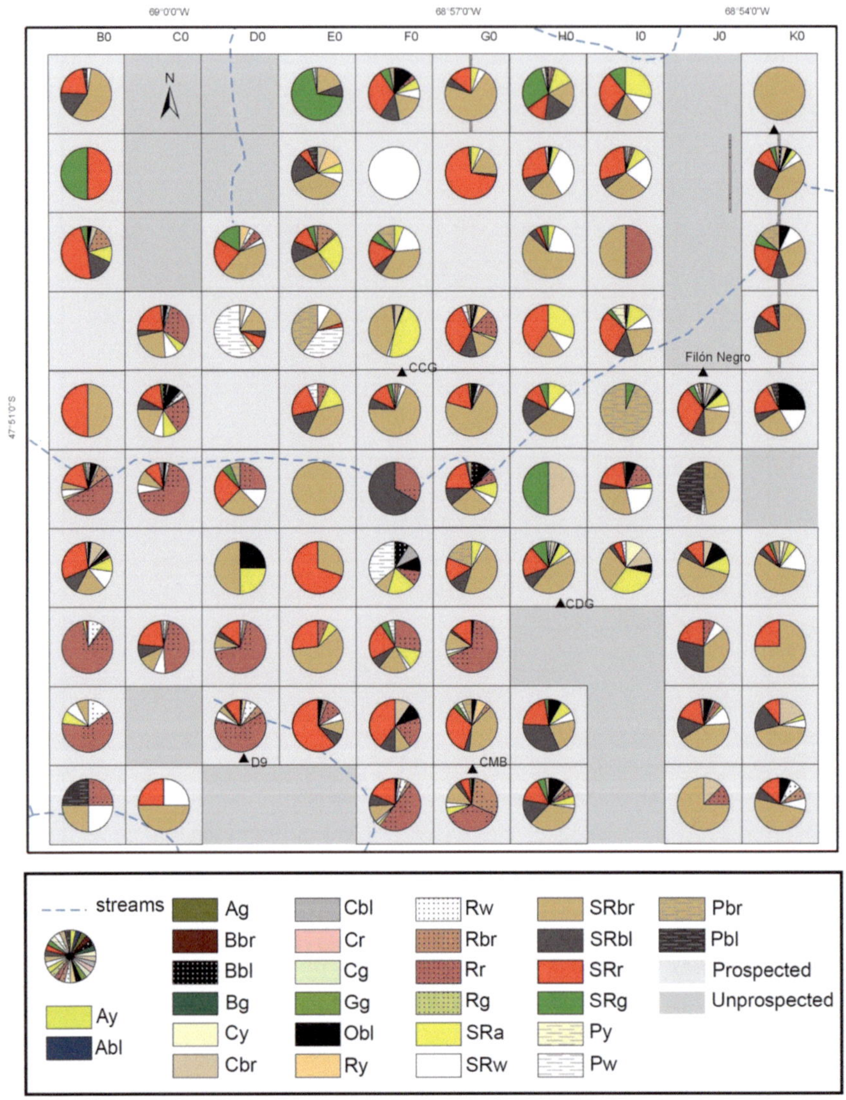

Fig. 7.8 Percentage of raw materials of the artifacts found in each transect section. The acronyms for the raw materials are composed of uppercase letters indicating the lithology: A: andesite; B: basalt; C: chalcedony; O: obsidian; P: petrified wood; R: rhyolite; and SR: silicified rock. The lowercase letters indicate the color: b: black; b: brown; g: green; r: red; y: yellow; and w: white. Percentage of raw materials of the artifacts found in each transect section. The acronyms for the raw materials are composed of uppercase letters indicating the lithology: A: andesite; B: basalt; C: chalcedony; O: obsidian; P: petrified wood; R: rhyolite; and SR: silicified rock. The lowercase letters indicate the color: b: black; b: brown; g: green; r: red; y: yellow; and w: white

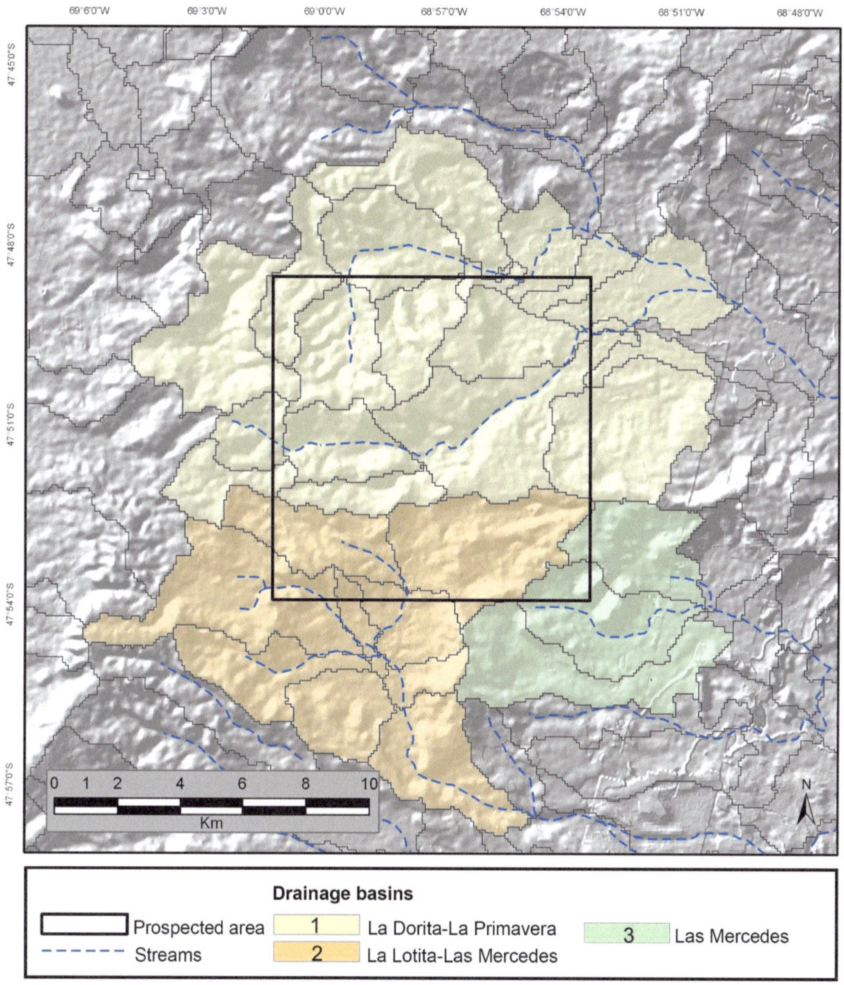

Fig. 7.9 Delimitation of rock catchment basins. The archaeological sites included in basin 1 are: La Primavera-La Lotita (basin 1); La Lotita-Las Mercedes (basin 2); and Las Mercedes (basin 3)

Distribution of raw materials used in each basin

A Chi-square test was used to analyze the distribution of raw materials in the basins. The rarer varieties were grouped together to prevent low or zero frequency values when applying the test. The groups are shown in Fig. 7.10 and Table 7.11. For the calculation, the frequency of raw material varieties was transformed into proportional values according to the sampled area of each basin (Shennan, 1992). The results are in Table 7.12.

As a result, an observed value of 17.26 and a critical value of 38.88 were obtained for a significance threshold of $\alpha = 0.05$ and Cramer's $V = 0.294$. Therefore, the

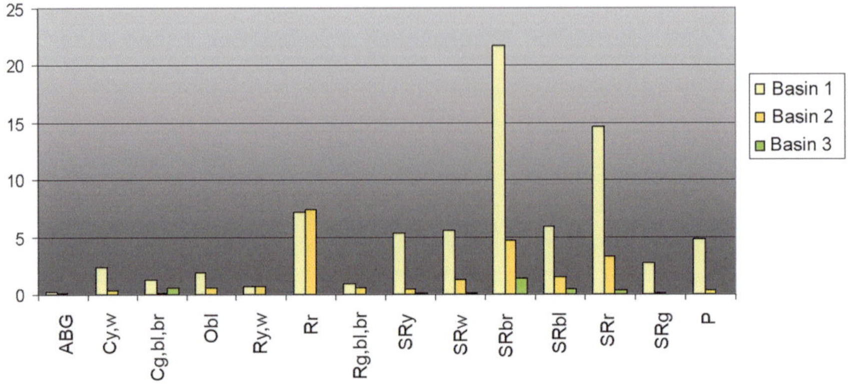

Fig. 7.10 This chart shows the distribution of the types of raw materials used to make artifacts in the basins. The values on the y-axis are proportional to the area surveyed by transects in each basin. The acronyms for the raw materials are composed of uppercase letters indicating the lithology: A: andesite; B: basalt; C: chalcedony; O: obsidian; P: petrified wood; R: rhyolite; and SR: silicified rock. Lowercase letters indicate color: b: black; b: brown; g: green; r: red; y: yellow; and w: white

Table 7.11 Distribution of varieties of raw materials used in each basin

Observed frequencies					Proportion in relation to the area sampled in each basin			
Raw material	Basin 1	Basin 2	Basin 3	Total	Basin 1	Basin 2	Basin 3	Total
ABG	39	7	0	20	0.26	0.07	0.00	0.34
Cy, w	123	28	1	152	2.33	0.34	0.02	2.69
Cg, bl, br, r	66	9	26	101	1.25	0.11	0.54	1.90
O	102	47	1	150	1.93	0.58	0.02	2.53
Ry, w	40	55	1	96	0.76	0.67	0.02	1.45
Rr	379	604	1	984	7.17	7.41	0.02	14.60
Rg, bl, br	48	48	1	97	0.91	0.59	0.02	1.52
Sry	282	34	4	320	5.34	0.42	0.08	5.84
SRw	290	103	8	401	5.49	1.26	0.17	6.92
SRbr	1149	388	67	1604	21.74	4.76	1.40	27.90
SRbl	312	122	25	460	5.92	1.50	0.52	7.94
SRr	774	271	17	1062	14.65	3.33	0.35	18.33
SRg	146	11	0	157	2.76	0.13	0.00	2.90
P	254	25	0	280	4.81	0.32	0.00	5.13
Total	3980	1752	152	5884	75.31	21.50	3.17	99.98

Indeterminate raw materials were excluded

Table 7.12 Significance and valus of the Chi-square test per cell for varieties of raw materials used in each basin

Basin 1	Basin 2	Basin 3	Raw material	Basin 1	Basin 2	Basin 3	Total
(+) NS	(−) NS	(−) ***	ABG	0.027	0.050	0.025	0.102
(+) NS	(−) NS	(−) NS	Cy, w	0.054	0.101	0.045	0.201
(−) NS	(−) NS	(+) NS	Cg, bl, br, r	0.014	0.240	3.896	4.150
(+) NS	(+) NS	(−) NS	O	0.000	0.003	0.042	0.045
(−) NS	(+) NS	(−) **	Ry, w	0.131	0.428	0.015	0.575
(−) **	(+) ***	(−) NS	Rr	1.683	5.857	0.436	7.976
(−) NS	(+) NS	(−) **	Rg, bl, br	0.063	0.219	0.016	0.299
(+) NS	(−) NS	(−) NS	Sry	0.258	0.607	0.049	0.914
(+) NS	(−) NS	(−) NS	SRw	0.017	0.034	0.011	0.062
(+) NS	(−) NS	(+) NS	SRbr	0.036	0.264	0.306	0.606
(−) NS	(−) NS	(+) NS	SRbl	0.000	0.025	0.289	0.315
(+) NS	(−) NS	(−) NS	SRr	0.059	0.097	0.081	0.237
(+) NS	(−) NS	(−) NS	SRg	0.197	0.418	0.085	0.700
(+) NS	(−) NS	(−) NS	P	0.300	0.623	0.151	1.073
			Total	2.840	8.967	5.448	17.256

X^2 per cell and contributions to x^2. Acronyms for the results of statistical test: (+): observed frequency higher than the theoretical frequency; (−): observed frequency lower than the theoretical frequency; NS: Chi-square test for non-significant cell at threshold alpha = 0.100; *: Chi-square test per significant cell at threshold alpha = 0.100; **: Chi-square test per significant cell at threshold alpha = 0.050; ***: Chi-square test per significant cell at threshold alpha = 0.010; DOF: degrees of freedom

null hypothesis cannot be rejected. The three basins have similar distributions of the raw materials used to manufacture the artifacts. This indicates the diversity of raw materials used and their homogeneous distribution in the three basins. However, as Fig. 7.10 shows, although the overall distributions do not differ, the individual raw material varieties do. To test this possibility and obtain information on the circulation of raw materials that would help characterize the observed settlement patterns, the contributions to X^2 were analyzed (Table 7.12). The results indicate a significant difference in the presence of red rhyolite in basin 2, which is present in greater proportions than expected.

On the other hand, the data suggest the following: A greater presence of andesites, basalts, and gabbro/diabase (A, B, G); chalcedonies of the yellow and white varieties (Cy, w); obsidian (O); siliceous rocks of the green, brown, red, yellow, and white varieties (SRg, br, r, y, w); and petrified wood (P) in basin 1. Chalcedony of the green, black, brown, and red varieties (Cg, bl, br, r) and black and brown silicified rocks (SRbl, br) were found in basin 3. These trends could be due to the natural distribution of raw materials in the basins. However, they could also respond to the differential use of raw material sources. An analysis of the classes and typological groups (Aschero, 1975, 1983) provides an alternative perspective.

Structure of artifact classes in the basins

In the next paragraphs we will continue to use drainage basins as a unit of analysis and develop a search for patterns in the artifactual classes and typological groups among the formatted artifacts (FA) and raw materials used. This analysis can provide information about how each basin was used differently. This will reveal the distribution of artifact classes in the different basins. The data are presented in Table 7.13 and Fig. 7.11.

The Chi-square goodness-of-fit analysis indicates non-significant differences between the expected and observed values in the distribution of artifact classes by basin (alpha $= 0.05$; $x^2 = 0.218$; critical value $= 15.507$; degrees of freedom $= 8$; Cramer's $V = 0.033$). However, Table 7.14 shows that the x^2 values per cell are significantly higher than expected for cores in basin 1 and a trend in higher than expected frequencies for FA and NFA, in that order. Basin 2 shows a trend of higher frequencies for strikers, NFA, and debitage. In basin 3, the frequency of the debitage class is higher than expected. The greater presence of cores in basin 1 may be related to their transport to this area from outcrops and secondary deposits. This suggests residential use of basin 1.

Table 7.13 Frequency distribution of artifact classes by basin

Observed frequencies					Proportion in relation to the area sampled in each basin			
Class	Basin 1	Basin 2	Basin 3	Total	Basin 1	Basin 2	Basin 3	Total
FA	326	130	12	468	5.85	1.87	0.24	7.97
NFA	619	274	13	906	11.11	3.95	0.26	15.32
Debitage	2965	1317	126	4408	53.21	18.98	2.54	74.74
Cores	85	25	1	111	1.53	0.36	0.02	1.91
Strikers	3	2	0	5	0.05	0.03	0.00	0.08
Total	3998	1748	152	5898	71.75	25.19	3.07	100.01

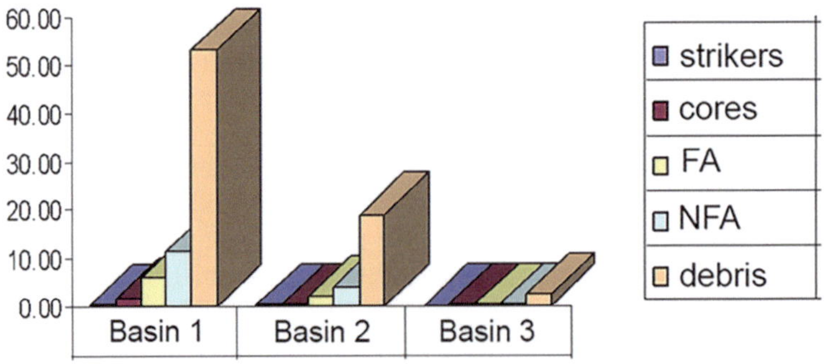

Fig. 7.11 Proportion frequency distribution of artifact classes by basin. "FA" formatted artifact; "NFA" non-formatted artifact

Table 7.14 Significance and values of the Chi-square test per cell for Artifact classes by basin

Basin 1	Basin 2	Basin 3	Class	Basin 1	Basin 2	Basin 3	Total
(+) NS	(−) NS	(−) NS	FA	0.242	0.546	0.000	0.788
(+) *	(−) NS	(−) NS	Cores	1.265	1.896	1.210	4.371
(+) NS	(+) NS	(−) **	NFA	0.038	0.112	4.587	4.738
(−) NS	(+) NS	(+) **	Debitage	0.177	0.086	1.353	1.616
(−) NS	(+) NS	(−) NS	Strikers	0.045	0.181	0.129	0.355
			Total	1.768	2.821	7.280	11.868

X^2 per cell and contributions to x^2. Acronyms for the results of statistical test: (+): observed frequency higher than the theoretical frequency; (−): observed frequency lower than the theoretical frequency; NS: Chi-square test for non-significant cell at threshold alpha = 0.100; *: Chi-square test per significant cell at threshold alpha = 0.100; **: Chi-square test per significant cell at threshold alpha = 0.050; ***: Chi-square test per significant cell at threshold alpha = 0.010; DOF: degrees of freedom

On the other hand, the frequency of cores in basin 2 is not as significant. Instead, the data suggests that strikers, NFA, and debris are more important. This trend could confirm that this sector was prioritized for the exploitation of lithic raw materials. The lower variability of raw materials in basin 2 reinforces the idea that more specific and concentrated activities were developed in that sector, specifically the extraction of red rhyolite. Basin 3 presents a predominance of debitage among the artifactual classes present.

Since the field-walked sector lacks outcrops and secondary deposits, the only available source of raw material is the isolated nodules recorded on the basaltic plateau covering much of the sector. The varieties of raw materials observed more frequently than expected (SRbr, SRbl, Cg, Cbl, and Cbr) may be naturally available as nodules or may have been transported from other sectors. Conversely, the higher relative presence of debitage in basin 3 suggests that it was used more expediently (Nelson, 1991).

Based on the data presented, in each basin, it is predominant the use of naturally available raw materials. However, not only the varieties located nearby were used. In fact, some varieties of raw materials that are available in large quantities have not been used significantly, such as rocks from the CDG and CP quarries in basin 2. Conversely, there is intense extraction activity of red rhyolite in the D9 quarry in the same basin.

7.2.5 Raw Materials Storage

Other archaeological evidence indicates the cultural distribution of lithic raw materials available for knapping in caches. Between basin 1 and basin 2, we found a small stone structure containing a brown siliceous core. Three more cores of brown siliceous rock were found outside the structure, together with unformatted artifacts within a radius of less than 3 m.

Macroscopic observations allow us to conclude that this core was not discarded. For example, it shows no signs of exhaustion and has flake blanks larger than the other artifacts found at the site. It has no serious defects, such as flakes with hinged ends, which could have caused it to be discarded. It does not have angles of 90° or more, nor is it too small to prevent flake extraction (Crivelli Montero & Fernández, 2004). Based on these observations and its location, we propose that the core was stored, not discarded. The stone structure containing the core is located in the sector dividing basins 1 and 2. It is far from the large secondary deposits at the bottom of the basins and from known outcrops (Fig. 7.12).

From an economic standpoint, this can be interpreted as a caching strategy for equipping locations. Such a strategy would be expected in an area with heterogeneous raw material availability (Kuhn, 1994). It implies detailed knowledge of the resource distribution in the area and indicates the depth of human occupation over time. At least, it suggests that the surface archaeological context is not related to the initial colonization of the territory (sensu Franco, 2002; Miotti, 1989). It also implies strategic thinking that denotes the intention to return to the site by planning for and anticipating future raw material needs. Visibility analysis indicates that the stones structure is highly visible from several sectors of the landscape. This allows us to argue that the stone structure itself could point out the lithic resource contained within

1

Fig. 7.12 1) Cache of siliceous rocks is located under a basaltic rock structure (ID_564). The lichen coverage indicates structural integrity; 2) Detail of the structure. The Sr core is indicated by the arrow. The basaltic rock covering was removed for the photo; 3) Discoidal core found within the structure; 4) Surface facing up. It has a layer of very fine sediment attached

2

3

Fig. 7.12 (continued)

4

Fig. 7.12 (continued)

it. In this sense, it does not hide the resource, but rather makes it more visible. The existence of caches shows that raw material availability was not homogeneous and that strategies other than residential mobility were used.

7.2.6 The MAL and the Quarries

If the location of the MAL had been significantly influenced by quarries, we would expect the raw materials used for artifacts at the MAL to coincide with those from the nearest quarry. Currently, we have all the information necessary to answer the question about the structure of lithic raw material sources in the study area.

Table 7.15 Multiple activity loci (MAL), predominant raw material, and the closest quarries (Q)

Basin	MAL	Closest quarry	Distance (minutes)	Rocks available in the closest quarry to MAL	Predominant rocks in the lithic assemblage (MAL)	Priority use of closest quarry
1	Cueva Maripe	Rocky quarry	6.78	Brown and red (Silicified ignimbrite)[2]	Red rhyolite (48.4%)[3]	no
1	Pico Blanco	Secondary deposit K3-K2	6	Varied	Brown silicified rock (35.78%) and black silicified rock (24.73%)	yes
1	El Descanso	Cantera F4	4.9	Brown silicified rock	Brown silicified rock (72.16%)	yes
2	El Mirador	Cantera Platense	8.8	Brown and white silicified rock (breccia-like structure)	Red rhyolite (52%)	no
2	D9	D9	0	Red rhyolite	Red rhyolite	yes
2	–	Cantera del Gris	–	Gray siliceous rocks (breccia-like structure)	No MAL were found nearby	–

Next, the six MALs with the lowest cost distance to the quarries are analyzed. Three of these MALs are located in basin 1, and the other three are located in basin 2 (Table 7.15).

7.2.6.1 Basin 1

The Cueva Maripe site (MAL ID_563) is located 6.78 min away from the Cantera Rocky site (Q ID_562) (see Fig. 7.8). The quarry is on the opposite side of the canyon from the cave and is 33 m above its base level. The cave offers excellent weather protection and is located near a source of siliceous rocks and water. These favorable environmental conditions make it an ideal location for MAL. The El Descanso surface site (MAL ID_22) is 4.98 min from the F4 siliceous rock quarry. The El Mirador site (MAL ID_44) is 8.8 min from the Platense quarry (Q ID_266). Finally, the Pico Blanco site (MAL ID_54) is close to the K3-K2 secondary deposit (Fig. 7.6).

---

[2] Some samples were determined by the petrographic section as silicified ignimbrite.

[3] Darío Hermo, personal communication.

7.2.6.2 Basin 2

In the case of Site D9 (MAL Site ID_78), the locations of a MAL and a quarry coincide. At the foot of a rhyolite outcrop of the formation, a high diversity and density of artifacts was recorded. Chon Aike, which extends 160 m by 50 m continuously (see Chap. 2, Fig. 2.9). The quantity and extent of the artifacts, which are mostly large-size primary flakes, allow us to consider this outcrop and the large anthropic deposit, a quarry for rhyolite extraction. Site D9 is also located near another potential source of raw materials: a cobble deposit in the bed of a temporary tributary located 12.11 min from Site D9 (Fig. 7.6).

The Gris quarry is an extensive outcrop of siliceous rocks (Q ID_729) with evidence of extraction, but no evidence of residential occupation within one kilometer (Fig. 7.6). The consistent quality of this siliceous rock for knapping likely led to less intensive use of the quarry. This, added to other factors, such as the lack of shelter and nearby water sources, has probably influenced the choice of this space for different purposes, but not for establishing residential camps.

Were quarries an important factor in the MAL's settlement strategy? One way to test this idea is to analyze whether each MAL used raw materials from the nearest quarry as a priority. A positive result would support the idea that quarries attracted MAL settlements. Conversely, if the quarries were not used more than other sources of raw material, and the rocks were collected from different places in the landscape, then the quarries would not be an important factor in the localization.

The results in the last column of Table 7.15 indicate that in basin 1, the raw material from the nearest quarry is used as a priority in two cases: Quarry F4, which is close to the Descanso Site, and the K3-K2 deposit, which is close to the Pico Blanco Site. In basin 2, this was verified for Quarry D9 and the MAL. This suggests that the MALs are located in consideration of this source. However, the opposite was found in two cases: MAL Cueva Maripe, near Cantera Rocky in basin 1, and MAL El Mirador, near Cantera Platense in basin 2. In these cases, the most frequently used raw material comes from more distant quarries. In a third case (CDG: Cantera del Gris), no MAL was located near a large rock outcrop. This variability in the use of raw materials allows us to conclude that the pattern of quarry locations is not solely derived from a strategy of minimizing distance. This is consistent with the mean cost distance between the MAL and the Q, which is 29 min. Evidence of the use and circulation of raw materials from other basins and regions indicates that different strategies were used to supply rock than those used for residential mobility toward the Q. For example, the red rhyolite primarily used in Cueva Maripe comes from at least 5.6 km away (D9 quarry), as indicated by the study of rock catchment basins. Based on studies of petrographic sections, this variety of raw material (ISG1) likely originates from Cantera del Rojo, which is located 11.5 km away in a straight line (Hermo, 2008) on the southern border of basin 2. The exploratory analysis indicates a sourcing strategy involving mobility within the home range for procuring raw materials. The use of gabbros and diabases to make boleadoras, which are less common than highly silicified rocks used for knapping, confirms that diverse lithic raw materials were moved to MAL rather than the opposite. Records of raw materials from non-local sources, such as obsidian, suggest that social networks were established with other groups to obtain them.

In summary, the two MALs in basin 1 (Pico Blanco and El Descanso) are located near the quarries that provide each site with the most frequent raw material (SRbr and SRbl). In contrast, the most frequently used raw material (Rr) is not naturally close to the Cueva Maripe site. The basin 1 sites contains a wide range of raw materials that are mainly available in basin 1's secondary deposits, such as petrified wood, siliceous rocks, and chalcedony. Some materials are not immediately available. The gabbros and diabases (G) come from an estimated distance of 10 km from the studied sector. Obsidian (O) is allochthonous and has an estimated provenance distance of more than 60 km (Fig. 7.13). In basin 1, there are no outcrops of red rhyolite. It was used in artifact manufacturing at a lower proportion than in basin 2, suggesting that the most commonly used raw material in Cueva Maripe was transported from sectors of the landscape between 5.6 and 11.5 km away. This affirms the idea of a procurement strategy involving movement to and from sectors further south (Hermo, 2008). Red rhyolite was most commonly used in basin 2, where several outcrops of this rock were identified. However, further data shows that this behavior is not solely a response to natural availability. A highly selective use of raw materials is evident. For example, the extensive outcrop of the Cantera del Gris quarry, which has siliceous rock, was not used. This is likely because, despite presenting a conchoidal fracture, this raw material is brittle due to its breccia-like structure. Another example of the selective use of raw materials is MAL El Mirador, which is close to the La Cantera Platense quarry. That site does not prioritize the use of brown and white siliceous rock from that quarry, but rather uses a greater proportion of red rhyolite. In fact, all of the assemblages in basin 2 feature red rhyolite as the most prevalent material. This evidence suggests that the use of raw materials was highly selective and that knowledge of the distribution of resources in the landscape was extensive. The LAM was not always located close to the most commonly used source of raw materials. Instead, we infer that other strategies were used, such as the movement of special purpose parties. Analyzing the raw materials in the MAL assemblages and local lithic resources allowed us to obtain further information (Fig. 7.13). MAL D9 is an exception because it is located directly next to a rhyolite quarry. This rock outcrop is the source of the raw material and provides shelter from the wind. The D9 MAL lithic assemblage includes evidence of projectile point replacement and the discard of processing artifacts, such as raclettes, scrapers, knives, and side scrapers. It also contains a large amount of NFA and debris. The analysis of the techno-morphological groups revealed that basin 1 is characterized by the presence of cores, FA, and NFA. Conversely, basin 2 is characterized by a greater presence of strikers, NFA, and debitage. Together with the results of the analysis of the distribution and use of raw materials and the location of the MALs, this analysis indicates that, within the prospected sector, basin 1 corresponds to a more pronounced residential use. The MALs are located close to water sources, rock outcrops, and secondary deposits. Conversely, basin 2 is characterized by the extraction of raw materials, mainly red rhyolite, and their transport to the MAL of basin 1. This is sustained in the low frequency of cortex in the rhyolite artifacts, which indicates their source is a rock outcrop, and in that, the only registered one in the study area is D9, in basin 2. On the other hand, in basin 1 artifacts present a greater proportion of cortex, more variability in raw materials, and there is a high frequency of cores, which indicates a strong use of the nodules available in the secondary deposits in this basin. Basin 3 represents a much smaller area analyzed, and no MAL sites are registered there, so at the moment, no inferences were made regarding its possible form of use. The only observation about basin 3 is the high proportion of debris and the lack of registered sources of raw materials. This indicates the occasional use of raw materials in the form of dispersed cobbles within the geological formations.

Finally, our exploratory analysis suggests that quarries did not substantially influence the location of residential campsites (MAL). Instead, other mechanisms for supplying raw materials were used, rather than settling campsites close to quarries. Immediate, locally available raw materials were obtained through the mobility of special task groups responsible for their supply and through fortuitous encounters with raw materials embedded in other activities. For more distant sources, exchanges most likely occurred through broad social networks (Miotti, 1989; Miotti & Salemme, 2004). All of the observed trends are summarized in Fig. 7.13, and the provisioning strategies are analyzed in more detail in other work (Magnin, 2015).

7.3 Activities Related to the Production of Rock Art (A)

7.3.1 Rock Art Sites

This type of evidence encompasses an environmental restriction regarding its location. It is dependent on the presence of adequate stone walls for rock art. While the locational analysis is valuable for comparison with other types of evidence, in the case of rock art, it is more useful to approach the study of its location differently. The landscape in the area has rock formations suitable for painting that were not used. Here, we focus our research on identifying factors that influenced people to create paintings in certain caves and not in others. During the surveys, we examined caves and rock shelters with and without art. We use this data to generate information about possible criteria important in choosing caves for this activity. In the following pages, we compare the locations of eight sites with art in the area to the locations of three caves without art.

7.3.2 The Caves and Rock Shelters in the Studied Sector

The eight rock shelters with art analyzed below are: Cueva Maripe (CM), La Primavera 1-2, 3, and 5 (LP1-LP2, LP3, and LP5), Cueva Mora (CMa), Cueva de la U (LM3), Cueva del Círculo (CC3), and Alero Lucifer (AL). Three caves without

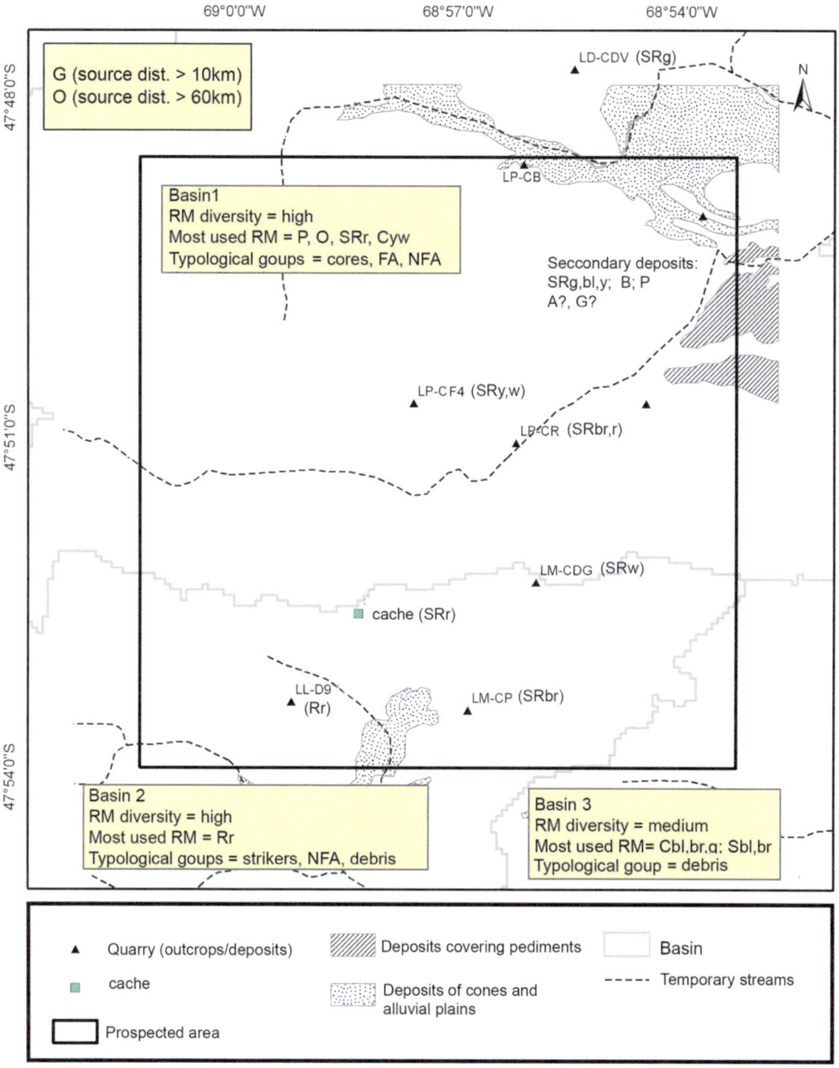

Fig. 7.13 Study area encompasses part of three rock catchment basins. The information on the use of raw materials and the structure of artifact classes summarized on this map provides an approximation of the use and circulation of raw materials, as well as a functional characterization of the area within each basin

art were also analyzed: Cueva del Valle Escondido, Cueva del Lagarto, and Cueva de la Subida. All of these locations are in the Chon Aike geological formation (see Chap. 2). As shown in Fig. 7.14, the art sites included are located in the prospected area delimited by the field survey design. Other sites with art beyond these limits are

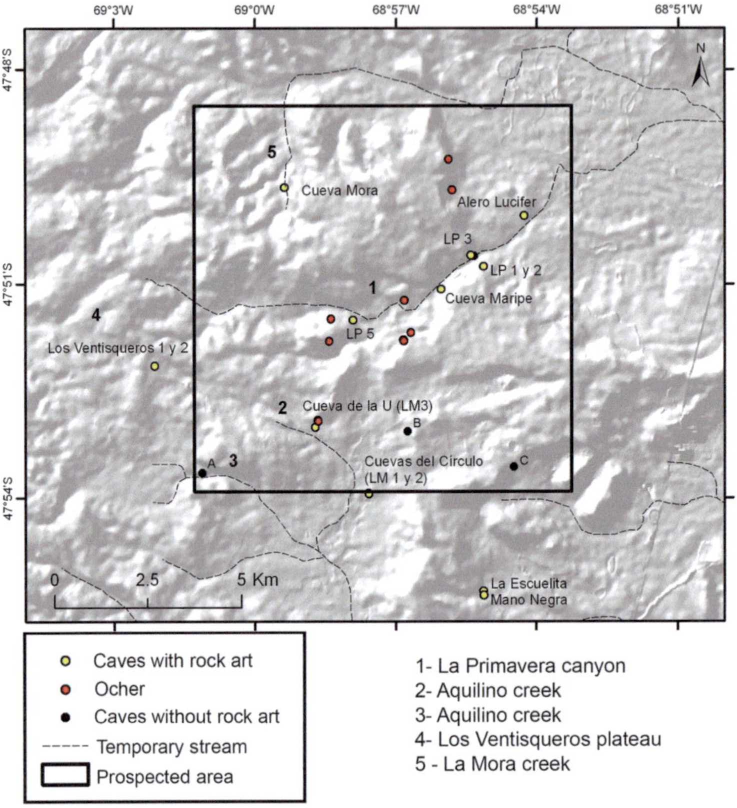

Fig. 7.14 Distribution of rock art in the study area and nearby sectors. Caves without rock art: A = Cueva del valle escondido, ID_232; B = Cueva del lagarto ID_17; C = Cueva de la subida, ID_552

not included in this analysis but are shown in the figure to provide contextual information. These sites with art in Los Ventisqueros and Las Mercedes (La Escuelita and Mano Negra Caves), which are located 1 km west and 2.5 km south of the transect layout limit (Fig. 7.14), as well as in other sites such as the engravings of Los Navarros (7 km northwest) and the Víbora Caves (7 km south), highlight the continuity of the physical and cultural landscape beyond the area delimited for this analysis.

7.3.2.1 La Primavera Canyon

The La Primavera Canyon contains six caves with rock art, as well as the Alero Lucifer rock shelter. A previous study (Carden, 2009) examined the rock art in all of these caves in depth. Maripe Cave is the most significant due to its size (Table 7.16) and the abundance of art on its walls and ceiling (Fig. 7.15). Due to its high complexity and deep sedimentary sequence, Maripe Cave (CM) is the site that has been most intensely studied. It has been defined as a locus of multiple activities with continuous reoccupations from the Early to Late Holocene (Hermo, 2008; Miotti et al., 2007b). CM is located in an intermediate position along the longitudinal axis of the canyon. The other caves and associated rock shelters are smaller (Table 7.16). They are located in three sectors: the upper sector closest to the springs (LP5); the intermediate sector, 1.2 km northeast of CM (Caves LP1, LP2, and LP3); and the sector of the affluent mouth outside the rocky canyon (AL).

Although no excavations have been carried out at these sites, we were able to propose more specific uses for the caves and rock shelters based on the surface

Table 7.16 Rock art in caves and rock shelters in the prospected sector and nearby areas

Site and archaeological locality	ID	Cave dimensions	Motifs and technique
La Primavera Cueva Maripe (CM)	563	Cave, 26 m in its mouth, by 24 m deep	Paint: adult and children hands, zoomorph and geometric motifs
La Primavera 1 (LP1)	699	Cave, 4 by 6 m deep	Paint: negative red and white hands, zoomorphs
La Primavera 2 (LP2)	699	Cave 8 by 12 m	Paint: hands (some children's), zoomorph and stains
La Primavera 3 (LP3)	702	Cave 3 by 10 m	Paint: negative red hands
La Primavera 4 (LP4)		Small cave 1.5 by 2 m	Paint: negative red children hands
La Primavera 5 (LP5)	731	Cave 3 by 10 m	Paint: polychrome painted motifs, hands stencils (half are children's) and complex geometric motifs
La Primavera Alero Lucifer (AL)	698	Rock shelter	Paint: red hands
La Primavera Cueva Mora (CMa)	18	Cave. North chamber 6 by 3 m, south chamber 3 by 3.50 m	Paint: orange circles stains and lines
Las Mercedes Cueva de la U (LM3)	697	Complex of rock shelters and caves	Paint: negative red and black hands
Las Mercedes Cueva del Círculo 2 (CC1)	70	Rock shelter	Paint: red hands negatives
Las Mercedes Cueva del Círculo 3 (CC3)	71	Cave: 6 m opening by 12 m deep	Paint: negative hands and circle of orange dots

Source Carden (2009)

materials and their contextual characteristics. The most common motif at all the sites is the hand stencil. In the smaller caves, LP5 features negative handprints, half of which are small, probably belonging to children between three and five years old. The rest can be assigned to children between six and twelve years old, with only 25% belonging to adults. LP5 is notable for its complex, polychrome painted motifs (Carden, 2009). These motifs are placed on the surface of a low ceiling, parallel to the floor, and 50 cm from the current surface, which provides poor lighting and allows only a few people at a time to appreciate them. This, combined with the cave's high elevation above the bottom of the canyon, suggests that it was used as a refuge rather than as a place for multiple activities (Carden, 2009). Cave LP3 has two entrances filled with sediment and sheep excrement. Little is known about its functionality, but it is probably a shelter.

LP 4 is a small shelter located in a lateral ditch of the ravine. Its dimensions indicate that it could not have been used for multiple activities. However, this cave is notable for the presence of small hand motifs (Carden, 2009). LP1 and LP2 are next to each other, located above the canyon, in the sector opposite LP3, in a high position on the slope. LP2 contains motifs of small hands similar to those mentioned, and LP1

1

Fig. 7.15 Some of the caves with painted rock art that are located within the prospected sector. 1) Interior of Cueva Maripe, panel painted above excavation unit B12; 2) Mouth of the cave located in the La Primavera canyon. Photograph B. Pianzola, 2006; 3) Mora Cave. Details of six orange color circles painted on the wall of the south chamber; 4) General view of the rock shelter and the south and north chambers. Photograph B. Mosquera 2008; 5) Cueva de la U. Hands of children and adults are painted on the rocky ceiling; 6) Entrance to the Cueva de la U; 7) Circle motif of painted dots; 8) Negative hands, both on dark walls of the Cueva del Círculo 3

2

3

Fig. 7.15 (continued)

4

5

Fig. 7.15 (continued)

6

7

Fig. 7.15 (continued)

contains motifs of small hands assigned to children aged 6–12. LP1 highlights the presence of a large zoomorphic motif, and LP2 highlights a feline footprint (Carden, 2009). The recurrence of painted children's hands in caves with high positions on the canyon slopes, which have good visibility of the access route, along with the aforementioned zoomorphic motifs, has led us to propose that these sites are related to activities controlling the passage of animals or people or to their possible importance in visual control or symbolic demarcation of the canyon entrance (Carden, 2008; Miotti, 2006).

Finally, AL has some poorly preserved negative handprints (Carden, 2009). This large rock shelter is located lower down and is sheltered by vegetation. It is located 50 m from a wet meadow. The presence of accumulated firewood indicates that the shelter is currently used by local settlers. Based on the intervisibility relationships between LP1 and LP2, which are opposite LP3 and the upper part of the rock formation in which the AL rock shelter opens, it has been proposed that these sites could have functioned as strategic hunting points within a system (Carden, 2009). On the other hand, the recurrence of children's hands in LP5 could have had symbolic importance in contexts far from multiple activities (Carden, 2008).

7.3.2.2 The Zanjón De La Mora

The Mora Cave (CMa), located in a south-north-oriented canyon, has two shallow chambers suitable for residential use. Rock art is present on its walls and ceiling (Fig. 7.15). The cave is located in a sheltered ditch with rock walls, where we recorded siliceous rock nodules formed in the tuff's cavities and fissures. These nodules have conchoidal fractures and few inclusions, making them good-quality raw material for knapping. Some of the rocks are more homogeneous and translucent and have been classified as chalcedony (see Sect. 7.2.2). Thirty-five hundred meters from the cave, wet meadows indicate the presence of permanent underground water. While there is not enough research to define the activities that took place at this site, radiocarbon dating indicates that the cave was used as early as 3678 $\pm$ 65 14C BP (see Chap. 8).

7.3.2.3 The Zanjón De Aquilino

In the area south of La Primavera Canyon, along one of the tributaries feeding the Aquilino Stream, there are caves and rock shelters carved into an ignimbrite outcrop. The outcrop's semicircular shape, which encloses an area of approximately one hectare, gives it the name "U Caves" (Cuevas de la U, LM3) (Fig. 7.15). Sixty meters south, a surface site near a rocky boulder yielded knapped artifacts. This site is called Sitio Mesa. Two hundred twenty meters to the south-southeast, the Alero del Balde also has archaeological remains on the surface. The complex has adequate conditions for residential use since an important spring is located 500 m upstream (where another site, Campamento Aguada de la U, was found).

Eighteen meters north is a source of ocher on a hill that is highly visible from various points in the locality due to its orange color. There is abundant evidence of knapped artifacts (including cores, scrapers, and knapping debris). Together with the variety of raw materials (obsidian, rhyolite, and siliceous rocks), evidence of thermal alteration in these materials, and the presence of rock art (three negative hands painted in red), these factors, along with the aforementioned natural context, allow us to propose this site as a probable TC or MAL. The site is also near significant surface lithic workshops, including site D9 (850 m southwest), El Ventarrón (350 m northeast), and site MAL ID_24 (El Piche Lento).

The El Círculo caves are located in the middle section of the same tributary of the Aquilino ditch, 2.2 km from Cuevas de la U. These caves consist of two adjacent rock shelters and one cave (CC1, CC2, and CC3, respectively). The caves are on a reddish rock outcrop of the Chon Aike formation that stands out from the surrounding sediments. All of them open to the south and offer a wide view of the open ground where the three tributaries of the ditch converge (this view is shared with Cueva de la U). Lithic artifacts (scrapers) and knapping debris were found at the entrance to CC1 and CC2. We recorded paintings in CC2 (seven red negative hands) and CC3 (a circle made with orange dots and negative hands, see Fig. 7.15). Above the caves, a currently dry meadow opens up. The functionality of both sectors is probably related to temporary camps associated with limited activities carried out in the surrounding area.

7.3.3 Description of the Caves Without Art

The caves without art that used here to contrast location patterns include Cueva del Lagarto (ID_17), Valle Escondido (ID_232), and Cueva de la Subida (ID_552). This is a small sample, but it allows us to explore possible explanations. These three caves have rocky roofs and walls that could support rock art. The caves with art are the ones mentioned above: CM, LP 1–5, CC 2 and 3, AL, CMa, and CU. The same variables as previously used are considered: the distances to ocher sources and the pigments used to make paints (Table 7.17 and Fig. 7.16).

Given the limited number of caves without art, we analyze each one individually and compare it to the average cave with art. The observed tendencies are useful hypotheses for rock art painting activity in the study area. Most caves without art, as well as more than half of those with art, are located in areas exposed to winds. As previously mentioned, this is due to the region's natural cave formation conditions (Panza, 2001). However, this did not prevent their use for specific or multiple activities in most of them. Of the caves without art, one (ID_17) shows evidence of occupation but not of having been painted. The remaining two caves (ID_232 and ID_552) have very steep slopes that make them difficult to access from the valley floor. This could be a factor in how the caves were used. Nevertheless, other caves and rock shelters with limited accessibility were painted. For instance, caves with paintings (LP1, LP2, and LP5) are located in elevated positions with slopes between 15 and 30 m in height.

Table 7.17 Average values of interest variables for caves with art compared to caves without rock art

Site Name	1-Shelter	2-Topography	3-Distance to springs	4-Distance to temporary water	5-Distance to quarries	6-Distance to mean paths	7-Distance to secondary paths	8-Distance to ocher	9-Percent viewshed (%)
Cueva Mora	E	Steep	10	3	57	40	44	86	3
Cueva Maripe	S	Steep	6	2	7	1	1	18	1
Cueva de la U	E	Gentle_low	5	5	17	0	3	5	2
Circulo	E	Gentle_low	2	2	24	9	11	38	5
Alero Lucifer	E	Gentle_low	16	4	15	5	2	33	8
La Primavera 1 y 2	S	Steep	15	10	34	9	1	46	11
La Primavera 3	E	Steep	9	5	29	6	10	41	6
La Primavera 5	S	Steep	8	1	40	4	7	136	1
Mean value for caves with Art			*9*	*4*	28	*9*	*10*	*50*	*5*
Cueva del Valle escondido. ID_232	E	Steep	4	3	53	39	37	73	1
Cueva del Lagarto.ID_17	E	Gentle_low	18	18	11	5	17	39	5

(continued)

Table 7.17 (continued)

Site Name	1-Shelter	2-Topography	3-Distance to springs	4-Distance to temporary water	5-Distance to quarries	6-Distance to mean paths	7-Distance to secondary paths	8-Distance to ocher	9-Percent viewshed (%)
Alero de la subida.ID_552	E	Steep	10	10	54	14	9	87	3
Mean value for caves without art			*9*	*5*	*30*	*11*	*12*	*53*	*4*

Due to the small number of cases, their individual values are presented in the table. 1) E = exposed to wind/S = sheltered; 2) topography; cost distance to 3) permanent water sources in minutes, 4) temporary water sources, 5) lithic supply quarries, 6) main paths network, 7) network of secondary paths, and 8) source of ocher supply; 9) amplitude of visual basins: percent visible from the theoric total

This is repeated in other areas of the Central Plateau. For example, in Aguada del Cuero, which is located about 15 km north of the prospected sector where the Cueva de la Hacienda is situated, the cave is located more than 70 m above the base level of the open depression in which it sits (Bajo del Avestruz). The same pattern is seen in the locality of Los Toldos, located 30 km north in the Zanjón del Pescado basin, where Cave LT-15 sits 60 m from the valley bottom. All of these sites have art. Based on this data, we can conclude that the difficulty or convenience of access alone has not prevented or favored the creation of art in other caves in the region.

In the case of Cave ID_232, the values obtained for the analyzed variables indicate that it is located a short distance from permanent and transitory water sources, but a long distance from lithic and ochre supply quarries, as well as the main and secondary access path networks between LAM. Furthermore, it is in a very steep canyon with restricted visibility; only 1% of the visible basin is visible. The cave lacks cultural remains. The closest artifact is a scraper located 200 m from the cave on the opposite side of the ditch.

The cave without art ID_552 presents unfavorable conditions regarding distance to resources, visibility, and access. It is particularly far from sources of ocher, and no archaeological evidence is associated with it. However, 500 m north is the MAL Oasis (ID_45), which is associated with a temporary shallow lake and has a high

1

Fig. 7.16 Los Navarros, a sector with petroglyphs, is located 7 km to the NW outside the micro-area of study. Engravings in sector "A" 1), and 2) sector "B". Photographs B. Mosquera, 2008

2

Fig. 7.16 (continued)

diversity and density of artifacts indicating intense use of the space. It is likely that the distance to the path network at MAL left this cave outside the interest of hunter-gatherers. This cave does not show signs of use. Finally, Cave ID_17 is close to sources of lithic raw materials and ocher and is located near paths. In this case, the cave has good conditions for habitation. It is also more enclosed than the previous caves and has enough space on its walls and ceiling for art. It measures 6 m in depth, 5 m at the mouth, 3.2 m at its maximum width inside, and 1.76 m in height from the current soil surface (see Fig. 2.6 in Chap. 2). Its viewshed is similar to that of other caves with art, and we found artifacts scattered around it.

7.3.4 Location of Rock Art Sites

By comparing the locations of caves with and without art, we can see that caves selected for rock paintings have shorter average distances to permanent water resources, sources of lithic and ocher resources, and paths between MAL. However, it is important to note that two caves without art, ID_232 and ID_552, are on the edge of the sampled area, so an edge effect may influence them. Any location on the edge of the study area is more likely to be outside the lines of communication between

MAL and registered resources. Caves without art could be restudied with more data through surveys of the surrounding areas. Based on the analysis of relationships with resources and mobility networks, we can conclude that, on average, caves chosen for painting are closer to resources and communication networks between MAL than other locations without rock art and have wider viewsheds.

Regarding communication networks and the relationship between rock art and the various uses of the landscape, the proximity to networks of paths that connect MAL, and sometimes their direct overlap, has implications for defining the function of rock art within the society that produced it. First, it affirms that rock art did not occur in a sphere separate from people's daily lives. For instance, rock art is present in Cueva Maripe, where evidence of prolonged occupation has been found. Based on lithic and archaeofaunal studies from stratigraphic contexts, this site was used for multiple activities (Hermo, 2008; Miotti et al., 2007b). Rock art is also present in caves and minor rock shelters that were probably temporarily used as AL, CU, CMa, and CC. Additionally, places that were likely used for hunting also have art, such as La Lotita, which is located to the west of the study area. However, the painted motives differ.

Rock art studies in the regional context of the Santa Cruz Central Plateau revealed recurrences between motifs and the functionality of locations (Carden, 2008; Miotti et al., 2007b). For instance, motifs of hands of various sizes that appear in high densities and with high degrees of overlap coincide with multiple activity sites that have been reoccupied throughout the history of occupation, as in CM and Caves 3 and 13 of Los Toldos Canyon. In some cases, when associated with limited activity sites, the motifs relate to specific activities. For instance, in the Los Ventisqueros sector (see Fig. 7.14), the depiction of a zoomorphic figure representing a guanaco in LP1 of the La Primavera Canyon points to the hunting function of the site. In other areas, such as the Pinturas River and the southern part of the Central Plateau, hunting scenes are also associated with these sites (Carden, 2008). On the other hand, it has been proposed that the rock engravings, including some animal footprints, coincide with hunting sites in Aguada del Cuero (Laguna, Cerro Bonete, and La Leonera), Piedra Museo, and Cerro del Doce (Carden, 2008; Miotti et al., 1999, 2007b). This relationship has also been proposed for other areas, such as La Meseta del Strobel (Belardi & Goñi, 2006; Belardi et al., 2005; Laguna del Faldeo Verde, Gradín, 1959–1960; Re et al., 2005,), where the engravings are associated with hunting blinds and residential parapets, similar to those found in the Alta Primavera and Los Ventisqueros sectors (Goñi et al., 2007). Within the study area, engravings were not recorded, but they were found 7 km away in the Los Navarros locality (Blanco, 2015) (Fig. 7.16).

The spatial analysis performed here expands our understanding of the relationship between rock art and functionality in this study area. Previous studies have suggested that rock art served to symbolically mark resources and territories and transmit information (Carden, 2004, 2008; Hartley, 1992; Ingold, 1986; Miotti et al., 1999b). If art was used to mark a resource, then it should be found near that resource. In this study, we found rock art in shelters, close to water sources, quarries, and access paths. Rock art also offers a better view of the surrounding landscape than other types of evidence. These tendencies persist when comparing caves with and without rock art. We also found that rock art is not only located in places where natural supports are available. It coincides with locations that concentrate human activity. Therefore, based on its spatial location, rock art may have been used to demarcate the symbolic appropriation of places with available natural resources, especially water. It may also have been used to demarcate places where resources were culturally concentrated through inhabitation and movement in the landscape. This appropriation would have occurred through cooperation rather than competition (Carden et al., 2009; Ingold, 1986; Miotti et al., 2007a). Regarding the association of caves with art and the network of access paths between base camps (MAL), the proximity of rock art to domestic spaces suggests its integral role in everyday life.

7.4 Hunting Activities

7.4.1 Hunting Stations (HS) and Hunting Grounds (HG)

This section jointly analyzes hunting stations (HS) and hunting grounds (HG) due to their complementary function within hunting activities. As seen in Chap. 6, HS are far from water sources and supply areas for lithic raw materials. These spaces, which are used for waiting and planning the hunt, occupy a specific place in the spatial structure of the landscape and are far from the secondary path network between MAL. The topographies most frequently used to locate HS are the high sectors. HG, where hunting occurs, tend to be close to the secondary path network in areas with limited visibility. These patterns suggest a general strategy for hunts, involving mobility from HS in high sectors to HG in areas with secondary paths.

Other archaeological studies have proposed the use of special topographies in hunting practices in different areas of the South Deseado Massif and the Central West Santa Cruz area. In these areas, hunting blinds are located in places with good visual dominance and close to bodies of water (Cassiodoro, 2011; Flores Coni et al., 2021; Franco et al., 2021). In this study area, however, HS and HG are not located near water sources. The only exceptions are the Oasis site and Alta Primavera Oeste (Fig. 7.17). The former is located in a closed depression that houses a temporary shallow lake. This site, classified as MAL, was probably also used as a hunting ground at another time. As expected in a mobile society, these sites were reoccupied for different purposes as the landscape was constantly restructured by movement (Binford, 1982). The Oasis site probably functioned in conjunction with the HS equipped with hunting blinds, located less than 1 km away. The use of bodies of water, which attract fauna, in hunting strategies has been documented in ethnography. For example, the well-known Yamnagoo, the Tehuelches' hunting paradise in Claraz (1988 [1865–1866]), and the chronicles of Musters (1997 [1871]), in which he mentions the use of special topographies in hunting practices.

In the Alta Primavera Oeste (APW) sector (Fig. 7.17), the hunting stations (HS) are equipped with hunting blinds (parapets 1), and their location allows for visual

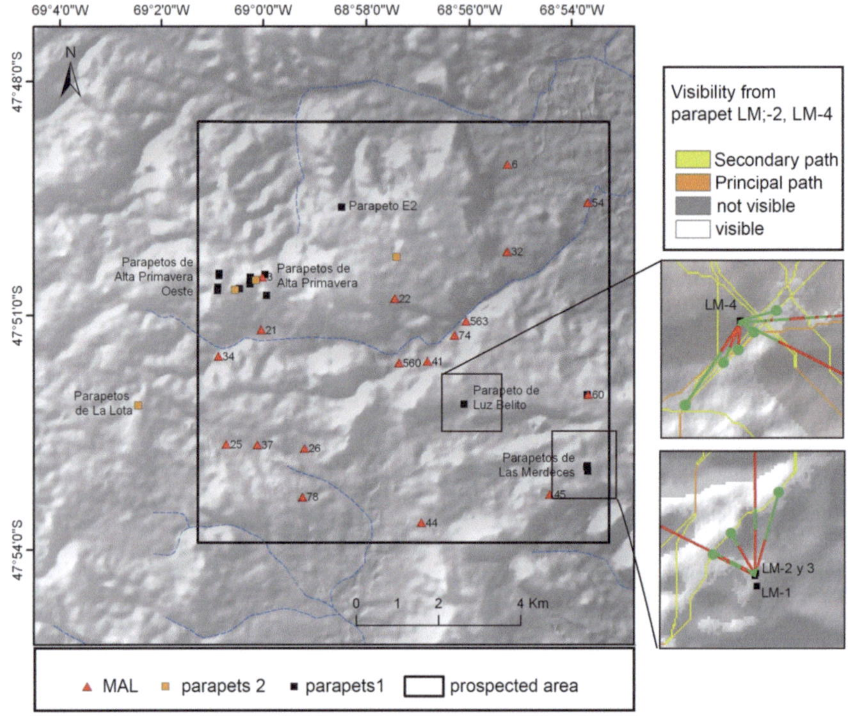

Fig. 7.17 This is a line of sight analysis from the hunting stands at the Luz Belito and Las Mercedes 2 sites. These sites have uninterrupted lines of sight (green lines in the figure) between the hunting blinds and the optimal paths. Parapets 1 are the hunting blinds and Parapets 2 are the residential parapets

control of small, temporary shallow lakes. Other hunting blinds that are interpreted as HS are located in Alta Primavera, east of the previous sector. These blinds do not seem to be associated with temporary shallow lakes. These stone structures and others in Las Mercedes show no relationship with water or shallow lakes and may have an alternative explanation. They were probably used to observe fauna movements and plan hunting strategies (Binford, 1982; Claraz, 1988; Franco et al., 2021; Krist & Brown, 1994; Martínez, 2003).

Regarding visibility, it is interesting to note that the viewshed analysis revealed complementary patterns between the hunting stations (HS) and the hunting grounds (HG). First, HSs have wide visibility, affirming their function as places for planning hunts and observing wildlife movements. Conversely, the HG have more restricted visibility. This is consistent with expectations for places where slaughter occurs. The lower visibility of these areas makes them ideal for ambushing prey. Regarding distance to access paths, HS and HG tend to be close to main paths. This suggests that the location of hunting blinds in these areas may be related to visual control of optimal access routes, where HG are frequently located. To test this theory, we overlaid the areas covered by visible basins, mobility paths, and HG and HS locations (Krist & Brown, 1994; Magnin et al., 2015). Figure 7.17 illustrates the distribution of hunting blinds (type 1 parapets) within the studied area. The smaller windows on the right show closer views of analyses performed in two sectors at the Las Mercedes locality to test visibility relationships between hunting stands (HS) and access roads. The analysis revealed uninterrupted visibility from hunting blind LM4 (Luz Belito) to a secondary path 130 m east and from hunting blinds LM2 and LM3 to a secondary path 450 m west. These results suggest how these spaces may have been used for hunting. The HS analyzed present locations from which the access paths can be controlled. The projectile point and boleadora ball registries found near these paths indicate where the hunting occurred (Magnin et al., 2015).

7.4.2 Limited Activities Loci

Another type of evidence possibly related to hunting activities is the limited activity loci (LALs) (see Chap. 6). These loci are located in high and low topographies, at relatively short distances from temporary water sources, paths, and quarries, but far from permanent water sources. The LAL class is comprehensive ($n = 138$), and because the artifacts we consider diagnostic (those with long edges, whether formatted or not) may indicate various activities, it is impossible to restrict LAL to a single activity context. For instance, processing tasks such as dismembering prey or extracting plant resources can involve the use of artifacts with very similar typologies (see Chap. 4). Thus, barking and cutting plant materials could leave archaeological evidence that we classify as LAL (Gamble, 1990). For this reason, we do not rule out the usefulness of distinguishing this class. However, analyzing particular cases will be more informative than searching for patterns common to all LAL.

7.5 Monument/Mortuary Activities

7.5.1 Location of the Chenques(Ch)—Tombs—In the Studied Sector

Based on the results of Chap 6, chenque structures can be characterized as located in sectors that are completely exposed to the prevailing winds, far from critical water resources (both permanent sources, such as springs, and temporary sources, such as affluents and shallow lakes), far from sources of lithic supplies, away from the main and secondary access path network between loci of multiple activities, and in sectors with panoramic views with a maximum amplitude of 18%, visible within a total theoretical basin. This information places the chenques in a class that appears spatially separate from the other analyzed classes of evidence. According to the K-S tests, the most significant difference is in the last of the aforementioned variables (see Chap. 6). This result led us to delve deeper into visibility patterns related to the location of the chenques (Magnin, 2009b).

The chenques are distributed throughout the studied sector. Their locations are shown on Fig. 7.18. As Table 7.18 and Fig. 7.19 show, they vary in degree of preservation and in the archaeological materials associated with them. These include knapped artifacts, faunistic bones, and other stone structures. The integrity of these probable mortuary structures varies and can be more clearly evaluated in cases with dense lichen coverage. For example, the distribution of lichens on the Chenque del Viento, located on the Meseta north of Maripe, showed that the rocks in the upper part of the chenque had been moved from their original location. This led us to consider it a looted chenque (Fig. 7.19). The indigenous origin of this structure is supported by the lithic artifact assemblage (Table 7.18) and two bones found among the stones of the structure. The bones were identified as a second phalanx of a juvenile *Lama guanicoe* and a first phalanx of *Ovis aries*. According to Beherensmeyer (1978), its weathering state is 4 in the first case and 5 in the second.

While most structures are isolated, we also observed groups of up to four chenques, as well as associations with smaller structures and habitational structures (type 2, see Chap. 6). For example, in the Cerro de las Estructuras sector, three chenques and three smaller structures were arranged 12 m along the edge of a steep plateau (see Fig. 7.19). In the La Lota sector, four chenque structures were found (Fig. 7.19). Two of these chenques (Arcos Bajos) are located on the slope of a hill with a western view. They are 14 m apart and covered in dense lichen. Both are associated with a small structure arranged down the slope, 1 m from each chenque and aligned with its longitudinal axis. Fifty meters from this complex, at the top of the hill, two more chenques (Arcos Altos) were found, 30 m apart. The structure located further north has a panoramic view of all the cardinal points. The association of chenques with smaller structures was also recorded in the case of chenque F2, with two of them found near it. Finally the Cabeceras del Cañadón Guanaco chenque is associated with a residential parapet (type 2 structure; see Chap. 6) (Fig. 7.19). In other cases, cairns that are currently used as visual signals by rural settlers were found in close proximity to chenques. For example, historical cairn structures were found 80 m away from the chenques at Chenque La Primavera 1 and Cerro de Las Estructuras (Fig. 7.18).

Regarding other recorded cultural evidence associated to chenques, in three cases: Del Xilopalo Chenque (on the Meseta south of Maripe), Chenque del Viento, and Chenque F2 (both on the plateau north of Maripe), they were associated with assemblages of knapped artifacts (see Table 7.18). Most of the other structures have little evidence of knapping, and some have no evidence at all (e.g., the chenques of Cerro de las Estructuras, the Loma Pelada structure, and the B10 chenque).

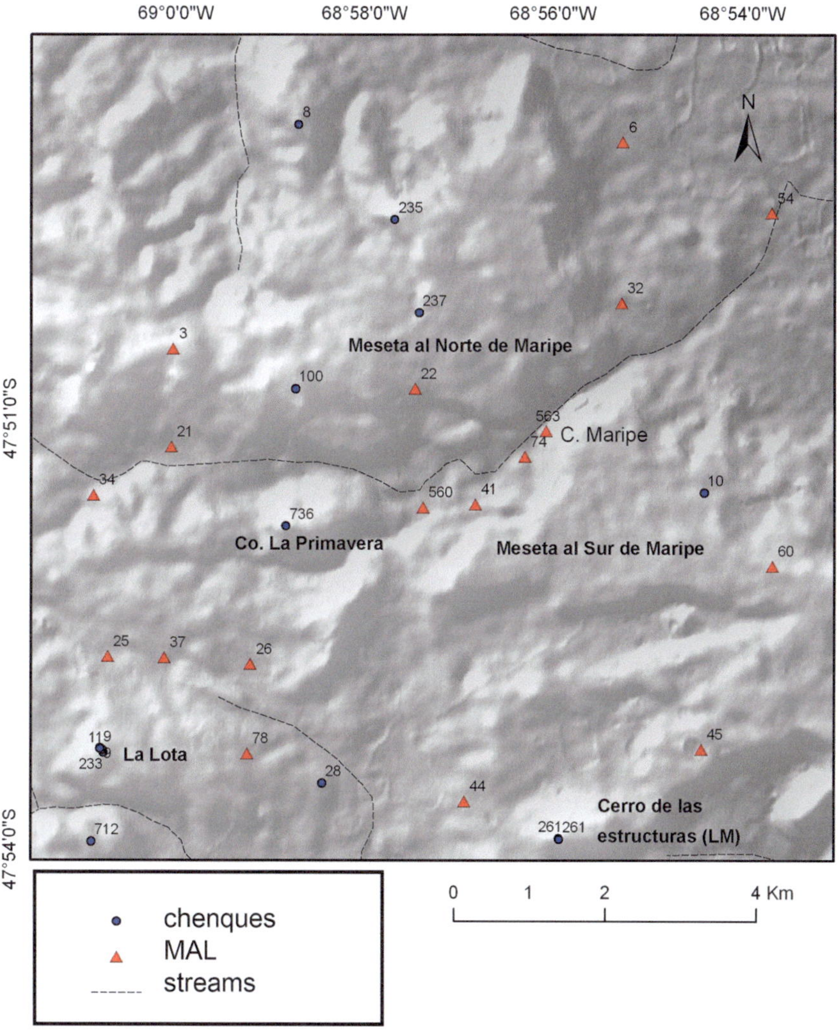

Fig. 7.18 Distribution of chenques in the studied area. The map also shows the MAL to appreciate the different location of both classes. The identification numbers of the chenques correspond to: 8—Chenque del viento; 11—Arcos Altos, N chenque; 235—Chenque F2; 233—Arcos Altos, S chenque; 237—Chenque Cabecera Cañadón Guanaco; 28—Chenque Loma Pelada; 100—Chenque La Primavera; 261—Cerro de las estructuras, chenque 5; 736—Chenque La Primavera 1; 261—Cerro de las estructuras, chenque 2; 232—Arcos Bajos, N chenque; 261—Cerro de las estructuras, chenque 1; 10—Chenque del xilópalo; 712—Chenque B 10; 9—Arcos Bajos, S chenque

Table 7.18 Summary of the characteristics of the analyzed chenques

ID	Name of the chenque	Lichens coverage	Integrity	Reuse evidences	Archaeological assemblage and other structures associated
8	Chenque del viento	Medium	Top stones repositioned	Yes, wiring rod nearby	2 bone specimen (*Lama guanicoe* and *Ovis aries*); Knapping sequence: 5 NFA 12 debitage 1 core
235	Chenque F2	High	Regular	–	1 scraper 1 retouched leaf 1 summary FA 1 undifferentiated FA fragment 1 core 4 debitage 1 smaller rock structure (ID_236)
237	Chenque Cabecera Cañadón Guanaco	Medium	–	–	1 basal projectile point fragment adjacent to a type 2 (residential) parapet structure
100	Chenque de Arriba	Medium	–	–	1 biface fragment 1 debitage 3 NFA
736	Chenque del Cerro La Primavera	High	Top stones repositioned	–	4 debitage
232	Arcos Bajos, North chenque	High	Good	–	1 smaller rock structure (ID_232b)
10	Chenque del xilópalo	Medium	–	–	6 UFA 57 debris 3 cores
9	Arcos Bajos, South chenque	High	Good	–	1 scraper 1 NFA 1 debitage 1 smaller rock structure (ID_9b)
11	Arcos Altos, North chenque	High	Good	–	1 big scraper 1 notch 1 artifacts with summary formatting 5 debitages
233	Arcos Altos, chenque S	High	Good	–	1 debitage
28	Chenque de la Loma Pelada	Medium	–	–	–

(continued)

Table 7.18 (continued)

ID	Name of the chenque	Lichens coverage	Integrity	Reuse evidences	Archaeological assemblage and other structures associated
261	Cerro de las estructuras—5	Medium	–	–	3 smaller rock structures (structures 3, 4, 6)
261	Cerro de las estructuras—2	Medium	–	–	–
261	Co. de las estructuras—1	Medium	–	–	–
261	Cerro de las estructuras—2	Medium	–	–	–
261	Cerro de las estructuras—1	Medium	–	–	–
712	Chenque B10	High	Good	–	–

"ID": identification code of the structure. "Integrity": it was evaluated whether the structure has a closed shape or whether the rocks that form it appear disarranged (see Chap. 6). NFA = non-formatted artifacts; FA = formatted artifacts

7.5.2 Amplitude of the Viewsheds from Chenques

The wide visibility pattern detected for these structures has been documented numerous times in ethnographic sources and archaeological research on aboriginal burial sites in the Patagonian region. These structures have been found at the top of small hills or on the edges of plateaus (Deodat, 1946; Moreno, 1879). This description coincides with the chenque structures of the present case study. However, regional archaeological studies have also recorded chenque burials in basin bottoms (Goñi & Barrientos, 2000, 2004) and on rocky outcrops near the seashore (Castro et al., 2000; Zilio et al., 2017). Despite the variety of locations on different geoforms, hunter-gatherer groups seem to have chosen their placement based on the dominance of wide panoramic views (see Deodat, 1946; Goñi & Barrientos, 2000, 2004; Martinić, 1995; Musters, 1871; Vignati, 1934).

The calculation of viewsheds in Chap. 6 considered the visibility of the landscape within a 10 km radius around each structure. The following section analyzes whether the same pattern emerges in chenques located in other sectors of the Central Plateau with different topographies, as well as how the pattern changes when considering a broader visibility radius. We included nine additional chenques in the macro-study area. These chenques are located in La Dorita ($n = 1$), Bajo Grande ($n = 4$), and Aguada del Cuero ($n = 1$); Bosques Petrificados de Jaramillo National Park ($n = 2$); and Piedra Museo ($n = 1$) (Table 1, Supplementary Material—Chap. 6). The landscape is more open in the last two localities, and the terrain altitude reaches 50 m above sea level. The parameters established for the analysis are as follows: (a) The observer's height was set to the same value as in the previous analysis (Chap. 6); (b) Two experiments were carried out: one with a maximum visibility radius of 10 km and another with a radius of 20 km. These radii were chosen based on good visibility conditions in the area, which allow viewers to see beyond 10 km when the air is free of dust and humidity (SMN, 2012). This allows us to test whether the amplitude of the viewshed remains consistent at each analyzed point or varies according to atmospheric conditions; (c) The base DEM used is the Shuttle Radar Topography Mission. Considering the scale of the analysis and computer requirements, the DEM with a spatial resolution of ~90 m was used (Table 5.1); (d) Finally, the reduction in visibility of distant objects due to the curvature of the Earth was considered in this calculation. This effect is approximately equivalent to a 7.86-m decrease in the Earth's surface for every 10 km of distance from the point of view (Connolly & Lake, 2006).

As a result, 22 viewsheds were obtained. Each was calculated for an individual chenque and is a binary raster image in which the values of the cells are 1 or 0, depending on whether the cell is visible or invisible from the observation point. The

1

Fig. 7.19 Chenques located in the study area. 1) Chenque del Viento (ID_8); 2) Chenque F2 (ID_235); 3) Chenque Cañadón Guanaco (ID_237); 4) Chenque La Primavera 2 (ID_100); 5) Chenque del xilópalo (ID_10); 6) Chenque La Primavera 1 (ID_736); 7) Cerro de las estructuras (ID_261); 8) Loma Pelada (ID_28); 9) Chenque B10 (ID_238); 10) Chenques de los Arcos Bajos south view (ID_232 and ID_9); 11) Chenques de los Arcos Bajos north view. From this last perspective, the lichen coverage is notable; 12) Small structure located downslope next to the Chenque de los Arcos Bajos South (ID_232) and 13) next to the North chenque (ID_9); 14) North (ID_11) and; 16) South Chenque; (ID_233) Chenques de los Arcos Altos; 16) Panoramic view of the fields of La Lotita from the southern Chenque

2

3

Fig. 7.19 (continued)

4

5

Fig. 7.19 (continued)

6

7

Fig. 7.19 (continued)

8

9

Fig. 7.19 (continued)

10

11

Fig. 7.19 (continued)

12

13

Fig. 7.19 (continued)

14

15

Fig. 7.19 (continued)

Fig. 7.19 (continued)

distribution of values in these calculated viewsheds was compared with that of an equal-size dataset ($n = 22$) calculated from observation control points to evaluate its statistical significance. These points were randomly selected within the negative evidence (NE) dataset. The landscape was stratified into three units with similar topography for the selection. This procedure guaranteed that differential landscape properties in the defined strata would not affect the size of the calculated viewsheds (Lake & Woodman, 2003). The first stratum corresponds to the broken terrain west of the macro-study area, in the headwaters and middle sector of the drainage basin. The main archaeological sites included in this sector are La Primavera, La Dorita, Bajo Grande, and Aguada del Cuero. The second stratum, located toward the east where the mouths of the ditches and the main temporary shallow lakes are found, includes the Piedra Museo locality, where one NE point was selected. The third stratum corresponds to the Jaramillo Petrified Forest Locality, where two NE points were randomly selected. Table 7.19 presents the viewshed amplitudes for the 22 random points and the 22 chenque locations. We compared these values using a Kolmogorov–Smirnov test for paired samples (Table 7.20 and Fig. 7.20). The results indicate that the width of the viewsheds calculated from the chenques is significantly greater than that of a random distribution.

In summary, the result is the same when considering lower (10 km) and higher (20 km) visibility ranges and varied topographic conditions (open topographies in

Table 7.19 Cells visible from chenques and a random set of NE in a circular area with a radius of 10 and 20 km

N°	N° of cells visible from			
	Chenques	Aleatory set from EN locations	*Chenques*	Aleatory set from EN locations
	In a 10 km radius	In a 10 km radius	In a 20 km radius	In a 20 km radius
1	3764	915	5021	915
2	3675	154	4937	154
3	3740	226	5927	237
4	9464	179	23,313	232
5	1089	174	8882	174
6	4332	75	5890	75
7	16,778	388	52,712	495
8	3571	286	4544	493
9	2185	680	3578	870
10	2401	114	4538	114
11	3646	193	7771	193
12	3246	275	5357	275
13	3975	345	4599	345
14	6153	76	9775	76
15	4706	159	5304	159
16	1665	191	2076	191
17	682	239	1831	214
18	4676	106	8801	106
19	4189	272	4929	314
20	5602	1589	17,085	1875
21	4268	8578	7157	13,083
22	2863	8521	3095	13,086
Total	96,670	23,735	197,122	33,676

Table 7.20 Kolmogorov–Smirnov test ($p < 0.05$) applied to the distributions of cells visible from chenques and random NE

Set 1	n	x (m^2)	Set 2	n	x (m^2)	D	P
NE 1	22	35,591,400	Ch (10 km)	22	8,739,900	0.8182	0.000
NE 2	22	72,576,000	Ch (20 km)	22	12,401,100	0.8636	0.000

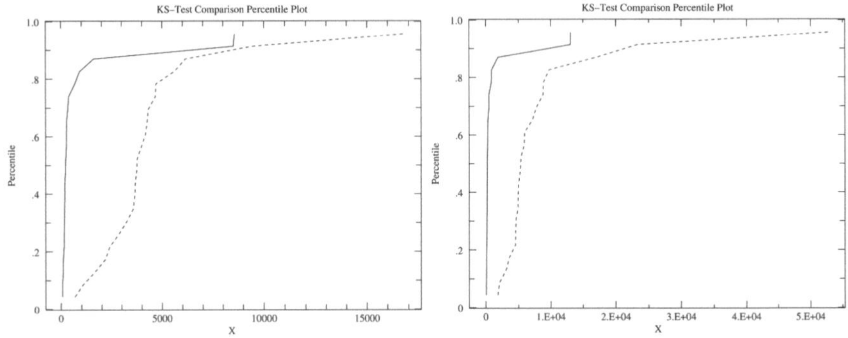

Fig. 7.20 Cumulative frequency diagrams showing the area visible from chenques (Ch) and random points (NE) within a radius of 1) 10 km; 2) 20 km. The results indicate that the Ch (continuous line) have significantly wider visual basins than the control sample (dotted line)

the eastern part of the macro-study area and broken topographies in the western sector): the viewsheds calculated from observation points coinciding with chenque locations are significantly wider than those calculated from random control points. This pattern is evident in the data, allowing us to propose that the visual landscape (García Sanjuán et al., 2006; Llobera, 2003) observed by people from the chosen chenque locations could have been an important locational factor (Magnin, 2009a).

7.5.3 Cumulative Viewsheds

Building on previous work, we became interested in analyzing which parts of the landscape are visible from these structures. Fundamentally, we are interested in testing whether chenques are more frequently visible from other chenques than would be expected by chance, as suggested by previous work (Magnin, 2009a). To complement this analysis, we carried out a cumulative viewshed analysis (CVA) (Conolly & Lake, 2006; Wheatley, 1995), which allows us to identify, in a digital terrain model of the microstudy area, the areas most frequently seen from the chenques. CVA analysis involves summing the values of individual viewsheds using map algebra. The result is a map in which the cells are classified according to the number of chenques visible from them, ranging from 0 to n. The theoretical maximum value for the cells is 14, as that is the total number of chenques in the sector. For this reason, the resulting map is also called a "times-seen" map (Wheatley & Gillings, 2002: 208) and can be considered a visual summary of how chenques relate to other landscape areas (Gaffney et al., 1996). Two similar maps were processed: one with 14 chenques and another with 14 randomly chosen negative evidence (NE) points. Specific queries on each cell (location of each chenque/NE) in the CVA map revealed how many other chenques/NEs are within their viewshed. The resulting distributions are presented in Table 7.21.

Table 7.21 Frequency of other chenques or random points (NE) visible from each location

ID	N of visible Ch	ID	N of visible NE
11	8	471	3
100	5	543	4
261a	7	544	2
736	3	436	2
261b	6	459	2
233	7	690	3
235	3	482	2
28	3	542	1
9 b	4	553	2
232	4	448	1
237	2	491	1
712	1	513	1
8	1	516	1
10	1	432	1

This information was obtained from accumulated viewshed maps. The values in this table include the observation point itself

Table 7.22 Kolmogorov–Smirnov test ($p < 0.05$) was applied to frequency distributions of other chenques/NE visible from each location

Set	n	x	Set 2	n	x	D	P
NE 2	14	1.857	Ch	14	4.071	0.5000	0.039

This information was obtained from accumulated viewshed maps

As seen in Table 7.21, the maximum value of the chenques is 8, while the maximum value of the NE set is 4.

To analyze these values, consider that the analysis includes the observation point in the account. For example, there are two chenques from which seven more can be seen. These are the Arcos Altos North chenque (ID_11) and the La Primavera 2 chenque (ID_100). For NE points, the maximum number of visible NE points is three (corresponding to NE ID_471, ID_543, and ID_544). A K-S test was performed on these, indicating a significant difference (Table 7.22 and Fig. 7.21).

7.5.4 Implication of These Patterns

The study of viewsheds indicates that the locations of the chenques in this sector of the Central Massif of Santa Cruz coincide with areas of panoramic views. There is a clear difference between the viewsheds obtained from the chenques and those

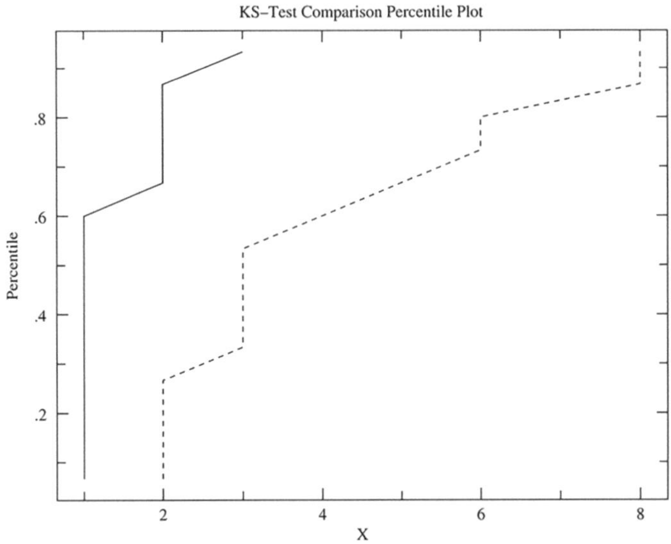

Fig. 7.21 Cumulative frequency diagram shows the frequency of chenques (dashed line) and NE (solid line) that are visible from the observation point. As the figure indicates, a significantly larger number of chenques can be seen from the chenque structures

obtained from the random control points. This difference is statistically significant at a level of 0.05 (K-S test of two paired samples). This pattern was detected when analyzing close visual radii (set at 10 km) and intermediate distances (20 km). Using the stratification method in landscape analysis (Woodman, 2000b), we determined that this result is not a topographic property of the landscape itself. These results support the hypothesis of intentionality in the selection of these locations and are consistent with the expected findings based on ethnographic information and field observations.

Additionally, there is a high degree of intervisibility between chenques. All the cases analyzed show visibility between other chenques, and two chenques have visibility with seven others simultaneously. This is an interesting result because it shows a statistically significant difference compared to the low degrees of intervisibility detected between randomly selected negative evidence points. As Table 7.21 shows, visibility frequencies from the NE locations are never higher than three. While this pattern partly reflects the grouping of chenques, it also suggests that direct visibility of other chenques was considered when placing new ones in the landscape. These results highlight two aspects of visibility from chenques. First, they confirm what is known about the region: the views from their locations are panoramic. Second, the visibility of other previously located chenque structures could have influenced the decision on the location of new chenque structures (Magnin, 2009a). If chenques are burial structures, then there are some interesting implications. The visibility relationships between funerary monuments and their surroundings have been analyzed using Geographic Information Systems (GIS) in other contexts, such as the Early

Neolithic period in England (Wheatley, 1995), the Bronze Age in Mull, Scotland (Woodman, 2000), and the Neolithic and Copper Age periods in the western Sierra Morena region of Spain (García Sanjuán et al., 2006). These studies demonstrate how the visual dimension of the landscape, specifically the location of burial sites, can be examined to understand the archaeological record in terms of perception and symbolism through GIS analysis. The patterns found in our study can be interpreted in different ways. First, the places chosen for burials may not have been arbitrary, but rather located in a previously structured landscape of ritual and routine cycles (Wheatley, 1996). Similar monuments have been interpreted in other studies as indicators of social cohesion, highlighting the idea that there may have been shared beliefs or meanings regarding the burial of the deceased (Wheatley, 1993). Conversely, burial sites may have served to establish territorial rights over an area's resources (Gaffney et al., 1996; Lobera, 2003; Wheatley, 1993,). The act of burial implies respect for the deceased and a belief in an afterlife. Because of this, funerary contexts can be considered sacred or culturally significant (Renfrew & Bahn, 1993). Therefore, it is reasonable to assume that the beliefs of these human groups influenced the selection of these structures' locations. Funerary structures built according to the same criteria over the years must have produced the pattern reflected in the archaeological record (Renfrew, 1982, as cited in Gaffney et al., 1996). This pattern relates to a highly ritualized context, such as a funerary or monument setting, and a physical aspect of the landscape, such as the visual. However, proving causality is more difficult. Based on direct ethnohistoric evidence, the amplitude of viewsheds is certainly a factor in locating funerary chenques. In contrast, ethnohistoric sources do not mention high intervisibility between funerary structures as a locational factor. Our finding is an interesting pattern that can be tested in other areas to evaluate its possible societal implications.

7.5.5 The Chenques on a Regional Scale. Chronology and Population Dynamics

In this region, aboriginal burial sites are arranged in three ways: concentrated in formal spaces, grouped in small numbers, or individually. Additionally, two types of burials belonging to different chronologies have been recorded: (1) burials in niches or under blocks, and (2) chenque burials (Berón et al., 2000; García Guráieb et al., 2007; Goñi, 2000,; Goñi & Barrientos, 2000). In the studied sector, only chenque burials were recorded. The chenque burial modality extends across regions, from the valleys connecting the steppe to the Pacific coast in the west (Reyes et al., 2006) to the Atlantic coast in the east (Castro et al., 2001) and north to La Pampa and southeastern Buenos Aires province (Berón et al., 2000). These regularities are probably a consequence of shared knowledge and beliefs among populations based on past mobility and exchange circuits that connected groups (Zilio, 2013).

Regarding their chronology, most of the chenque burials at Salitroso Lake, in the western part of Santa Cruz Province, are dated between 1500 and 300 years BP (Guráieb et al., 2007; Zilio, 2013; see Chap. 8 for more information). Lago Salitroso, along with Puerto Ibáñez, and the Atlantic Coast, is one of the significant concentrations of formal burial sites at the regional level. In Lago Salitroso, at least 60 chenques were recorded in an area measuring about 10 by 20 km, with others scattered across the steppe (Cassiodoro, 2008; Goñi et al., 2004). Studies indicate a predominance of women's and children's skeletons, and several constitute multiple burials. Another interesting finding is that two of the excavated chenques did not contain human remains (Cassiodoro, 2008). In the Puerto Ibáñez sector of the Chilean foothills, 14 grouped chenques and other isolated ones have also been identified as formal burial sites (Mena & Lucero, 2004). Finally, to the east of the study area on the Atlantic coast, mainly in Punta Medanosa, numerous structures formed by accumulated stone piles were recorded ($n = 274$). These structures are approximately 2 m in diameter and often acquire a circular shape due to looting, a common practice among the region's residents and visitors since the beginning of European settlement (Deodat, 1946; Vignati, 1934). In these chenque fields (Castro & Moreno, 2000; Castro et al., 2001), the chenques were found both in isolation and in groups of up to 40, along the coastal strip of rocky outcrops, a few meters from the current beach (Castro & Moreno, 2000).

According to Goñi and Barrientos (2000) and Goñi (2000, 2000/2002), the form of disposal (in formal spaces or cemeteries with multiple burials) and the reuse of chenques have important implications regarding societal organization, group mobility, and environmental use. The burials at Lago Salitroso, along with surface residential sites (MALs SAC 3, 5, and 6), have been proposed to indicate restricted mobility or stable residence (Goñi, 2000). The co-occurrence of base camps with formal cemetery sites, especially those containing burials of sub-adult and female individuals, indicates residential areas. Since these population sub-groups are always inside the base camps, their presence in the majority of burials indicates the residential character of the space. High infant mortality is also interpreted as reflecting situations of population aggregation, where the likelihood of disease contagion is greater (Goñi et al., 2004). The factors that explain this settlement pattern are complex. They are based on climatic variations, which during the period called Medieval Climate Anomaly (MCA) between 1200 and 600 years BP determined a higher temperature and aridity than the current climate (Morales et al., 2009). The settlement pattern would also have been accentuated with the introduction of the horse upon the arrival of the Europeans (Goñi, 2000), by the end of the sixteenth century (Martinić, 1995). This pattern of population concentration in the main pre-Andean lake basins has been contrasted with the pattern of logistical use of the high plateaus around these basins (Goñi, 2000; Goñi & Barrientos, 2000). The high sectors such as the steppe areas of Lake Posadas/Pueyrredón, and Ghío, as well as in the highlands of Pampa del Asador and Perito Moreno National Park (PMNP) have only found isolated funerary structures (Goñi et al., 2004). These are interpreted as occasionally, sporadically, or seasonally occupied environments. Based on this model, and given the presence of isolated burial sites or in small groups, the case studied here would be closer to these environments of occasional or seasonal occupation. The particular settlement pattern proposed here about the funerary structures indicates a continuity in occupation or seasonal reoccupation by the same group, or at least by people who share the same mortuary customs.

On the other hand, the pattern from which we started at the beginning of the chapter, according to which the burial sites are separated from the domestic campsites and from the mobility paths between them, seems opposite to the pattern in the northwest of Santa Cruz province. Other areas at the regional scale could be analyzed using the presented methodology to determine whether the settlement and visibility patterns are similar or different. The Patagonian region presents numerous cases of isolated funerary structures similar to those found here, which have not been the subject of studies or publications. The large formal burial sites mentioned in the north Santa Cruz coast and in Salitroso Lake were subject of excavation and systematic research, but the isolated structures or in small groups in the region have not been examined to determine if they are arranged in a specific intervisibility pattern, as found here.

Chapter 8
Chronology of the Occupations

Abstract This chapter presents arguments that provide a chronological framework for the spatial distribution study. It presents the absolute dates obtained from samples taken from Cueva Maripe, Cueva Mora, and La Quinta. La Quinta is of special interest because it is the first open-air site dated in the area. The chapter discusses the possibilities and limitations of using existing information to establish a temporal context for patterns found at sites, concentrations, and isolated surface finds.

Keywords Absolute and relative chronology · Chronology hiatus · Early to Late Holocene

8.1 Absolute Chronology

8.1.1 Cueva Maripe Site

This work places a strong emphasis on searching for and analyzing patterns in spatial distributions within the study area. As previously mentioned, the objective of this study is not to reconstruct settlement patterns. Such approach assumes that assemblages derived from simultaneous occupations can be clearly distinguished within a settlement system. However, this is not feasible in practice (Binford, 1982; Dewar & McBride, 1992; Gamble, 1990). Instead, we expect to find patterns in archaeological distributions derived from long-term settlement systems (Gamble, 1990). Consequently, estimating the chronology of occupations is a matter that must be considered carefully. In this chapter, we present the range of radiocarbon dates that serve as the temporal anchor for the archaeological locality under study. Most of these dates were obtained from archaeological material from Cueva Maripe and two other sites: Cueva Mora and La Quinta. The latter is the first open-air date in the locality. Here, we also evaluate the use of relative chronology indicators, which are useful for surface assemblages. The main results of the investigations carried out at these sites are presented below (Figs. 8.1, 8.2 and 8.3).

L. A. Magnin, *The Landscape and Mobility of Hunter-Gatherer Societies*, The Latin American Studies Book Series, https://doi.org/10.1007/978-3-032-08127-8_8

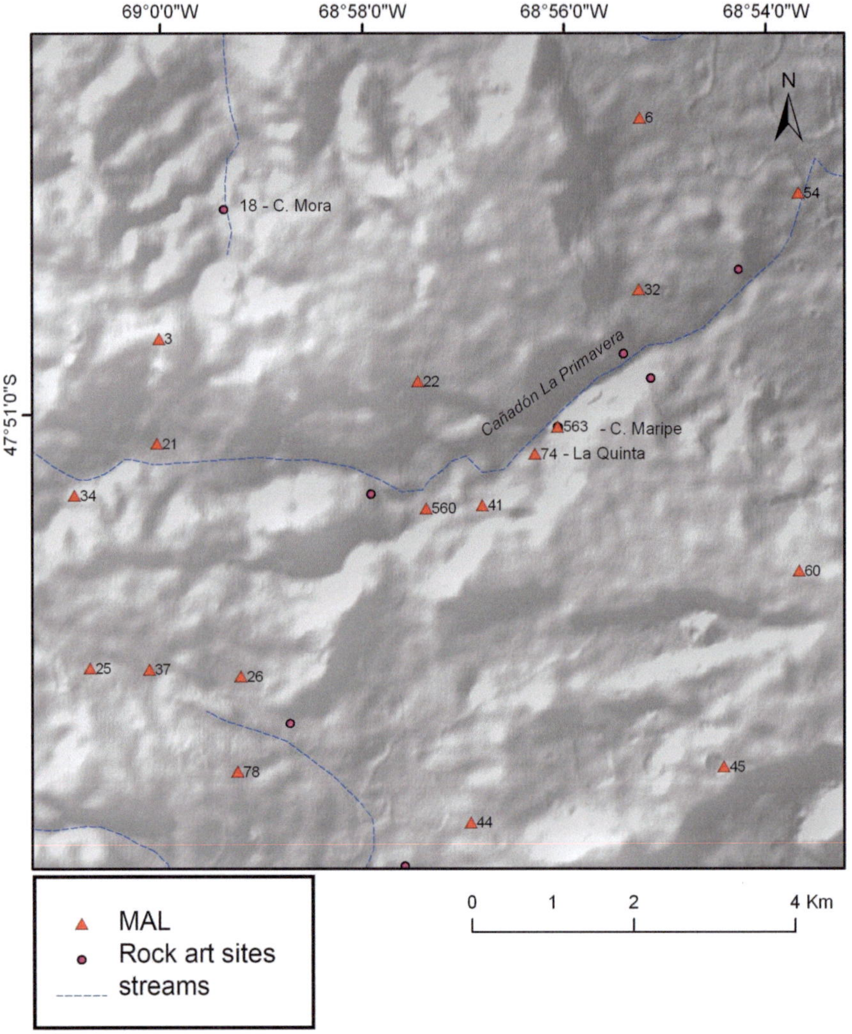

Fig. 8.1 Location of multiple activiries sites (MAL) in the study area: 3—Alta Primavera; 6—Camino Clandestino; 18—Cueva Mora; 21—El Denticulado; 22—El Descanso; 25—El Pájaro Muerto; 26—El Piche Lento; 34—GB; 37—La Herradura; 41—Loma Blanca Bajo; 44—Mirador; 45—Oasis; 54—Pico Blanco; 60—Sitio del Paso; 74—La Quinta; 78—D9; 560—La Playita; 563—Cueva Maripe

The archaeological contexts are presented within a heterogeneous stratigraphic deposit. As the excavations progressed, significant differences were noted between the two chambers in terms of the characteristics of the sedimentary matrix, the thickness of the layers identified and correlated between grids, and the conservation, frequency, and distribution of artifacts and archaeofauna. This heterogeneity was also evident in the distribution and preservation of rock art in both chambers (Miotti

1

Fig. 8.2 1) View of the Maripe Cave on the southern wall of volcanic tuffs, La Primavera canyon; 2) Excavation carried out in Cueva Maripe during 2004. Grids C5 and D5, North Chamber

et al., 2007b). Therefore, the analyses were structured as a comparison between the two chambers. The main results of the Cueva Maripe studies are summarized below.

8.1.1.1 The Historic Occupation

Maripe Cave was inhabited until the beginning of the twentieth century by a countryman of probable aboriginal ancestry (Echeverría Baleta, 1982; Miotti et al., 2009). In the south chamber, which is more sheltered from the wind, remains of a stone pile from this occupation are preserved (in grid B12). Historical materials, such as metal nails and glass, have been extracted from the associated grids (A12 and AA12). Later, the cave was used to protect livestock. For this reason, the current surface consists of aeolian material, clasts from the collapse of the bedrock, and a significant proportion of cattle and sheep manure. This layer acts as a seal, placing itself above the strata with archaeological content. In the South chamber, this layer is approximately 20 cm thick and in the North chamber, it reaches a maximum thickness of 70 cm (Miotti et al., 2007, 2009).

2

Fig. 8.2 (continued)

8.1.1.2 Previous Occupations

The cave is an important sediment trap that receives contributions from both outside and inside sources. Wind processes play a crucial role in depositing particles from outside the cave, primarily from the colluvium slope at its base and on the opposite side of the canyon. Wind also actively erodes the cave's walls and ceiling (Miotti et al., 2007). Wind and water filtering from various points inside the cave actively undermine and deepen its current limits. Evidence of this can be seen in the collapses on the north wall of the north chamber and at the bottom of the south chamber. The importance of these agents in the cave's formation is evident in the entrance's orientation toward the west, from which the dominant winds originate (Panza, 2001). Since the cave is 4 m above the canyon's bottom, it is unlikely to receive input from the temporary tributary. Other exogenous sediments that contribute sediments to the cave include biological material left by humans and animals. Examples include guano deposited by birds and bats, layers of cow and sheep dung, litters of straw found in sedimentary layers, the bones of animals that also occupied the cave, and construction materials from the pirca, as well as archaeological artifacts. Due to the complexity of the sedimentary matrix, layers or continuous strata with clear lateral and horizontal limits could not be defined during the excavation (Farrand, 1985), so artificial excavation levels were established (Miotti et al., 2009). The northern chamber has a

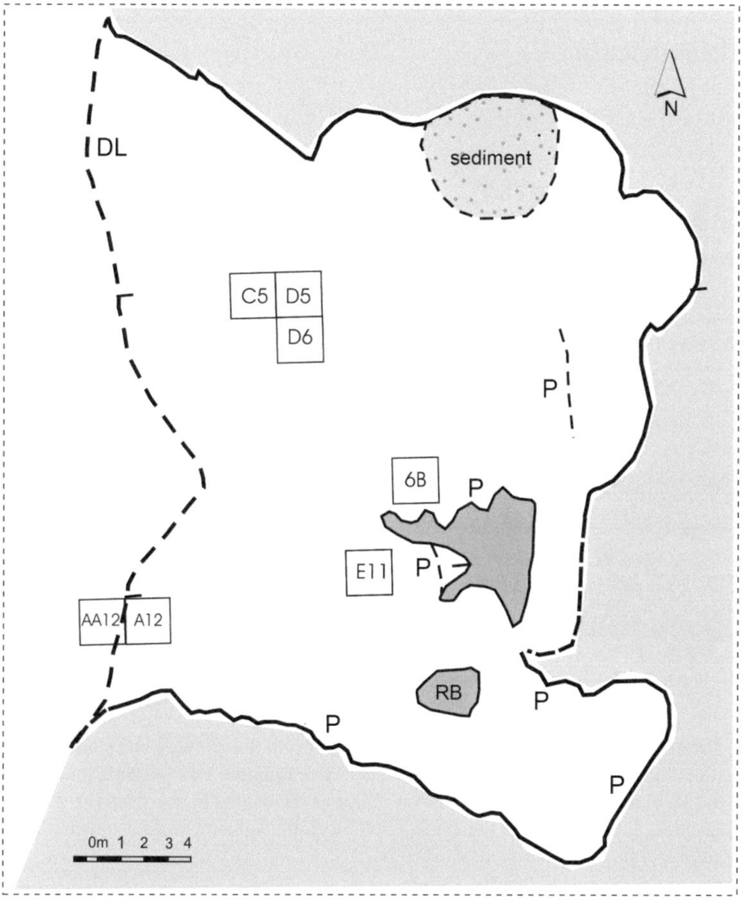

Fig. 8.3 Plan view of Maripe Cave. The grids excavated in 2003 and 2004 are named C5, D5, E11, A12, and AA12; those excavated in 2005 are D6 and B12; and 6B in 2006. Planimetric survey carried out by Santiago Medel in 2003

greater sedimentary thickness, reaching 129 cm of archaeologically fertile sediment. In contrast, the sedimentary depth in the southern chamber only reaches 60 cm, which is why it is considered a compressed sequence. A three-dimensional plot of materials allowed us to visually analyze and distinguish a discontinuity of approximately 20 cm in the northern chamber. Based on this, we differentiated two components (Fig. 8.4).

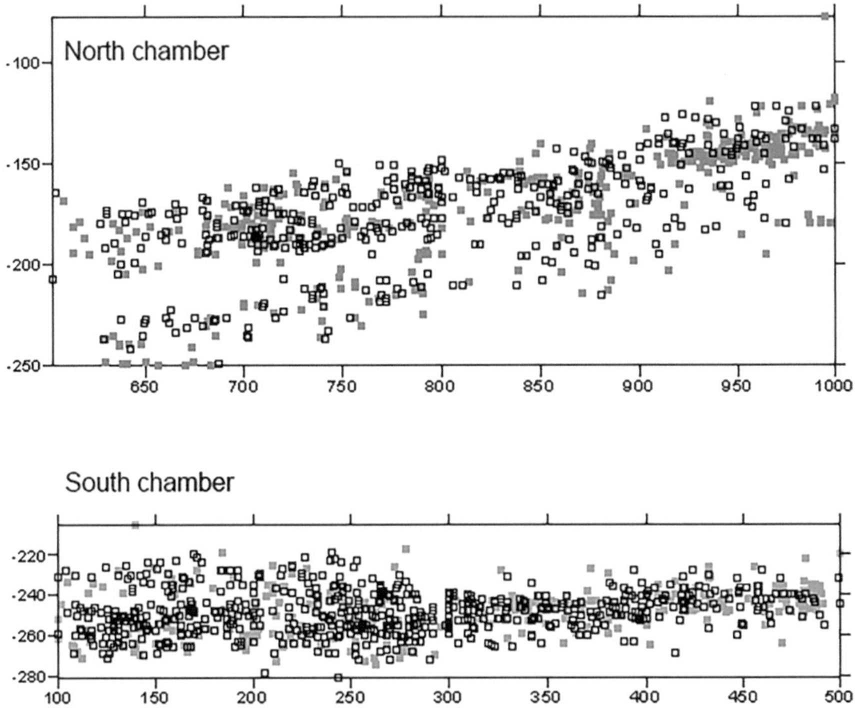

Fig. 8.4 Profiles showing the bone and lithic distributions in the northern and southern chambers. E–W sections for the north and the south chamber. The ordinate axis presents the distance data in cm in the W–E direction. The abscissa axis indicates the depths in cm, from the general datum of the excavation. Lithic materials are symbolized by empty squares, and bone materials by filled squares (graph published in Miotti et al., 2007)

8.1.1.3 Archaeological Assemblages Recovered

From a sample of the materials analyzed up to 2006 consisting of 2184 lithic artifacts and 606 bone materials in three dimensions, the following trends could be established in the data (Miotti et al., 2007):

1. The percentage frequency of total lithic artifacts is 72.49% for the south chamber ($n = 1393$) and 27.51% for the north chamber ($n = 791$), which is equivalent to an average of 203.49 artifacts per m^3 for the south chamber and 49.44 artifacts per m^3 for the north chamber, based on the volume excavated in each chamber (9.6 m^3 and 16 m^3, respectively).
2. The distribution of bone materials between the chambers shows few differences: 51% in the north chamber and 48% in the south chamber. However, the greatest difference in deposition is obtained by evaluating the frequency of bone materials per cubic meter. In the northern chamber, it is 1.49, while in the southern chamber, it is 3.81 (Table 8.1). Therefore, the southern chamber has a greater artifact

Table 8.1 Comparison between the densities of lithic artifacts and bone materials in each chamber of Cueva Maripe

	North chamber	South chamber
Lithic artifacts (n/m^3)	49.44	208.49
Bone specimens (n/m^3)	19.25	31.18

density, which is always greater than twice that of the northern chamber. This high variability of the deposit is quite common in cave sites (Kornfeld et al., 2007; Miotti, 2006). Thus, taphonomic and sedimentary rates were calculated.

8.1.1.4 Taphonomic Studies and Sedimentary Rates

Dating results from charcoal samples taken from nine hearth structures allowed us to determine that the site was occupied between the Pleistocene/Early Holocene transition and the Late Holocene. These results also verified that there were no stratigraphic inversions (see Table 8.2).

Studies were carried out to interpret and compare the contexts in the north and south chambers. These studies allowed for direct and indirect estimation of the prevailing fill accumulation rates for both chambers. The indirect method determines the rate based on the location of the dates at the top and bottom of the deposit. The interval between the dates obtained allows us to estimate, on average, whether the accumulation rate was high or low. This method is based on the premise that there are no stratigraphic inversions and that the incorporation of materials was relatively constant (Favier-Dubois, 1997; Ferring, 1986). However, the obtained estimate is preliminary, as rates based on only two dates can give the impression of a continuous process, which is a simplification of the actual process (Stein et al., 2003). Two of the nine dated samples were selected, one from the upper sector and one from the lower

Table 8.2 Radiocarbon dates were obtained from charcoal samples in the stratigraphic contexts of the Cueva Maripe site. With the exception of LP1497, these dates were obtained using AMS (Miotti et al., 2007b)

Period	Chamber	Grid unit	Lab code	Radiocarbon dating (14C years BP)
Pleistocene/Holocene transition	S	A12	AA65175	9518 ± 64
Early	N	D5	AA65179	8992 ± 65
	N	C5	AA65178	8762 ± 50
	S	A12	AA65174	8333 ± 63
Middle Holocene	S	B12	AA65177	7703 ± 47
	N	D6	AA65173	5084 ± 49
	S	A12	AA65181	4113 ± 39
Late Holocene	N	C5	LP1497	3210 ± 60
	S	B12	AA65176	1078 ± 40

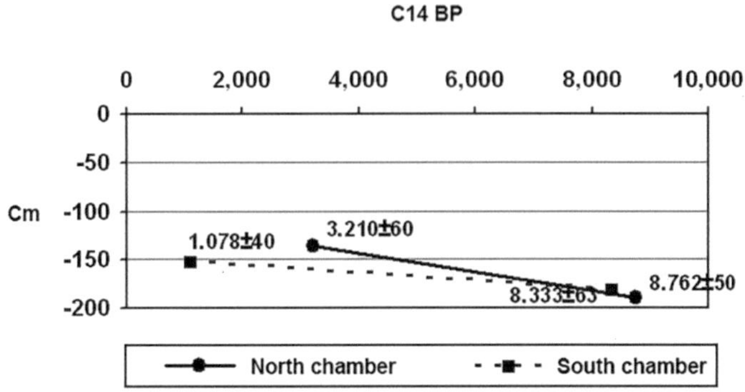

Fig. 8.5 Fill accumulation rates in Cueva Maripe. The lowest filling rate in the south chamber is represented in this graph by a lower slope in the dotted line, indicating that there is a greater compressed time in the sedimentary sequence in that sector of the Maripe cave than in the north chamber, with positive accumulation rate (graphed here with a solid line)

sector of the sedimentary sequence in each chamber. The criterion for selection was that the two samples from each chamber be superimposed in the same vertical column with the greatest possible depth difference to achieve the most accurate estimate.

In the North Chamber, the difference in depth between the two selected samples is 54 cm. In this thickness, 5552 years 14C BP are represented, resulting in a sedimentation rate of 0.0097 cm/year. The distance in depth between the South Chamber samples is 29 cm, representing 7255 years 14C BP and resulting in a sedimentation rate of 0.0039 cm/year (Fig. 8.5). These results indicate higher rates of fill accumulation in the North Chamber than in the South Chamber. The lower rate in the South Chamber suggests the possibility of temporary hiatuses due to erosion processes. Among the possible influencing factors, the endogenous action of an active spring at the bottom of the cave and the action of the drip line, which affects the grids of the South Chamber more directly than the North Chamber, are proposed. However, both rates are generally low (Ferring, 1986; Miotti et al., 2007b).

In summary, analysis of recovered archaeological materials revealed a lower rate of artifact deposition and a higher rate of fill accumulation in the northern chamber than in the southern chamber.

The evaluation of archaeological resolution and integrity also includes the analysis of the stages of weathering and fragmentation of bone materials and the analysis of reassembly of lithic artifacts (Miotti & Marchionni, 2009; Miotti et al., 2007b). The bone weathering profiles in both chambers present different trends, with a marked slope toward the lower stages (stage 2) in the northern chamber and greater weathering of the specimens (stage 3) in the southern chamber. This indicates that it is likely that the bone assemblage from the south chamber has been exposed to atmospheric agents longer than the assemblage from the northern chamber, but it is also likely that the runoff of water from the spring (still active) on the roof of the southern chamber has generated a less conducive environment for the conservation of materials from this sector. Regarding the bone refits, they all occurred with pieces from the same level in the northern chamber and the same component in the southern chamber. The horizontal distances between the pieces that comprise the refits were not significant. There are currently no assembly relationships between pieces extracted from different stratigraphic units, indicating the integrity of the assemblages (Miotti et al., 2007).

8.1.1.5 Interpretation of the Activities Developed in the Site

In summary, the analysis of the recovered archaeological materials revealed a lower artifact deposition rate and a higher fill accumulation rate in the northern chamber. Evaluating archaeological resolution and integrity includes analyzing the stages of bone material weathering and fragmentation, as well as the reassembling lithic artifacts (Miotti & Marchionni, 2009; Miotti et al., 2007b). The bone weathering profiles in both chambers exhibit distinct trends, with a steeper decline toward lower stages (stage 2). Studies indicate that the southern chamber is best represented by lithic and bone manufacturing, which may correspond to various domestic activities (Hermo, 2008; Miotti & Marchionni, 2009; Miotti et al., 2007, 2009). This suggests that the southern chamber was used for habitation. The historical occupation of the site is also interpreted in this way, since the remains of a stone wall in the southern chamber indicate its use as a dwelling place. The degrees of resolution in this chamber are lower and involve a greater variety of post-depositional processes.

The northern chamber could have been used for processing guanaco carcasses and the initial stages of artifact manufacturing, but not for habitation in the traditional sense (Miotti et al., 2007b). Based on these differences and an alternation between the dates in both chambers, Miotti and Marchionni (2010) propose that the occupation of these cave spaces occurred alternately throughout the cave's period of occupation. Regarding the distribution of rock art, differences are observed in the cave's various sectors (Carden, 2008; Miotti et al., 2007b). These differences are due in part to differential preservation processes, which mainly affected the paintings in the entrance and intermediate areas of the cave by preventing the preservation of paintings exposed to light and other elements. However, this cannot explain the differences in spatial distribution of different motifs in the cave. In the most visible areas of the cave, the hands of adults and children were painted. In the least visible areas, only small hands were painted, presumably those of children. Zoomorphic and dotted motifs were painted inside natural cavities or niches in the rock. Finally, differential colors were used for negative handprints according to age group (a higher proportion of children's hands were painted pink and adult hands red than would be expected by random chance). The differential distribution and use of different colors in the recorded motifs suggest formalism and traditionalism among different groups of images, which could explain their different symbolic values. Since these are aspects that define ritual behaviors, these patterns probably indicate ritualized uses of the cave (Carden, 2008).

8.1.1.6 Occupancy Continuity/Discontinuity Indicators

The sequence in the southern chamber at this site spans all periods of human occupation in the region, from the Early to the Late Holocene. However, there are still many gaps in the available information, making it difficult to determine whether this space was occupied continuously or discontinuously. Based on the existing radiocarbon dates and their standard deviation values, it can be concluded that the undated intervals range from 520 years for the shortest period (between dates 2 and 3) to 3504 years (Table 8.3).

Based on the current state of knowledge, the extensive stratigraphic sequence of the northern chamber of Cueva Maripe dates back to the Early, Middle, and Late Holocene (Table 8.4). As previously mentioned, an absence of archaeological materials was identified in a 20-cm-thick sedimentary layer that separates the sequence into an Early Holocene lower component and a Middle and Late Holocene upper component. The absence of organic materials in that layer could be explained by preservation issues under certain physicochemical conditions. However, it is difficult to explain the lack of inorganic archaeological evidence in this way. Therefore, a possible hypothesis is that the cave, or at least its southern chamber, was not used during a certain period (Miotti & Marchionni, 2009; Miotti et al., 2007b).

Assuming constant filling rates over time, the period represented by the 20 cm discontinuity between the upper and lower components of the northern chamber is estimated to be 2056 years. However, this estimate is hypothetical and requires micro stratigraphic studies to distinguish the contributions of different sediments. Another possibility is that this stratum without archaeological evidence represents a very short period, which would not allow for the cave to have been empty for such a

Table 8.3 Radiocarbon dating sequence for the southern chamber, Cueva Maripe

South chamber Datings (14C years BP)		*No. dating*	*Difference* in 14C years BP	*Period*
9518	– 64	1	9454	Pl/EH
8333	+ 63	2	8396	EH
			Difference = **1058**	
8333	– 63	2	8270	EH
7703	+ 47	3	7750	MH
			Difference = **520**	
7703	– 47	3	7656	MH
4113	+ 39	4	4152	MH
			Difference = **3504**	
4113	– 39	4	4074	LH
1078	+ 40	5	1118	LH
			Difference = **2956**	

Table 8.4 Radiocarbon dating sequence for the northern chamber, Cueva Maripe

North chamber Datings (14C years BP)		*No. dating*	*Difference in* 14C years BP	*Period*
8992	– 65	1	8927	EH
8762	+ 50	2	8812	EH
			Difference = **115**	
8762	– 50	2	8712	EH
Sterile layer open gravel				
5084	+ 49	3	5133	MH
			Difference = **3579**	
5084	– 49	3	5035	MH
3210	+ 60	4	3270	LH
			Difference = **1765**	

long time. In this sense, one or several accretion events with high sedimentation rates could have occurred without implying that the space was abandoned for such a long period. Geoarchaeological studies by Dr. Rabassa have made significant contributions to understanding this issue. The occupational voids in the northern chamber are probably due to water runoff from active springs inside the cave. This could generate lenses and gravel strata (*open gravel*), representing periods when that part of the cave was unavailable for occupation. The previously presented idea of alternating use of both chambers is now supported by dating (Miotti & Marchionni, 2009) and these microlocal sedimentation episodes inside the cave.

8.1.2 Cueva Mora Site

Cueva Mora (ID_18) is located in the Zanjón de La Mora creek, which runs in an S–N direction. The creek is a tributary of the Zanjón Blanco, like the La Primavera Canyon (Fig. 8.1). The cave opens onto the Chon Aike Formation. It consists of continental volcanic rock made up of ignimbrites, agglomerates, and rhyolitic tuffs, with a few tuffites and rhyolitic porphyries (Panza, 2001). The entrance opens to the west, where strong prevailing winds blow. The rock wall forms an overhang extending 21 m from north to south with two deepening chambers. The north chamber has a maximum depth of 6 m and is 3 m wide at the entrance. The south chamber is 3.5 m deep and has similar dimensions at the entrance (Fig. 8.6). Cave painted motifs were recorded on the ceiling of this same chamber: a negative hand made with white paint and nine red circles (see Chap. 7). The site is named after the Mora bush (*Senecio filagionides*) that grows there.

A survey was conducted in this cave to identify archaeological materials within the stratigraphy and to collect samples for dating purposes. Excavation took place at

1

Fig. 8.6 1) The sector where the survey was carried out was in the northern chamber of the Cueva Mora site. 2) Manufactured bone flakes were collected at the 60–70 cm level in Cueva Mora. The bone flake identified with D409 was dated

the northern end of the entrance to the northern chamber. The 50 cm × 50 cm grid rested on the rock wall that constituted its northern limit. The excavation followed natural levels, which were identified by changes in the sediment's color and texture. The sedimentary thickness from the surface to the base rock was 70 cm. All sediment was screened using a 3 mm mesh. Cultural evidence was continuously registered from the base of the borehole to 16 cm from the surface. The lithic archaeological materials recovered include knapping debris, NFA, and an FA (notch). Charcoal spicules, bone flakes from *Lama guanicoe* and other species, as well as microfauna bones, were also recorded. Between 16 cm and the soil surface, there is a layer of compacted sheep guano that does not contain any archaeological evidence. The recorded sequence is briefly described according to levels defined by changes in the sediment's color and texture. From the base rock to the surface, these levels are:

Level 60–70 cm: Fine brown sandy sediment. The sample for dating was extracted from this level (Fig. 8.6).

Level 50–60 cm: Fine brown sandy sediment (artificially established layer).

Level 39–50 cm: Angular tuff clasts. Below them, a large root was recorded.

2

Fig. 8.6 (continued)

Level 33–39 cm: The fine brown, sandy sediment is interrupted by a thin, black layer containing charcoal spicules.

Level 21–33 cm: Angular tuff clasts.

Level 16–21 cm: Loose soil with roots corresponding to an A horizon.

Level 0–16 cm: Guano (Table 8.5).

The sample used for dating is a human-made bone flake that presents a bulb of percussion, a curved profile, and a thin-edge termination. It was found at the deepest

Table 8.5 Radiocarbon dating of bone material procedent from site Cueva Mora

Cueva Mora site	GPS ID: 2108
Lab code	AA85461
Weight	2gr
Material	Bone flake on guanaco metapodium
Field label	LP'08 D # 409
d13 C	– 20.1
F	0.6326 ± 0.0051
14C years BP	3678 ± 65

level, 60–70 cm deep, before reaching the base rock. It was found in spatial association with two other bone flakes, three bone fragments, two lithic flakes, and rodent bones. The sample's state of methodization is low (stage 1, according to Behrensmeyer, 1978). The expected age was calculated to be 3500 years BP, a date close to the actual result. This calculation was based on sedimentation rates established for the northern chamber of Maripe Cave (Miotti et al., 2007b). The depth at which the sample was taken was used to estimate the age. Once the absolute date was obtained, the fill accumulation rate for the site was calculated. This calculation uses two known dates in a stratigraphic column and the thickness of the sediment between them (Ferring, 1986). In this case, however, we only have one date and the surface taken as a measure of the current date. Since these caves have been used by livestock for shelter, there is generally a layer of manure of varying thickness. In this case, the layer is 16 cm thick. Guano deposition began after sheep were introduced to the area and acted as a stratigraphic seal. This layer was excluded from the calculation. Therefore, the sedimentary thickness was measured below this seal and was computed to be 44 cm (the 60 cm of the dated sample minus the 16 cm of the guano layer). The surface dating where the guano layer begins was estimated at 1930. This estimate considers that the first cattle drives occurred in 1884 and that the 1937 agricultural census indicated the Deseado area had reached the historical maximum for the entire province (see Chap. 2). Considering these parameters, the calculated accumulation rate is 0.01200 cm/year.

8.1.3 La Quinta Site

The La Quinta site was detected in 2003. Located in the middle sector of La Primavera Canyon, it sits 430 m southwest of the Cueva Maripe site (ID_74 in Fig. 8.1). The site is situated at the point where the slope changes from the bottom of the canyon to its southern slope. Though the substrate is sandy and stony with loose materials, the site is arranged around a wet meadow, both upslope and downslope. This meadow is an important factor in sediment retention (J. Rabassa, pers. comm.). The meadow is fed by four active springs that maintain a small forest of *Schinus sp.* (molle), *Colliguaja sp.* (peach bush), *Junellia sp.* (black bush), and *Senecio sp.* (mata mora). One hundred meters from this site is an orchard of fruit trees that gives the site its name. The survey of this site was conducted as an initial assessment of the type of archaeological evidence present at open-air sites in this area and to obtain materials for radiocarbon dating. A relatively flat area protected by vegetation and above the wet meadow was chosen for excavation (Fig. 8.7). This allows us to rule out water's action as a degrading agent of organic matter (at the suggestion of Dr. Patricia Fanning). We excavated a grid of 1 by 1 m in artificial levels of 5 cm (Fig. 8.8) and recorded a gray to light brown, sandy sediment. From the bedrock to the surface, the levels are as follows:

The level is 50–60 cm. Brown, sandy sediment.

1

Fig. 8.7 La Quinta Site 1) It is located about 20 m uphill and 30 m downhill from the meadow and extends 150 m longitudinally, 2) The survey was conducted at a point above the level of the plain

Level: 45–50 cm. Brown, sandy sediment.

Level 40–45 cm: Brown, sandy sediment with roots.

Level: 35–40 cm. Brown sandy sediment; the dated sample comes from this level.

Level: 30–35 cm. Brown, sandy sediment with charcoal spicules.

Level 25–30 cm: Gray sandy sediment with volcanic ash.

Leveling was carried out at 25 cm.

All levels presented archaeological evidence, including lithic and bone materials.

In this case, since none of the bones had cut marks or were bone flakes, we selected the bones in the best state of preservation that were spatially associated with lithic artifacts at the deepest level, before reaching the bedrock. The expected age was 3000 years before present. This estimation was performed using the sedimentation rates of the northern chamber of Cueva Maripe as a reference. However, the obtained date is younger than expected (see Table 8.6), indicating that sedimentation rates at this surface site are higher than in cave environments. The prevailing sedimentary and anthropogenic accretion rate was calculated to be approximately 0.04046 cm/year; that is, one centimeter of sediment represents a temporal extension of about 24.71 years. The only radiocarbon dating available and the sedimentary layer thickness were used to carry out the calculation. In this case, it was necessary to correct

2

Fig. 8.7 (continued)

the depth measurements. Given that the different excavation units were defined by artificial levels to simplify measurements, to obtain the real sedimentary thickness it was necessary to adjust a difference in unevenness with the terrain (which had lower altitudes toward the northern sector where the sample was taken). The sedimentary thickness decreases by 5 cm toward that sector given the natural slope.

8.1.4 Radiocarbon Dating for the La Primavera Locality

The new dates for this location, along with the previously published dates, support the idea that this area of the Deseado Massif has a long history of occupation. Table 8.7 and Fig. 8.9 present all existing dates for the La Primavera locality. Based on the complete sequence, the major hiatuses in the chronological data are 2523 radiocarbon years (between datings No. 5 and 6 in the Middle Holocene) and 2032 radiocarbon years (between datings No. 9 and 10 in the Late Holocene). The first hiatus coincides with a regional trend proposed for the entire central Deseado Massif, where an archaeological silence exists between 7000 and 6000 years before present (BP)

1

Fig. 8.8 Archaeological sample used to date the La Quinta site. 1) The sample comes from the southeast sector of the excavation; 2) a guanaco cuneiform was dated. The sample exhibited a weathering stage between 1 and 2 (Behrensmeyer 1978). Dr. Laura Marchionni personal communication

2

Fig. 8.8 (continued)

Table 8.6 Dating of bone material from the La Quinta site

La Quinta site	GPS ID: 2217
Labcode	AA85460
Weight	4 g
Material	Guanaco cuneiform bone
d13 C	– 20.9*
F	0.8897 ± 0.0065
14C years BP	939 ± 59

(Hermo, 2008; Mosquera, 2016, 2018; Salemme & Miotti, 2008). Possible explanations for this gap in the archaeological evidence include environmental phenomena, such as:

(a) catastrophic events, such as eruptions of mountain volcanoes and tectonic events, which have been detected at many sites in the central Deseado Massif, including Piedra Museo, Los Toldos, La Martita, El Ceibo, Cueva de la Mesada, and Cerro Tres Tetas (Salemme and Miotti, 2008); (b) extreme environmental drying; (c) increased seasonality, which prevented winter occupations; (d) preservation problems; (e) lack of clear records due to very brief occupations, or (f) sampling problems, as caves were sampled more than open-air sites. Regarding the second hiatus

Table 8.7 Radiocarbon dates in the La Primavera locality (Miotti et al., 2007a); (*) unpublished

No. in the Fig. 8.9	Siti	14C years BP	Period	Lab code	Material
1	Cueva Maripe south chamber	9518 ± 64	Pl/H	AA65175	Charcoal
2	Cueva Maripe north chamber	8992 ± 65	EH	AA65179	Charcoal
3	Cueva Maripe north chamber	8762 ± 50	EH	AA65178	Charcoal
4	Cueva Maripe south chamber	8333 ± 63	EH	AA65174	Charcoal
5	Cueva Maripe south chamber	7703 ± 47	MH	AA65177	Charcoal
6	Cueva Maripe north chamber	5084 ± 49	MH	AA65173	Charcoal
7	Cueva Maripe south chamber	4113 ± 39	MH	AA65181	Charcoal
8	Cueva Mora	3678 ± 65*	LH	AA85461	Bone
9	Cueva Maripe north chamber	3210 ± 60	LH	LP1497	Charcoal
10	Cueva Maripe south chamber	1078 ± 40	LH	AA65176	Charcoal
11	La Quinta	939 ± 59*	LH	AA85460	Bone

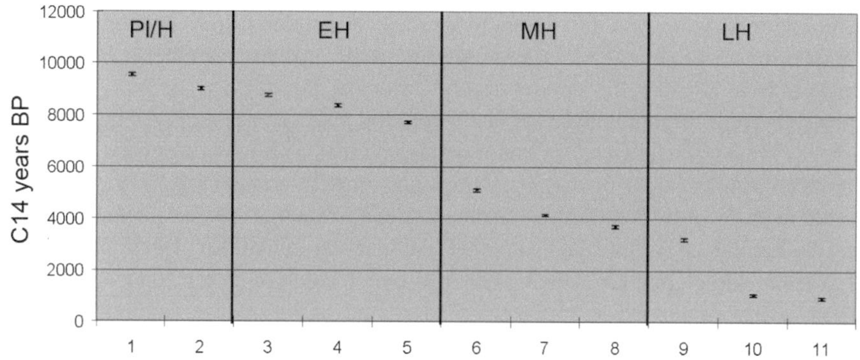

Fig. 8.9 Radiocarbon dates with their standard deviations (1 sigma)

in local dates corresponding to the Late Holocene, however, there is no correlation elsewhere in the Central Plateau. In Los Toldos (Miotti, 1989), La Martita (Aguerre, 1987), Cerro Tres Tetas (Paunero, 1994), Cueva Moreno, and La Hacienda (Miotti et al., 2005), dates were recorded within these ranges (see Chap. 3). However, the second hiatus in local dates corresponding to the Late Holocene does not correlate with the rest of the Central Plateau. In Los Toldos (Miotti, 1989), La Martita (Aguerre, 1987), Cerro Tres Tetas (Paunero, 1994), Cueva Moreno, and La Hacienda (Miotti et al., 2005), dates were recorded within these ranges (see Chap. 3).

Regarding the other sites dated in the La Primavera loality, they both correspond to the Late Holocene. The dating of Cueva Mora does not overlap with the dating of the other sites. It is chronologically located between two dates obtained at Cueva Maripe. Throughout this period, the caves have been continuously occupied since their initial settlement. The dating obtained at the open-air site is the most recent in the area. It is presented separately from the youngest dating obtained at Cueva Maripe by a 40-year 14C hiatus. Currently, based on these data, we cannot confirm the contemporaneity of the cave's occupation and that of this site. However, since the dating comes from the lower level of the La Quinta site sequence, it is possible to propose other, later open-air occupations.

8.1.5 Average Filling Rate (Cave Sites) and Sedimentary Accretion Rate (Open-Air Site) of the La Primavera Locality

Table 8.8 shows information from archaeological contexts dated, in the locality; Fig. 8.10 plots the infill (or sedimentary accretion) rates for the Cueva Maripe, Cueva Mora, and La Quinta sites. This allows us to see that the rates in the North Chamber of Cueva Maripe and Cueva Mora are very similar, suggesting that they formed in a similar way. Conversely, the South Chamber of Cueva Maripe exhibits different conditions, possibly due to reduced sediment deposition, greater sequence compaction, or erosion processes not present in the northern chamber a few meters away, at least not with the same intensity (Miotti et al., 2007b). The La Quinta site shows an opposite trend: a greater sedimentation package accumulated in a shorter time than in previous cases. It is important to note that infill rates calculated for caves and accretion rates for open-air sites are not directly comparable. The processes acting in both contexts differ, and their complexity requires specific geoarchaeological studies (Dr. Adriana Blasi, personal communication; Bútzer, 1982). The graph shows that some open-air sites, such as this one, located in protected environments where sediment deposition processes predominate and that have features allowing entrapment, such as stabilization surfaces due to nearby water sources promoting abundant vegetation growth, have a chance of presenting good archaeological resolution. In these cases, they do not form surface palimpsests of an accretionary nature (Binford, 1980; Lyman, 1994). A wide range of conditions that conform to the local dynamics will be studied in future between the sedimentary rate calculated for the La Quinta site and the sectors where there are no accumulated sediments (as has been observed on the rock outcrops and in high sectors of the basaltic plateau).

In addition to the survey of the La Quinta site, we have information from other surface assemblages. During the 2008 prospecting work, 75 one-square-meter excavation quadrats were carried out at sites with sediment coverage. Due to the limited time available for the fieldwalking survey, the soundings reached a depth of 5 cm. To recover small artifacts, all sediment was sifted using a 3 mm mesh. Of the total, only 9.33% (seven tests) yielded negative results, meaning they did not contain archaeological material. This indicates that at least 90.67% of cases with materials on the surface have buried artifacts in the subsurface. Mixing by trampling and other processes is possible in these contexts. However, many other sites are located in

Table 8.8 Radiocarbon dates used to generate the graph in Fig. 8.10

Sitie	14C years BP	Grid	Field label	Sediment thickness (cm)	Represented time span	Sediment rate (cm/year)	Represented time span/cm)
Cueva Maripe north chamber	8762 ± 50 3210 ± 60	C5 (2 by 2 m quadrat)	MA C5 se 83 MA C5 nw	54	5552	0.00972	103
Cueva Maripe south chamber	8333 ± 63 1078 ± 40	A12 (2 by 2 m quadrat)	MA A12 se 217 MA B12 sw 305	29	7255	0.00399	250
Cueva Mora	3678 ± 65 1937	1 by 1 quadrat	LP'08 D409	44	3665	0.00120	83
La Quinta	939 ± 59 2008	1 by 1 quadrat	LP'08 LQ 47	38	997	0.03811	26

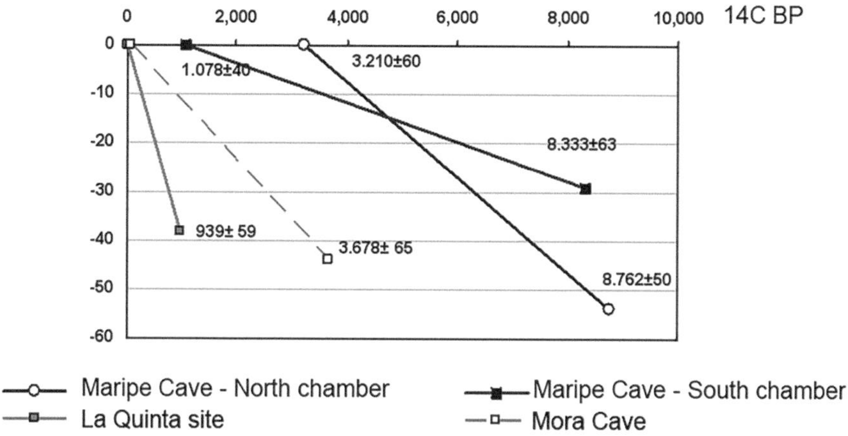

Fig. 8.10 Comparison of the thickness of sediment layers deposited per year in the dated sites at La Primavera Locality

environments where erosion processes predominate and where there is sometimes no sedimentary matrix (or where it has been lost due to the action of transport agents). In these cases, artifacts rest on rocky outcrops or erosion pavements.

From the observations, it is clear that the diversity of environmental conditions is key to studying the temporal dimension of surface assemblages. Although it is likely impossible to date the assemblages, knowing which sectors have similar environmental conditions allows us to begin comparing assemblages with similar information in the temporal dimension. Given the current state of research, we can distinguish assemblages that exhibit signs of being palimpsests from those that appear to have a coherent integrity. Currently, the most we can do to estimate a chronology of surface sites is to use the maximum possible date obtained for the La Quinta site (939 ± 59 years 14C BP) and analyze the techno-morphological indicators of the lithic artifacts collected, as well as other indicators.

8.1.6 Other Chronological Indicators in the Cueva Maripe Sequence

In addition to the radiocarbon date sequence, other chronological trends were studied in Cueva Maripe (Hermo, 2008). These trends were based on the proportions of obsidian and siliceous rocks among the raw materials found at different depths within the cave's stratigraphy, as well as the proportion of bifacial and unifacial knapping technologies used. In the northern chamber, two components from different chronologies could be separated. The upper component contains more obsidian (89% of the obsidian in the northern chamber is in the upper component). In the southern chamber, however, 91.4% of the obsidian is found in agglomerated levels between

150 and 180 cm deep. Despite the cave's complex stratigraphy, an analysis of the spatial relationships with the dated hearths revealed that these distributions indicate an increase in obsidian use during the Middle Holocene compared to previous occupations (Hermo, 2008). Considering the regional record, this finding aligns with trends in the use of obsidian, which is absent in the lower component of Piedra Museo (Cattáneo, 2002) and Level 11 of Los Toldos Cave 2 (Cardich & Paunero, 1992), but is present in later assemblages. An exception is the initial component of Cerro Tres Tetas (C3T1) (Paunero, 2000), which has five knapping debitage pieces of this material (Hermo, 2008).

8.2 Relative Chronology Indicators in Open Air-Sites

As shown in Table 8.9, obsidian is widely distributed in the surveyed sector of the surface sites. It was determined that 8% (60 out of 712) of the analyzed assemblages and isolated finds present obsidian. Furthermore, 3% of the total artifacts found on the surface were made from this material. However, the mixture of artifacts of different chronologies in surface sites makes it impossible to define the proportions of obsidian in relation to other materials, which prevents us from using it as a specific chronological indicator. The mere presence of obsidian does not indicate a particular chronology. However, we can estimate the latest date it was used, assuming it entered the region around 8000 years before present (BP) (Hermo & Magnin, 2012; Hermo, 2008).

Similarly, 32 of these assemblages contain evidence of lithic projectile points, bifaces, foliate pieces, and bifacial knapping flakes among the debitage pieces in assemblages lacking formatted artifacts. However, due to their possible accretionary origin, the proportions of bifacial pieces were not measured within the assemblages.

Table 8.9 Frequency of surface assemblages that present different lithic raw materials

Raw material	No. assemblages with raw material present	%	No. artifacts	%
Andesite	6	1	6	0.1
Basalt	16	2	37	1
Chalcedony	64	9	255	4
Gabro/Diabase	2	0	3	0.0
Obsidian	60	8	154	3
Other/Undetermined	9	1	84	1
Rhyolite	99	14	1190	20
Siliceous rock	379	53	4029	67
Petrified wood	77	11	280	5
Total	712	100	6038	100

At this point, it is important to introduce techno-typological indicators of relative chronology. These indicators are often the only way to determine the chronology of an isolated discovery or a surface lithic assemblage. However, the procedure is criticized for its association with an essentialist perspective. The use of these indicators to provide chronologies relating to surface assemblages has been widely debated. As Martínez (2003) points out, some authors, such as Orquera and Piana (1986), propose using typologies in a more neutral way, as a tool to make comparisons between different sets of artifacts. However, typological classifications are widely used for chronological purposes in contexts where other time indicators are unavailable (Martínez, 2003). Types (and morphological specimens, as defined by Aschero, 1975) derived from classifying and comparing artifact attributes are defined in relation to artifact assemblages in stratigraphy. Their presence in partial samples (soundings) or surface sites can be used to determine the distribution of types in research areas (Aschero, 1975:45). Without delving into the ontological status and meaning of the aforementioned types (there are numerous points of view in the theoretical and methodological discussion, see Aschero, 1975; Sackett, 1985; Dobres & Hoffman, 1994), we consider their use justified as a chronology estimator, supported by carefully recorded chronostratigraphic correlations. In this sense, we present the master stratigraphic sequence for the region, which has been used as an axis of comparison with the archaeological sequences derived from research in the Central Massif.

O. Menghin defined the first stratigraphic sequence of the region in Los Toldos in 1951 and 1952. Excavations there provided evidence of three occupations characterized by different types of artifacts. The chronology was established relatively, with the oldest occupation being called Industria Toldense, followed by Industria Casapederense, and then Tehuelchense (Menghin, 1952). In the 1970s, A. Cardich continued studying this and other caves in Cañadón de Los Toldos (Cardich et al., 1973). As a result, Cardich defined an earlier archaeological component, Industry Level 11, which predated those studied by Menghin. The application of radiocarbon techniques for the first time in Patagonia yielded a date of 12,600 ± 600 BP, stimulating debate about the antiquity of human occupation in the region. This finding supports the hypothesis proposed by J. Bird in the 1930s at the Fell and Palli Aike sites in Chile (Cardich, 1985). Level 11 Industry is associated with extinct fauna and lacks bifacial instruments. It is succeeded by two levels called Toldenses (levels 10 and 9), which are located chronologically between 10,600 years BP (the limit established by relative chronology) and 8750 years BP (the limit defined by radiocarbon dating). The first of these levels is the Typical Toldense (Cardich et al., 1973), characterized by thin, sub-triangular, non-stemmed bifacial projectile points. This level is earlier and includes projectile points, discoidal pieces, and extinct fauna. These indicators are absent in Toldense II, also known as the late Toldense (Cardich, 1987). Continuing the stratigraphic description, after a 1500-year occupational hiatus (level 8), the Casapedrense industry (levels 7 and 6) would appear. It is dated at 7260 ± 350 years BP and is characterized by the absence of projectile points, the predominance of leaves, and the presence of stone balls. Cardich correlated a layer of volcanic ash that overlies these deposits (level 5) with an eruption of the Hudson volcano dated to 4850 BP. Finally, the industries of levels 1, 2, and 3 follow one another. Cardich mentions that flakes, leaves, scrapers, side scrapers, engraved stone plaques, bifacial pieces, and points appear in these levels. However, the small projectile points and ceramics that would indicate very recent pre-Hispanic Patagonian archaeological contexts are absent (Cardich, 1987). Theoretical aspects of sequences like this have been overcome. Most notably, the use of artifactual types interpreted as industries from a normative perspective has been abandoned (see Borrero, 1995 for a discussion). As studies of the region advanced and new research emerged from other sites throughout the Deseado Central Massif, new lines of analysis were employed. This led to the proposal of the coexistence of knapping technologies that, in the master sequence of Los Toldos, exist in separate contexts. Cardich interpreted these as cultures carrying industries that can clearly be distinguished from each other. These cultures are the Level 11 industry, which is characterized by unifacial technology; the Toldense culture, which is characterized by bifacial triangular stone heads; the Casapedrense culture, which is characterized by the replacement of bifacial artifacts by blade and boleadora technology; and the return to bifaciality in Industries 1, 2, and 3, which are characterized by stemmed projectile points. Beyond its theoretical connotations, extrapolating it to the rest of the region formed the basis for subsequent interpretations. For instance, in the early levels of Piedra Museo (occupation levels 6 and 5/4, with dates between 13,000 and 9300 years BP), both technologies appeared simultaneously during the Pleistocene transition/Holocene. Miotti (1996) proposed that Level 11 of Los Toldos and El Ceibo, as well as the lower component of Piedra Museo, share the same "technological conception" as defined by Nami (1987). This idea was then reaffirmed in several subsequent works with different analyses (Cattaneo, 2005; Miotti & Cattaneo, 1997; Miotti, 2003), which led to the proposal that the coexistence of technologies, along with the variety of raw materials used, is the result of technological decisions related to expedient or curated strategies for solving particular problems at different times and places (Nelson, 1991; Torrence, 1983).

In Cueva Maripe, the trends identified by Hermo (2008) both support and contradict the regional sequence proposed by Cardich et al. (1973) and previously by Menghín (1952) for the Middle Holocene. While Hermo agrees that blade technology is predominant and identifies an important standardization in its production, he also found that boleadora balls are associated with bifacial production flakes. This finding does not coincide with the "Casapedrense" model. In the northwest sector of the province of Santa Cruz, the sequence defined by Gradín (1980) in the Pinturas River shows coexistence between Toldense and Casapedrense industries and a transitional step toward Patagoniense industries. Similar conclusions were reached in the Belgrano River–Lake Posadas Aschero area when analyzing components dated from 7000 BC (see Hermo 2008 for a detailed discussion of lithic technology, mainly with respect to Casapedrenses assemblages in sites throughout the Massif Central).

In the Deseado area, the dividing line between Level 11 and the Toldense tradition, as well as between the Toldense and Casapedrense traditions, is blurred based solely on the presence or absence of bifacial technology.

On the other hand, there are discrepancies regarding the expected replacements of the later industries based on Bird's (1938) and Menghín's (1952) master sequences for the most diagnostic pieces of the assemblages: the lithic projectile points. The triangular lithic projectile points with a wide stem (Fell IV) were expected to be replaced by the triangular ones with a narrow stem (Fell V) later in the stratigraphic sequence (after 4500 BP). However, as more radiocarbon dates were obtained, it was noticed that the two types appeared simultaneously in some archaeological sites (Borrero, 1989, 1994–95; Gómez Otero, 1986–87). Studies by Ratto (1994) in the Gallegos and Chico River basins show that the variety of point designs may be the result of different hunting strategies. Projectile points with wide stems would have been used to propel throwing spears with muscular force, while those with narrow stems would have been hafted arrowheads propelled by a bow. These strategies coexisted in the Gallegos and Chico river basins between 3600 and 740 years BP, according to dating of the sites. Finally, unstemmed lithic points would have constituted non-throwing hand weapons (Franco et al., 2009; Ratto, 1994).

Taking these considerations into account, investigations that include surface sites currently use these types as diagnostic indicators of relative chronology. This is always done in the context of broader evidence and with the support of stratigraphic information (Borrero & Nami, 1996). The only techno-typological difference among projectile points used as indicators of chronology is the transition from non-stemmed designs to stemmed designs. However, the chronology of this transition varies by area. For instance, triangular stemmed projectile points from Pali Aike, Tom Gould, Fell, and other sites (see Borrero, 1994–1995) characterize Bird's (1938) Period IV and are dated to over 4500 years BP (Borrero, 1994–1995). However, these types appear in Perito Moreno National Park in contexts dating to after 2500 years BP (Aschero et al., 1992–1993). These authors also argue that the surface sites Campo Río Roble 1 and 3 (CRR1 and CRR3) and isolated finds of triangular stemmed projectile points constitute the earliest manifestations in the area (Aschero et al., 2005).

Belardi et al. (2005) studied the distribution of projectile points on surface sites in the basins of Lakes Cardiel and Strobel. Based on stratigraphic information from a rock shelter (Manuk 1) and a dune site (GSLN 2), as well as regional information (e.g., the hunting blinds of Cerro Pampa and the open-air residential and funerary contexts of the Salitroso Lake basin), they associated the stemmed projectile points with radiocarbon dates no older than 2500 years BP. Goñi et al. (2007) consider the morphologies of the projectile points found on the Strobel Lake Plateau to be homologous to late designs, along with the presence of ceramics at the Don Edmundo site, to support late chronologies. Cassiodoro (2004) interprets the absence of non-stemmed projectile points throughout the Salitroso-Posadas Lake Basin as an indicator of late occupation. Additionally, based on excavated burials with radiocarbon dates, a technological change in projectile points occurred around 380 BP, when triangular points with small stems (< 1 cm) replaced points with stemmed triangular limbs or wider stems (> 1 cm). This suggests that the technology of throwing spears and/or hand weapons, used earlier, was replaced by the use of the bow and arrow (Aschero et al., 1992–1993; Ratto, 1991). Finally, Velásquez et al. (2007) assigned the APPELEG 1 open-air site in Alto Río Cisnes, Chile, a late chronology (between 1000 and 1500 years BP) based on diagnostic indicators within the assemblages. This was later confirmed by absolute dating.

About the triangular non-stemmed design of weapons, recent studies of regional archaeology assign them to the early and middle Holocene (Cassiodoro et al., 2020, Magnin et al., 2017).

8.2.1 Projectile Points Designs in La Primavera Surface Contexts

A comparison of the stratigraphic record and investigations of the Lago Salitroso-Posadas area reveals that the sector under study here has a reduced number of projectile points. These points are found in the South Chamber of Cueva Maripe (three fragments) and the North Chamber (one basal fragment). They are dated to the Middle Holocene (ca. 3500–7500 BP) (Hermo, 2008). One point is non-stemmed, and two have stems greater than 1 cm wide. Absolute chronologies to which different head designs can be assigned are under review. Regarding the surface record in the entire La Primavera Locality, of the 41 projectile points and fragments found in 30 different locations, three are triangular and non-stemmed, 21 have stems wider than 1 cm, and four have stems narrower than 8 mm. Non-stemmed projectile points were not found with other types of points but rather with debitage and NFA of various raw materials. Notably, a non-stemmed triangular point made of red rhyolite was found with its apex fractured, possibly by impact (Magnin et al., 2015). This point was found next to an obsidian blade core and an obsidian blade that fits onto it, as well as brown siliceous rock and white rhyolite debris. The stemmed points had no association with other types of heads in eight cases; were next to heads of the same type in two cases (they were projectile points with stems greater than 1 cm); and next to heads of different stem widths in two cases.

8.2.2 Chenques

There are other types of chronology-indicating evidence recorded in the sector, such as burials in chenques. The chenques in La Primavera Locality have not been dated. However, at a regional scale, chenques are considered indicators of late Holocene chronologies (see Mena & Reyes, 2001). In the Central Plateau, there is one radiocarbon date for a chenque. It is the El Sargento chenque, located in the Piedra Museo Locality, and dates to 727 ± 48 years BP (Miotti, 2006). To the west of this area, in Puerto Ibáñez, Chile, open-air burials in chenques were dated to between 360 and 570 years BP. In the northwest of Santa Cruz, chenque groupings received dates ranging from 350 to 1480 years BP. Burials in chenques at sites such as El Rodeo, located at the foot of walls or rock shelters, have been dated to between 1400 and 5000 years BP (Gradín & Aguerre, 1994) (but see Mena & Reyes, 2001). These

Fig. 8.11 Lithic projectile points designs. 1 and 2) Triangular non-stemmed; 3 and 4) With wide stems (> 1 cm); 5 and 6) with narrow stems (< 1 cm)

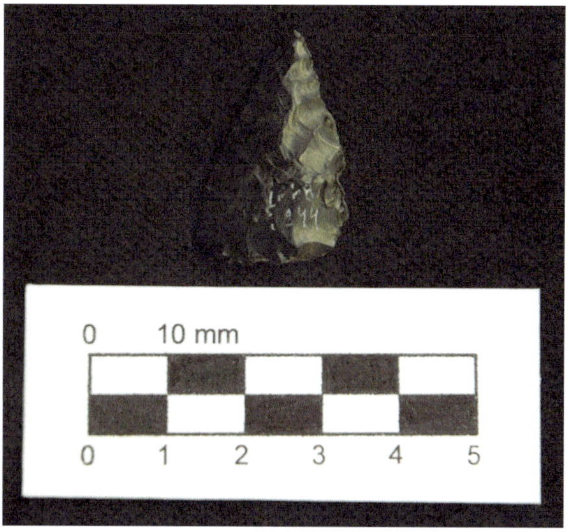

1

authors also differentiate between the construction of chenques in central Patagonia, where this practice emerged earlier than in southern Patagonia.

The chronology of this funerary practice differs from that of niche burials, which are dated between 2500 and 2600 years BP, and burials under boulders, which are dated between 2200 and 2500 years BP (García Guráieb, 2011; García Guráieb et al., 2015; Goñi, 2000–2002). The earliest chronologies recorded for burials in the region are for stone stacks similar to chenques found inside a cave at the Oreja de Burro 1 site in the Pali Aike region of Santa Cruz Province, Argentina, and at the Baño Nuevo-1 site in the XI Region of Chile. These were dated to 3500 and 9000 years BP, respectively (L'Heureux & Barberena, 2008; Zilio 2013).

Therefore, in the case of La Primavera, where there are no dates for the open-air chenque structures, we hypothesize that they probably belong to the late Holocene (Fig. 8.11).

8.2.3 *Stone Structures and Petroglyphs*

The presence of stone structures in the Patagonian region is also an indicator of late chronology. Among the described stone structures in Patagonia, those with a quadrangular layout correspond to residential structures from the Equestrian period of the Historic Tehuelches (see Gradín, 1971 for more information). The chronology of the hunting blinds on the Strobel Lake, Somuncurá, and Buenos Aires plateaus was first established based on the archaeological contexts recovered at Cerro Redondo on the Buenos Aires plateau (Gradín, 1976). In that archaeological site, 17 habitation structures were found. Gradín established their chronology based on the excavation of one

Fig. 8.11 (continued)

2

Fig. 8.11 (continued)

3

Fig. 8.11 (continued)

4

Fig. 8.11 (continued)

5

Fig. 8.11 (continued)

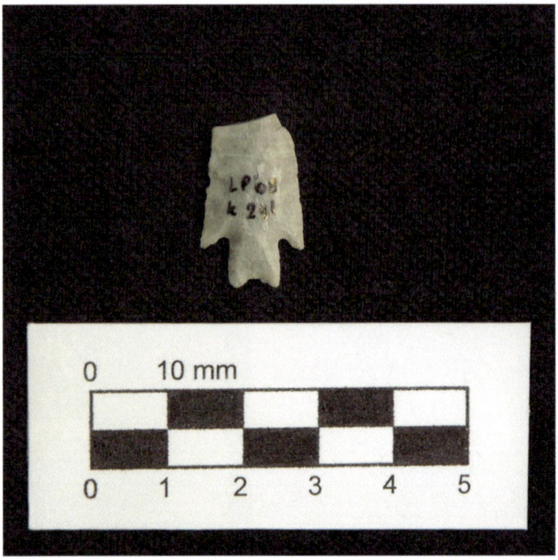

6

of the structures and related them to the Patagonian II or ceramic period (Menghín, 1952). Gradín emphasized the importance of lithic projectile points as chronological indicators in his work. The recovered projectile points were all triangular and stemmed; however, the points in the lower layer of the excavation (10–15 cm in excavation unit A) are the "type A" points. These projectile points are larger and have blades with slightly convex edges, barbs extended at right angles, and a long stem with a strong notch. "Type B" points are smaller with an elongated triangular blade, straight edges, and pronounced barbs at an acute angle. They have a short or very short, not very low-cut stem, and they appear in the upper stratigraphic levels, the latest of which has ceramics.

The habitation structures in the Lago Buenos Aires Plateau are associated with rock art manifestations characterized by engraving techniques (Gradín, 1976). The rock art motifs in the Buenos Aires Lake Plateau are engraved on basaltic walls surrounding some of the shallow lakes (sites Puesto Blanco Slick, Puesto Filín González, Laguna Honda, and Grabados de las Bardas Bajas) (Gradín, 1959–1960). Late chronologies are proposed for the Strobel Plateau (Belardi & Goñi, 2002, 2006; Gradín, 1959–60), the Pinturas area, and the Lake Buenos Aires Plateau. These chronologies are also proposed for the petroglyphs of the Central Santa Cruz Massif, based on regional archaeological information (Gradín, 1976). In the Strobel Plateau, Belardi and Goñi (2002, 2006) found that evidence of human populations primarily dates to after ca. 2500 years BP. Engraved rock art occurred alongside the presence of pottery and late Holocene projectile point designs. However, these authors propose an earlier chronology for the petroglyphs based on a date of approximately 6700 years BP obtained in the lower canyons and the presence of zoomorphic motifs that could correspond to the middle Holocene (Belardi & Goñi, 2002; Goñi et al., 2004). For a summary of the chronological arguments, see Carden (2008).

Most absolute chronologies of habitational stone parapets in the province of Santa Cruz originate from the mountain foothills in central western Santa Cruz. Radiocarbon dates obtained for repaired spaces, such as caves and shelters, in the Cardiel Lake sector (Goñi et al., 2004, 2014b), suggest human occupation during the Middle and Late Holocene. However, radiocarbon dates obtained for the stone parapets in Pampa del Asador, as well as on the Asador, Guitarra, and Strobel plateaus, indicate that these structures date back only to the Late Holocene (Espinosa & Goñi, 1999; Goñi, 2000–2002; Rindel et al., 2007; Re et al., 2017). Flores Coni et al. (2021) recently studied the 24 existing radiocarbon dates for rock structures in Pampa del Asador, as well as the Plateaus of Lakes Guitarra and Strobel. The structures' varied sizes and morphologies, as well as the associated archaeological evidence, suggest that they were used as hunting blinds for procuring prey and as residential camps. The analysis enabled the authors to conclude that the structures were used in the Late Holocene, beginning 2000 years BP, with the majority of dates falling between 1300 and 300 BP, and with periods of reoccupation of some structures (Flores Coni, 2019; Flores Coni et al., 2021). About the Deseado Massif, recent information shows that the technology of stone structures was used in the interior of this region some centuries before. The Parapeto Meseta de la Horqueta in the Bosques Petrificados de Jaramillo National Park is the earliest recorded stone structure (possibly a residential parapet) in Southern Patagonia, was dated to 2430 ± 30 years BP (Magnin et al., 2023). Other findings in that protected area documented in recent years are unique to the region. Sites in the park, such as Grabados de Seccional Cerro Horqueta (GCH) (Blanco & Magnin, 2023) and Cerro Horqueta 5 (CH005) (Magnin & Blanco, 2019), contain artificially arranged rocks that are part of stone structures located next to basaltic outcrops with petroglyphs. While no absolute chronology is currently known for these sites, the presence of both engraved rock art and stone parapets provides a relative indicator of Late Holocene chronology.

8.2.4 Ceramics

Ceramic records are scarce. In the province of Santa Cruz, the most important ceramic registry is in the Salitroso-Posadas Basin of Pueyrredón. A total of 1099 potsherds corresponding to 24 containers were recovered (Chaile et al., 2019). Based on microscopic analyses and compositional aspects of the pastes and clays of the ceramic sherds, the area has been proposed as a possible manufacturing site for ceramics (Cassiodoro & Tchilinguirian, 2007). In the rest of the province, only scattered sherds have been documented. In this case study, only two sherds were recorded in the upper levels of the south chamber of Cueva Maripe. Regarding the absolute chronology of this technology, the datings in Lake Salitroso range from 960 to 390 years BP (Cassiodoro, 2008). Between the Deseado Massif and the Santa Cruz River, residues adhering to the interior of sherds were dated to between 420 ± 20 and 140 ± 20 years BP (15th–twentieth century) (Cirigliano, 2019).

8.3 Conclusion

In summary, based on the information presented in this chapter, it is possible to partially assign chronologies to open air and surface assemblages in the La Primavera study area. The stratigraphic evidence from Cueva Maripe and Cueva Mora indicates that the analyzed occupations may span from the Early to the Late Holocene. However, at the open-air La Quinta site, absolute dating at 939 ± 59 BP and relative indicators show a strong human presence assignable to the late Holocene. Additionally, stone structures in La Primavera Locality are similar to those in neighboring areas dated after 1000 and 2000 years BP (chenques and hunting blinds/habitational parapets, respectively). The presence of rock art produced by engraving can also be interpreted as an indicator of late chronologies. Engraved rock art was registered in the Los Navarros site, located 7 km from the study area. Finally, based on the known chronological information for the province, the ceramic sherds found in Cueva Maripe may predate the Hispanic-indigenous contact; however, this indicator is scarce and not associated with open-air sites.

Chapter 9
Conclusions

Abstract This chapter synthesizes the interpretations and ideas that emerged from the archaeological study of La Primavera Canyon and its surroundings. The analysis revealed spatial patterns specific to each class of archaeological evidence. These different locational tendencies constitute new information interpreted based on natural and cultural factors, including human decisions during the area's hunter-gatherer occupation period.

Keywords Archaeological silences · Enculturated landscape · Mobility dinamics · Regional connections · Residential decisions · Sourcing strategies

This research applies an approach that considers all archaeological materials important and potentially informative, whether they are found on the surface (sites, concentrations, or isolated finds) or in stratigraphy (Borrero, 1995). Even the absence of evidence plays an important role in contrasting spatial patterns. Studying archaeological evidence across the landscape allowed us to detect contrasts in how its various sectors and resources were used (Belardi & Borrero, 1999; Borrero et al., 1992; Foley, 1981). Additionally, the archaeological record was not directly interpreted as a product of human use of space. Rather, the observed patterns are the result of natural and cultural processes, including the activities of human groups throughout a region's history of occupation, the degree to which spaces are reoccupied, and the differential preservation of evidence (Belardi, 1992; Butzer 1982; Dewar & McBride, 1992; Zvelebil et al., 1992). The ways the landscape was inhabited, perceived and signified is also considered when analyzing human spatiality (Ingold, 1993).

Before searching for patterns, we presented the theoretical tools that allow us to link archaeological and taphonomic indicators with inferences about the types of activities that could have occurred at different locations in the landscape (see Chap. 4). Formalizing the analytical procedure is a powerful tool for re-studying the same case and comparing it with other geographical sectors in future. These interpretations may be revised based on new data and different lines of study. We stored, managed, acquired, and analyzed the data through the application of Geographic Information Systems (GIS), which is explained step by step in the work. Therefore,

L. A. Magnin, *The Landscape and Mobility of Hunter-Gatherer Societies*, The Latin American Studies Book Series, https://doi.org/10.1007/978-3-032-08127-8_9

it is a valid methodology for application to other archaeological contexts. GIS is a fundamentally important mechanism for obtaining data and recognizing complex patterns. We agree with Kvamme (1988) that the success of any archaeological effort using this tool hinges on interpreting the patterns' meaning. In this regard, a key point in the search for patterns is selecting the relevant variables. In this study, environmental variables (natural and cultural) relevant to explaining location patterns were selected based on the theoretical framework (Chap. 4) and the restrictions imposed by the availability and adequacy of existing data (Chap. 5) (Magnin, 2013a). Next, we analyzed the distributions of the archaeological evidence concerning the locational variables, such as distance to water (a critical resource), distance to quarries supplying lithic raw materials, location in sheltered or exposed sectors, location in different topographies, amplitude of viewsheds from each location, and distance to main and secondary access paths. We understand that the information obtained from the locational analysis is more than just a map illustrating the fieldwork findings. The main patterns found are presented below and integrated with information from chronology, paleoenvironment, and regional archaeology.

9.1 Locational Patterning Al La Primavera Canyon and the Residential Sites

Multiple activity loci (MALs) are often found near water sources, indicating that they played a significant role in shaping residential occupation patterns in the past. In this landscape, the most reliable water sources are permanent springs fed by snow and precipitations. The water is captured by extensive basaltic plateaus that act as reservoirs, which then pour water into specific areas, forming springs that feed meadow vegetation and ecosystems known as "mallines," or wet meadows. The presence of MALs in close proximity to these sources highlights their importance for human occupation in this area. For example, the dating results from the La Quinta open-air site (939 $\pm$ 65 BP) suggest that the area was inhabited during the period of increased aridity known as the Medieval Climatic Anomaly. This is consistent with its proximity to a permanent spring. This data confirms the importance of these water sources for site location, especially during arid periods. Additionally, the continuity of archaeological materials at the same site affirms successive reoccupations of the area, associated with the resource's reliability and availability over the last ca. 1000 years. Some MALs are located close to temporary water sources. The placement of residential sites next to temporary creeks and shallow lakes must have coincided with periods of water availability. In periods of strong desiccation, such as the current one, water is only available during the rainy season in winter and early spring. This information is important for understanding the seasonal patterns of occupation in the area.

In the case of Alta Primavera and the Oasis sites, assemblages classified as MAL were located in high sectors of the plateau, far from permanent springs. Due to their topographic position, these locations indicate that occupation could have occurred during the warmer year season or climatic periods.

At the Oasis site (and similar sites, such as Laguna Cósmica), the arrangement of the archaeological evidence outside the limits of a temporary shallow lake indicates that occupation occurred when the shallow lake contained water. Two possible situations can be proposed: (1) the occupation occurred during periods of greater humidity, such as the *Optimum Climaticum* between 6000 and 7000 years BP or the Little Ice Age between 500 and 100 years BP (see Chap. 2), or (2) the occupation occurred during periods of aridity like the current one, specifically after rains or thaws, which suggests occasional and less prolonged occupations. These findings underscore the importance of conducting thorough geoarchaeological and paleoenvironmental studies in the area.

The Alta Primavera site is the exception to the locational pattern of MAL close to water sources in the study area. It is located far from them. It was probably used as a campsite, involving the transport of water from permanent springs located 200 m downslope. Based on the size of the site and the stone structures present, it is proposed that this area was occupied repeatedly over time. Besides the artifactual structure of the lithic assemblage, the variations in shape and size of two stone structures suggest they could be associated with residential use. Near this site, in a similar topographic position (in the sector called Alta Primavera Oeste), the other stone structures registered near the MAL were assigned a different function. Due to the associated indicators and by analogy with other case studies (Gradín, 1959–1960), they were classified as hunting stands (HS). Excavating the MAL and HS in the Alta Primavera and Alta Primavera Oeste sectors (Chap. 7) will allow us to characterize the use of this space and explain the patterns found, which, in some respects, differ from what was expected. The location of the Alta Primavera (MAL) site and the nearby HS aligns with the general pattern of hunting stands (HS) in the area, such as the La Mercedes parapets near the Oasis site and the Luz Belito parapet. These locations offer high visibility of the surroundings and are near access areas used as paths by humans and animals.

Another locational pattern of the MAL that emerged from this study is that they are located closer to lithic supply quarries than other types of evidence. This pattern was unexpected in a landscape where the distribution of raw materials for knapping was assumed to be homogeneous and abundant. It motivated an exploratory analysis of the distribution of raw materials and the artifactual structure at different scales. Analyzing the structure of lithic resources in the study area made it possible to identify trends in the location of MALs, analyze the origin and use of raw materials, determine sourcing strategies, and characterize how different sectors of the landscape were used in terms of technological organization. The raw materials in the La Primavera locality are abundant and varied. This is well-known throughout the Deseado region; however, the level of detail with which the fieldwork was carried out here allowed us to document the heterogeneity in the availability of raw materials. The main sources of supply are rocky outcrops and secondary deposits occupying the lowest landscape sectors. Raw materials are also available in minor deposits in dry stream beds and in isolated nodules scattered throughout plateau sectors. Regarding the MALs, their proximity to quarries seemed to suggest that this was a criterion for their placement. However, the analysis of the most commonly used raw materials did not always confirm this, as indicated by the comparison of the raw materials from the closest quarries to the MAL with those most commonly used by the MAL for manufacturing artifacts (Chap. 7). This suggests that the observed spatial pattern responds to factors other than reducing distances to quarries. The rock catchment analysis allowed us to study the natural topography of the landscape and how it affects the distribution of lithic resources. In a landscape with abundant raw materials, this analysis is useful for studying the natural distribution of rocks. Combined with other techniques, such as petrographic section analysis and characterization of smples with the naked eye, it provides information about sourcing strategies. The results for Basin 1 indicate that, in one case, the closest raw material to the MAL was preferred (at the El Descanso site, the most commonly used raw materials are brown silicified rocks from the F4 outcrop quarry). At the Pico Blanco site, however, diverse rocks were used, and they were mainly supplied from secondary deposits at the bottom of the basin (for example, petrified wood and a variety of siliceous rocks come from these deposits). Conversely, the lithic assemblage from the stratigraphic levels of the Cueva Maripe site shows that the most commonly used raw material, red rhyolite, comes from a quarry that is not the closest to the site (D9, or Del Rojo quarry). In this basin, the use of allochthonous materials, such as gabbro, diabase, and obsidian, stands out. Obsidian is sourced from Pampa del Asador, located more than 60 km away. Basin 2, on the other hand, presents a great diversity of raw materials, but the use of red rhyolite is predominant. In this basin, in one case -D9-, a campsite was set directly on a red rhyolite quarry. In the El Mirador site, the most used raw material was not sourced at the closest quarry. In this case, the strategy used was the transport of this lithic raw material to the campsite from quarries located at greater distance. The register of a lithic raw material reservoir or *cache* of cores in the water divide sector between basins 1 and 2 responds to the heterogeneity in the natural distribution of raw materials in the locality, and constitutes a strategy to resolve it. This strategy of provisioning places consists in equipping a place that one probably expects to visit again (Kuhn, 1994). The stone structure was constructed in a highly visible point of the landscape, suggesting an intentionality to make this visible, and not hide it from others.

The presented data supports the idea that past populations had very detailed knowledge of the distribution of raw materials in the landscape. This allows us to study how information related to lithic resources circulated. The study of the sourcing of raw materials confirmed that ancient inhabitants of the locality used a sourcing strategy to access raw materials located more than 5.6 km from campsites, such as red rhyolite and gabbros. They also disposed of high-quality raw materials for knapping in *caches* located in sectors of the landscape where these materials are not naturally available. In the case of obsidian, these strategies involved long-distance mobility or social exchange networks.

9.2 Extra Local Materials: Obsidian and Malacofauna

Considering the information obtained in the macro study area (other localities of the Deseado Massif), the presence of allochthonous elements (obsidian, bivalves, and marine snails) suggests the consolidation of broad social networks. These networks allowed access through exchange or high mobility, enabling groups in the La Primavera locality to supply more than 80 km away in the pre-Andean cordillera sectors (Miotti, 1989, 1995; Miotti & Salemme, 2004; Stern, 1999). The widespread use of obsidian in the study sector is evidenced by its distribution across the locality's surface sites, rather than being concentrated in a few places. This is consistent with the idea that this raw material entered the Central Deseado Massif in the Early Holocene, and its use was well-established with widespread distribution in the Middle Holocene (Hermo, 2008).

9.3 Questions About Year-Round and Seasonal Occupations

Both MAL and temporary camps (TC) are indicators of residential activity and key to understanding the general use of the La Primavera locality. It is probably not possible to distinguish transient occupations (TC) from longer-term MALs, as both have likely contributed discarded, deposited, or lost artifacts to these same places. In fact, the TC that are recognizable on the surface as separate classes of evidence are those that do not coincide with the MAL in their locational patterns. The main difference in the placement of these two types of evidence is that TCs are located in a greater proportion than MALs in sectors of the landscape at a higher altitude and exposed to low winter temperatures. This location tendency raises questions about seasonal use of space. For example, this trend may indicate that the temporary residential settlements were occupied in the spring or summer. Furthermore, based on previous observations of water availability in high-altitude landscapes, the most probable season for temporary campsites is early spring, when temporary water sources are available in shallow lakes. In contrast, the MAL tend to be located in lower, more sheltered sectors. Further insolation studies could help characterize the occupation of the lower, most repaired sectors. Were the residential campsites (MAL) located in sunny areas of these low, sheltered terrains, suggesting winter use? If the summer and winter residential use of the study area is confirmed, as represented by the TC and the MAL, respectively, then the hypothesis of year-round use of the La Primavera locality could be proposed. This does not imply a low mobility since the locality could have been used occasionally as part of longer mobility cycles. At the regional level, the annual range of action (sensu Borrero & Nami, 1996; Foley, 1981) of the groups that occupied the sector could be expanded to include other areas. This idea is supported by the additional local materials mentioned previously.

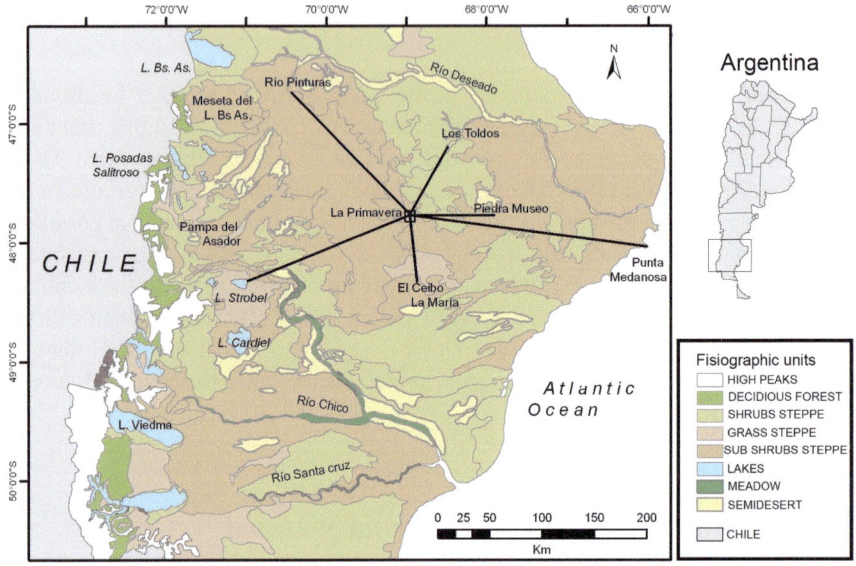

Fig. 9.1 Santa Cruz province map, with the location of the study area and other sectors of the regional space that could be included in long-distance movements

The environments that complemented these movements could be other sectors of the Central Plateau or more distant areas (Fig. 9.1).

9.4 Regional Mobility

Some of the possible locations related to long-distance movements in the Central Plateau include Los Toldos, which is located 71 linear kilometers north-northeast of Cueva Maripe (centrally located in the study area), as well as Piedra Museo and Bosques Petrificados de Jaramillo National Park, which are located 82 km east. Other possible locations include the localities of El Ceibo, Cerro Tres Tetas, and La María, which are located 68 km southeast (Carden, 2009; Hermo, 2008; Miotti, 1989). Groups of people could have moved to these localities by following the course of streams or touring different environments. In Los Toldos, the age-at-death structure of the faunal assemblages associated to the earliest occupations (between ca. 10,600 and 8750 BP) of Cave 3 (Miotti, 1989) was used to infer year/round occupations. Therefore, Los Toldos is similar to La Primavera in that both have large caves and rock shelters that provide reliable refuge and concentrated resources in canyons. The main difference is that Los Toldos is located at a lower latitude and altitude above sea level. In La Primavera, the heights range from 400 to 800 m above sea level (from the mouth to the upper sectors of the canyon), while in Los Toldos, they range from

450 to 500 m. The ecological implications of these differences in terms of human habitability are the subject of current research.

Two basins with significant bodies of water west of the Central Plateau and toward the foothills have been considered attractors of human populations in the late Holocene, specifically during the Medieval Climate Anomaly (ca. 900 BP) (Belardi & Goñi, 2006; Cassiodoro, 2011; Goñi, 2000–2002; Morales et al., 2009). These basins are Lake Posadas-Salitroso, located 200 km west, and Lake Strobel, located 180 km southwest. According to the regional model for this area, the range of distances for population attraction exerted by these basins is between 60 km (between the Pinturas River area and Lake Posadas-Salitroso) and 150 km (between Lake Argentino and Lake Cardiel). Could they also have functioned as attractors for the populations that inhabited the area under study here? As previously mentioned, there is evidence of contact with the Andean sector from the middle Holocene due to the continuous use of obsidian from Pampa del Asador and Cerro Bayo (Belardi et al., 2006; Hermo, 2008; Miotti, 1989; Miotti & Salemme, 2004). Social contacts between the Strobel Plateau and the Central Plateau were also proposed through another means of analysis: the similarity of the rock art motifs of both sectors (Carden, 2008). In this regard, the study of rock engravings recently found very close to the north of the La Primavera locality (Los Navarros) (Chap. 7) (Blanco, 2015), in addition to those studied on a regional scale (e.g., Carden, 2009; Guichón & Re, 2014), will shed light on regional movements.

The presence of stone structures is another piece of evidence that suggests connections between the Andean region and La Primavera. These structures are similar to those found at Cardiel Lake, Strobel, the Águila Plateau, and Cerro Pampa (see Chap. 6) (Flores Coni, 2019). This reinforces the idea of high mobility or information exchange with these groups in the foothills sector. Stone structures are also present in other sectors of the Santa Cruz Central Massif interior, such as the Bosques Petrificados de Jaramillo National Park (near Piedra Museo) and Los Toldos (Hermo et al., 2021; Magnin et al., 2018).

Another attractive area for human groups, either for seasonal occupation or longer-term movements, is the Atlantic coast. It is currently located 200 km east of the studied sector (Salemme & Miotti, 2008). At the Cave 2 site, Los Toldos Locality, necklace beads made from marine mollusks were found in layers assigned to the Middle Holocene (Cardich & Paunero, 1991–1992). These beads suggest contact with the coastal sector. As in the foothills, contact with the coast could have occurred through the movement of human groups or exchanges between groups (Miotti, 1989, 2006).

Regardless of the mobility scheme or its changes over time, the extent of the area analyzed in this study can be covered in a day's walk by people with a hunter-gatherer lifestyle (10 by 10 km). Nearby localities within 50 km could be linked by path networks for long-distance logistical hunting and gathering. These expeditions perhaps involved the entire population (Albert & Le Torneau, 2007; Binford, 1982). About the annual ranges, they may encompass areas of approximately 22,500 km^2. Thus, annual movements could develop within the Central Plateau, which has a similar areal extent, and could also include movements between the Central Plateau and the coast or the pre-mountain range. Similar annual ranges are observed in current studies of hunter-gatherers. For example, the Nunamiut, the G/wi, and the Australian desert aborigines can cover 25,000 km^2 in an annual cycle (Binford, 1983). Archaeological studies of hunter-gatherers in the Pampas region propose similar ranges (ca. 22,000 km^2; Martínez, 1999). However, mobility ranges are neither fixed nor easy to define, as they are expected to vary according to ecological, social, and symbolic conditions (Politis, 2006).

Topographic modeling of access routes to and from La Primavera at a regional scale using GIS cost-distance techniques showed the theoretical mobility routes used for seasonal or longer-term transfers to the Andean foothills. Starting in La Primavera, these routes head south to the Massif limit, then west to the headwaters of the Chico River and the basins of Lakes Cardiel and Strobel (Magnin, 2013b). Further analysis of large-scale movement is needed, including modeling snow cover during the winter season (for this methodology see Pallo, 2012).

9.5 Chronology

The chronological information for the dated open-air site La Quinta (939 ± 59 years BP) is relevant to the models of spatial occupation and current regional mobility. These models propose that human occupations in Patagonia underwent significant changes during the last 1000 years of the Holocene. This period was marked by demographic concentration in areas around the pre-cordilleran lakes, including Cardiel and Salitroso, as well as population relocation in Argentine Lake. In the Central Massif, human activity has been documented in both open-air sites and caves. The most recent open-air dating was obtained at the El Sargento chenque, which was dated to 727 ± 48 years BP (Miotti, 2006). In a cave context, the youngest dates are 1078 ± 40 years BP in Cueva Maripe (Miotti et al., 2007) and 830 ± 60 years BP in Cerro Tres Tetas (Paunero, 2003). Due to the limited available data, it remains challenging to assess the intensity of human occupation in the region and evaluate hypotheses regarding the heterogeneous distribution of populations in the steppe during the Medieval Climatic Anomaly (MCA) (Morales et al., 2009).

About Cueva Maripe, the chronological information indicates prolonged occupation from the Pleistocene/Early Holocene transition through the Late Holocene. Since the Middle Holocene, the occupation has been continuous (Hermo, 2008; Miotti et al., 2007). Based on the current state of research, it is not possible to confirm or refute the occurrence of human withdrawal events in the area. However, synthesis studies of the Deseado Massif indicate periods of archaeological silence spanning 1394 years in the Middle Holocene and 295 years in the Late Holocene (Mosquera, 2016, 2018). It is likely that hunter-gatherer populations adapted to environmental changes by altering their locational patterns and mobility.

The most represented occupations, as evidenced by surface archaeology, belong to the Late Holocene. This is confirmed by the radiocarbon dating results from the open-air site La Quinta and is consistent with other indicators of a later chronology, such as stone structures, rock engravings, and burials in chenques. However, given the site formation processes that contribute to the sector's high archaeological visibility (see Chap. 2) and the presence of techno-morphological indicators in the assemblages (see Chap. 8), we propose that the surface evidence, which was formed throughout the sector's history of occupation, also encompasses earlier periods. For example, the non-stemmed heads can be assigned to chronologies predating ca. 3500 BP in this sector (Hermo, 2008) and have been associated with dates of 8333 ± 63 years BP in Cueva Maripe. In other areas of the Central Plateau, this same design of projectile points has been associated with dates from around 7700 years BP in Piedra Museo (Cattáneo, 2002) and chronologies between 8750 and 10,600 years BP in Los Toldos (Cardich et al., 1973; Menghin, 1952).

9.6 The Location and Role of Rock Art

The rock art in the studied area is found in places where daily activities are carried out. It is located in residential areas and hunting grounds, and its motifs highlight this functional variability (Carden, 2008). Our study comparing caves with and without art provided insight into past decisions. Caves close to natural ochre sources and the network of paths between residential base camps (MAL) were chosen for art creation. Rock art is not located directly in places where resources are found, such as lithic quarries. Rather, it is found in caves and rock shelters where campsites (MAL) were established or near the roads that connect them. Residential campsites are places that structure human movement and where resources are brought. The network of paths between them is spatially associated with evidence of temporary camps (TC) and places of limited activity (LAL). This suggests that rock art coincides with the most frequently used spaces and is associated with residential activities. Considering its potential function of signaling the presence of a resource (Carden, 2004; Miotti et al., 1999), the main resource it would point out is the shelter provided by the caves. These sites also have a concentration of other resources as a result of provisioning activities, as well as ample viewsheds.

9.7 The Chenques in Regional Archaeology

Our studies showed that the chenque sites (Ch) are dispersed throughout the area and located in elevated areas of the landscape with wide, panoramic views. GIS analysis revealed that the viewsheds obtained from the chenques are significantly wider than those obtained from other types of evidence and control points. These results confirm expectations based on ethnographic records, chronicles, and other archaeological studies of burials. Additionally, a pattern of reciprocal visibility between the chenques was verified (Chap. 7). We propose that the location of these structures was defined by taking into account the existence of previous structures. This suggests the existence of a symbolic territorial claim to the space through the arrangement of monuments or mortuary structures in sectors of wide visual dominance, as well as a use of the landscape that involves a certain temporality (Tilley, 1994). Assuming these chenques are human burials, their arrangement likely refers to generations repeating the practice of burying the deceased in specific locations chosen based on the visibility of their ancestors' burials. Their location in sectors far from critical resources (water, shelter, quarries, and mobility paths) distinguishes them from other recorded evidence belonging to domestic environments. However, this presents an exception since hunting stations (HS) share a common pattern of being located far from resources and having wide visual basins. In this case, based on the visibility analysis, we have proposed that the HS are part of a spatially structured hunting landscape. They are the places where the hunt begins and are related to the hunting grounds, where the hunting ends. The main difference in the location of CH and HS is how each class relates to access paths. While HS are located near the main access path network between MAL, HS are far from any access paths. This suggests a clear separation between domestic and sacred spaces in the locality (Miotti, 2006).

Given their significance in regional archaeology, as discussed in Chap. 8, chenque structures may indicate late chronologies (after 1100 BP, according to Goñi et al., 2004). In the studied area, the fourteen chenque-type structures are isolated and arranged in small groups of two to four structures. In other study areas, chenques with funerary functions have also been found in small, isolated groups. Examples include the Jenimei Valley, near Lake General Carrera/Buenos Aires (Mena & Lucero, 2004), and Pampa del Asador and Perito Moreno National Park (Goñi et al., 2004), where single chenques have been documented. Conversely, larger groups of chenques interpreted as formal burial spaces (cemeteries) have been documented in Lago Salitroso, Puerto Ibáñez, and along the Atlantic coast.

Based on the archaeological and climatic records of the past 1000 years in northwestern Santa Cruz, Goñi and Barrientos (2000) proposed a model of reduced mobility and spatial use. Assuming that people are buried where they die, the presence of cemeteries was considered an indicator of a more sedentary lifestyle and greater population density (Goñi et al., 2004). The presence of cemeteries in low-lying areas such as the Salitroso and Cardiel Lake basins during the Medieval Climate Anomaly (MCA) suggests that these spaces attracted populations due to their concentration of critical resources such as water, firewood, and shelter. These basins would have served residential and logistical functions, similar to the high plateaus in Argentinean Patagonia (Meseta del Lago Buenos Aires, Pampa del Asador, and Meseta del Strobel), as well as the ecotonal and forested environments in the upper regions of Chilean valleys, such as the upper Chacabuco (Belardi et al., 2005; Gradín, 1959–1960; Goñi, 2000; Goñi & Barrientos, 2000; Goñi et al., 2004; Mena & Lucero, 2004). Based on this model, it is worth asking what interpretation can be made of the late Holocene registration in the La Primavera sector.

In terms of the space's functionality, a dispersed pattern of burials like the one recorded would indicate a logistically occupied space. However, in the case of the La Primavera locality, there are no nearby residential areas, which, according to the above-presented model, should be indicated by the concentration of chenques in formal cemeteries. These indicators are located 150–200 km away from La Primavera in a straight line, one in the Cordillera and the other on the coast.

Regarding environmental restrictions, and considering the climate data available, around 1000 BP this sector could have been unoccupied and populations concentrated in the pre-cordilleran sectors. However, evidence from caves and open-air sites in the La Primavera canyon dated to around 1000 AD shows that the area was habitated. Still, this evidence is insufficient to evaluate population density. The chronological information presented for La Quinta is consistent with the use of other sectors of the Central Massif during the same period. This is indicated by De la Ostia, a site in the BPJNP dated to 1096 $\pm$ 25 (Magnin et al., 2023), as well as the chenque in the Piedra Museo Locality, dated to 727 $\pm$ 48 years BP (Miotti, 2006). Thus, two scenarios can be proposed for the occupation of this sector in the last 1000 years BP: (a) This space was only used for short, transitory occupations during long-distance movements of passage toward sectors with strong residential occupation in times of great drought; (b) The resources provided by the sector (permanent springs, shelter in ravines and caves, lithic resources, and fauna) were not drastically reduced during the MCA. These resources were sufficient to allow the population to continue their usual mobility circuits and to use this locality for more prolonged stays.

Considering that chenques may have a funerary function and analyzing their distribution as an indicator of residential use when they are in great numbers or logistical use when they are isolated (Goñi, 2000; Goñi & Barrientos, 2000), makes it difficult to interpret the evidence found in this study area. Although the chenques in the La Primavera study area are dispersed, their location pattern, which offers wide panoramic views that include other chenques, suggests a structured funerary space that was constructed and repeated through generations.

9.8 Final Words

The evidence presented in this study allows us to describe the landscape of the La Primavera Locality as non-homogeneous with respect to the available natural resources and the recognized patterns of use by hunter-gatherer societies. These observant and knowledgeable inhabitants surely knew the landscape and its natural cycles very well, including the diversity in the distribution of water, rocks for knapping, sheltered areas, firewood, animals, and sectors with special views. These resources were searched for, distributed, administered, and shared.

The presence of goods from long distances since the Early Holocene and more frequently since the Middle Holocene allows us to conclude that large-scale movements toward the foothill regions or exchanges with people from those areas were established mechanisms very early on (Miotti, 1989; Miotti & Salemme, 2004). This is surprising given that, even though there are provincial routes and local roads, the area is currently still considered "the hills far from everything" or "the place where there is nothing" (Miotti & Carden, 2006). The ancient people who inhabited this landscape were highly mobile. Nevertheless, they left numerous marks suggesting that the places were known and revisited. The rock art and stone structures preserved in today's landscape evoke a strong sense of awe when recognized. They indicate a modified, enculturated landscape (Zvelebil, 2003). One surprising pattern indicative of this is the arrangement of chenques, which are probably mortuary monuments, in a pattern of visual interconnection that denotes a cyclical return to the landscape. The same people, or others with the same beliefs and customs of burying their dead or monumentalizing their landscape, must have carried out these practices generation after generation. In each repetition, the structures previously placed were antecedents for the funerary practice. Taking an ancestor into account when burying the deceased affirmed the unity of their present about the past. Through this act, they asserted their connection to and authority over the landscape. All of this leads us to believe that this was not a marginal area of hunter-gatherer occupation. Although the intensity of occupation probably fluctuated over time, evidence of human presence is visible throughout the landscape.

Chapter 10
Recent Advances and Perspectives

Abstract This chapter restudies the locational patterns at the La Primavera site. A GIS analysis was performed using free, open-source software. New explanatory variables were included, and an advanced statistical modeling technique, generalized linear modeling (GLM), was applied using the statistical programming tool R. The results revealed that the indigenous people's activities were more spatially stable when they manufactured lithic artifacts in workshops, painted art in caves and rock shelters, constructed chenques for burying the dead and/or as monuments, and stalked prey. Residential activities (base camps and temporary camps) were comparatively less stable. These patterns demonstrate a recurrence of spatial practices throughout the hunter-gatherer dwelling process, indicating its temporal depth. This spatial structuring in the La Primavera locality supports the idea of a consolidated occupation of the locality by high mobile societies. The locational patterns explored in Chaps. 6 and 7 at the La Primavera locality are re-examined in light of new questions arising from the exploratory analysis. This further work contrasts the previous results with more sophisticated multivariate statistics. Previously, the behavior of each archaeological class was examined in relation to the locational variables. Here, we analyze each functional class to construct multiple locational models and determine the best model for each class. We use the previous exploration to determine which variables are relevant to include in the modeling process. We agree with Gillings et al. (2020: 13) that "the application of quantified, formal methods can be used in an exploratory way—not least for developing more nuanced understandings of space and spatiality—and that any results produced are not 'the answer' but rather the starting point for a process of interpreting and understanding the past." Along these lines, we believe that this restudy is important for gaining greater confidence in the results obtained. This new application of other analytical methods, including different variables, scrutinizing the data, and adding controls for specific formation processes allows to a better understanding of locational patterns and how they respond to specific spatially and temporally situated modes of living.

Keywords Archaeological silences · Enculturated landscape · Mobility dinamics · Regional connections · Residential decisions · Sourcing strategies

L. A. Magnin, *The Landscape and Mobility of Hunter-Gatherer Societies*, The Latin American Studies Book Series, https://doi.org/10.1007/978-3-032-08127-8_10

10.1 GIS and Statistical Analysis

Locational modeling is performed in a raster-based Geographic Information Systems (GIS) environment using GRASS GIS v. 7.8.3 (GRASS Development Team, 2016) and the statistical software R (R Core Team, 2016). The goal of this modeling is to identify environmental and cultural variables that can potentially distinguish between locations of different types of archaeological evidence and control points (negative evidence, NE) (Fig. 10.1). The binomial generalized linear modeling technique is used to analyze the presence and absence of archaeological evidence (the different classes defined in Chap. 6). Numerous methodological antecedents of locational analysis exist in other regions of the world (e.g., Connolly & Lake, 2006; Warren, 1990; Warren & Asch, 2000; Wescott & Brandon, 2000), and the approach is current in world archaeology. However, it is a recent approach for Santa Cruz and Tierra del Fuego (Magnin, 2010, 2016; Negre et al., 2018).

10.1.1 Archaeological Evidences Classification: The Response Variables

The distribution of control points (negative evidence) is presented in Table 10.1. The analysis proposed here is performed on the same set of archaeological evidence

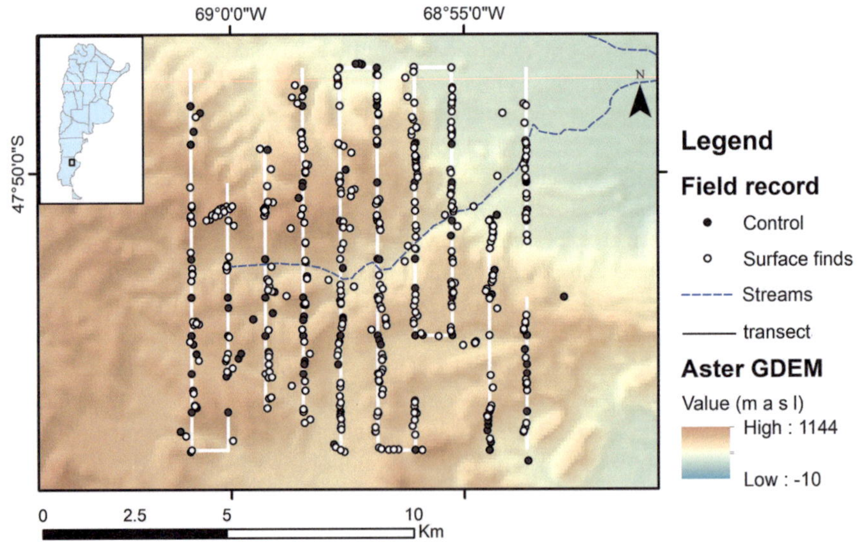

Fig. 10.1 Distribution of control points (negative evidence) and the 411 archaeological registers in the La Primavera locality (the different classes are not symbolized on this map). This figure's base map is an SRTM V3.0 DEM with 60% transparent ground shading coverage

Table 10.1 Archaeological evidence from the 2008 fieldwork surface survey at La Primavera was assigned to ten functional or activity classes

Class of archaeological evidence	*n*
A (rock art)	*9*
Ch (stone structures of "*chenque*" type)	*14*
HG (hunting ground)	*10*
HS (hunting stand with/without associated hunting blind)	*16*
TC (transitory camp)	*24*
LAL (limited activity locus)	*134*
LAM (multiple activity locus, residential base camp)	*18*
LW (lithic workshop)	*5*
OK (occasional knapping)	*171*
Q (lithic quarry)	*10*
NE (record of negative evidence)	*110*

Nine of these classes are the response variables studied here. Class NE is the control sample

analyzed in the original work (Magnin, 2010) and in this book previous chapters. The classes of evidence are presented in Table 10.1. Each "class" refers to archaeological evidence that can be assigned to general types of use. This classification is based on ethnographic (Martinić, 1995), ethnoarchaeological (Binford, 1982), and archaeological (Borrero, 1994–95; Miotti, 1998; Miotti & Salemme, 1999; Salemme & Miotti, 2008) information. Although the classes are not functionally specific, they distinguish inter-assemblage variability and locational tendencies in this landscape. The response variables are rock art (A), chenques (Ch), multiple activity loci (MAL), quarries (Q), lithic workshops (LW), transitory camps (TC), limited activity loci (LAL), hunting grounds (HG), and occasional knapping (OK). Each class is assumed to respond to different location logics; therefore, they are analyzed separately. There are 411 archaeological registers in the La Primavera locality (the different classes are not symbolized in this map).

10.1.2 The Explanatory Variables

As stated in the previous study, the variables included in the modeling process allow us to study formation processes and characterize location decisions. Past decisions were influenced by the characteristics of the natural environment as well as social and symbolic factors (Warren, 1990). Our models must include information about past environments and resources to assess human–land relations and social aspects to assess human–human relations (Roper, 1979). At the same time, the modeling considers today's environmental processes, such as the detectability and preservation of archaeological evidence on the ground surface (Gómez & Magnin, 2009;

Marchionni et al., 2020). Once relations between classes of response and explanatory variables are detected, a focus on the different dimensions of the data and their inherent limitations is needed for a complex approach to inhabited places (Ingold, 1993). Table 10.2 presents metadata important for interpreting the results and for future studies. The modeling process includes twenty-three explanatory variables (Magnin, 2023). Their inclusion in the present study responds to the general bibliography of hunter-gatherers, especially in Patagonia (Chap. 5). Their representation in GIS considers the possibilities and limitations throughout the modeling process and when interpreting the results (Magnin, 2013a). The selection of variables also considers the results obtained in previous work (Chap. 7). Some describe the natural characteristics of the analyzed location's landscape, while others approach the relationship between classes of archaeological evidence and landscape characteristics involved in mobility and visibility. These characteristics point to possible relations between patterning in archaeological evidence and past landscape use. Many of the geoprocessing methods used to model the explanatory variables are derived from the Shuttle Radar Topography Mission (SRTM) Digital Elevation Model V3.0, which has a resolution of 1 arc second and covers nearly the entire globe. Released in 2015, these data showed fewer anomalies and artifacts in the study area than the previously used ASTER GDEM V1. The geoprocessing details for each GIS layer are presented in Table 10.2. The variables "altitude," "slope," "aspect," and "geomorphology" characterize the terrain surface. In previous GIS analyses, we used a coverage that described the wind-exposed and sheltered sectors of the landscape. This coverage combined different aspect values in relation to the wind direction (Chap. 6). For this study, we included the aspect data directly to analyze the variability of the archaeological evidence locations. The variable "shading" explores the landscape's seasonal use by examining the different insolation conditions in winter and summer. Other variables included in the statistical modeling refer to the visual aspects of sites and their surroundings. "Visibility from each location" (within a 10 km radius) is not a unique GIS coverage, but rather the result of a viewshed analysis performed on each location. The previous analysis used a subset of data (viewsheds were analyzed for chenques and rock art sites, Chap. 7); this analysis includes the total set of archaeological locations (Table 10.2). Additionally, we included a map of "general visibility" of the terrain, calculated from accumulated viewsheds, to analyze the most visible sectors of the landscape and evaluate whether the "visibility of" certain places was a variable considered in locational decisions.

Evidence about natural resource distribution was described by the variables "distance to water" and "distance to rock supply sources." The modeling process also considered the distances to other classes of archaeological evidence and to the modeled path network between residential campsites (multiple activity loci, MAL). These variables explore the locational importance of other evidence classes that provide information about other uses of the landscape.

Finally, to study the landscape's general accessibility, an MADO analysis was performed using starting and ending points around the study area (Fábrega & Parcero, 2007; Llobera et al. 2011). The distance to primary and secondary paths between

Table 10.2 Metadata for the raster information layers that model the explanatory variables used in the GIS environment

Variable group	Name of raster map for explanatory variables (EVs)	Unit of measurement	Minimum value	Maximum value	GRASS GIS modules and data used to generate the raster layer
Characterization of the ground surface	1. Altitude (numeric)	Meters above sea level (masl)	132	1127	SRTM DEM v.3.0 (~ 30 m spatial resolution) (1)
	2. Slope (angular)	Terrain slope (°)	0	53	Data derived from map 1 (r.slope.aspect)
	3. Aspect (angular)	Terrain angle respect to North (°)	0	360	Data derived from map 1 (r.slope.aspect)
	4. Geomorphology (categorical)	Geomorphological unit	1—Basaltic plateau 2—Alluvial plain 3—Mass-removal area (slopes) 4—Serrania 5—Terrace: aggradation plain		Geomorphology map 1:50,000 (Gómez & Magnin, 2009)
	5 and **6.** Shading of the land in winter and summer	Shading in summer Shading in winter (high values indicate that the area receives more solar radiation)	134 107	254 254	Data derived from map 1 (r.relief) Map 5: Day: 10/2/2008, 12:00 AM Map 6: Day: 10/8/2008, 12:00 AM

(continued)

Table 10.2 (continued)

Variable group	Name of raster map for explanatory variables (EVs)	Unit of measurement	Minimum value	Maximum value	GRASS GIS modules and data used to generate the raster layer
Characterization of accessibility and visual properties	**7.** Natural accessibility of the terrain	Accessibility measure (‘)	17,437	74,286	The accumulative cost-distance map* was calculated using the GRASS Module (r.walk, option-k). As input, we used a 20 points separated every 3 km surrounding the perimeter of the study area; and a slope-based friction surface. Accumulated flow surfaces (r.accumulate) were then calculated and summed to generate the MADO analysis (Fábrega, 2006) (2)
	8. General terrain visibility (visual prominence)	Number of landscape locations (from a random set of $N = 500$), from which each map cell is visible	0	176	Viewsheds generated from the 500 random locations (r.viewshed) are summed to generate a Cumulative Viewsheds surface (r.mapcalc)
	9. Viewshed from each site	Visible area from each location (N of cells)	101	117,519	Viewshed extent from each of the 521 locations within a 10 km radius (r.viewshed)

(continued)

Table 10.2 (continued)

Variable group	Name of raster map for explanatory variables (EVs)	Unit of measurement	Minimum value	Maximum value	GRASS GIS modules and data used to generate the raster layer
Distance to resources	**10.** Distance to permanent water sources (springs)	Walking minutes (‘)	0	2976	Accumulative cost distance to springs. The location points of the springs were recorded with GPS during fieldwork and later complemented by digital processing of satellite images
	11. Distance to temporary water sources (creeks, shallow lakes)	Walking minutes (‘)	0	1188	Accumulative cost distance to temporary tributaries. Temporary tributaries were modeled using a hydrological analysis tool (r.stream.extract; using threshold = 300, course length = 300) and supplemented with hydrology vector data from IGN SIG:250
	12. Distance to lithic provisioning sources	Walking minutes (‘)	0	2918	Accumulative cost distance to sources of lithic raw materials. The location of these sources was recorded with GPS during fieldwork. Five quarries were located in secondary deposits and five in rocky outcrops

(continued)

Table 10.2 (continued)

Variable group	Name of raster map for explanatory variables (EVs)	Unit of measurement	Minimum value	Maximum value	GRASS GIS modules and data used to generate the raster layer
Settlement pattern	**13 and 14** Distance to the network of main and secondary paths	Walking minutes (‘) to main and secondary roads	0 0	2936 2936	Map 13: Accumulative cost distance to the path network between MALs. Principal paths are defined by the overlapping of three or more optimal paths between MALs Map 14: Secondary paths are defined by non-overlapping or two paths overlapping (MADO Analysis: Fábrega & Parcero, 2007, Llobera et al., 2011)
	15. Distance to A	Walking minutes (‘)	0	1614	Map 13: Accumulative cost distance to A
	16. Distance to HS	Walking minutes (‘)	0	3056	Map 14: Accumulative cost distance to HS
	17. Distance to HG	Walking minutes (‘)	0	3646	Map 15: Accumulative cost distance to HG
	18. Distance to Ch	Walking minutes (‘)	0	3484	Map 16: Accumulative cost distance to Ch
	19. Distance to TC	Walking minutes (‘)	0	2988	Map 17: Accumulative cost distance to TC
	20. Distance to LAS	Walking minutes (‘)	0	2928	Map 18: Accumulative cost distance to LAL
	21. Distance to MAL	Walking minutes (‘)	0	2936	Map 19: Accumulative cost distance to MAL

(continued)

Table 10.2 (continued)

Variable group	Name of raster map for explanatory variables (EVs)	Unit of measurement	Minimum value	Maximum value	GRASS GIS modules and data used to generate the raster layer
	22. Distance to W	Walking minutes (‘)	0	3488	Map 20: Accumulative cost distance to LW
	23. Distance to OK	Walking minutes (‘)	0	2916	Map 21: Accumulative cost distance to OK
Control	Distance to NE	Walking minutes (‘)	0	640	Map 22: Accumulative cost distance to NE

(1) Digital elevation model (SRTM, or Shuttle Radar Topography Mission); (2) MADO (Optimum Displacement Accumulation Model, Llobera et al., 2011); (3) IGN: National Geographic Institute of Argentina. A: rock art site; HS: hunting stand; HG: hunting ground; Ch: chenque structure; TC: transitory campsite; LAL: limited activity locus; MAL: multiple activity locus; LW: lithic workshop; OK: occasional knapping; and NE: negative evidence

MADO was used as an indicator of spatial association with domestic campsites and landscape mobility vectors.

10.1.3 Preparation of the Dataset in GIS

All variables were represented by georeferenced layers of information. The response variable layers were in point vector format, while the explanatory variables were in raster format. The raster layers extend between the following coordinates (in meters): N: 4,710,021.79570504, S: 4,690,139.27513844, E: 2,513,019.52945935, O: 2,494,034.9291764. The coordinate reference system is POSGAR 2007 Argentina 2 (EPSG code 5344). The layers consist of 775 rows and 740 columns of pixels, with a spatial resolution of 25.65 m, and cover an area of 377.31 km^2. The values of the explanatory variables were extracted to populate the data columns in the attribute table of the response variable layer. Then, the values were exported to a .dbf table used in R statistical software (SI_1_data.xlsx in Magnin, 2023).

10.1.4 Generalized Linear Modeling in R

To evaluate the relationship between the location of the archaeological evidence and the response variables, we created generalized linear models (GLMs) using R software (Nelder & Wedderburn, 1972). For each class of evidence, we selected models based on the Akaike information criterion corrected for small samples (AICc). This criterion combines the level of support of the data with the complexity of the model, which is measured by the number of parameters (Burnham & Anderson, 2002). GLMs with a binomial error distribution and a logit link function were adjusted in all cases, using the presence of a certain class of evidence (coded as 1) and control points (coded as 0) as the response variable. The variables were standardized with a mean of 0 and a standard deviation of 1 to facilitate comparisons between predictors and prevent convergence problems. To avoid collinearity problems between the predictors, we used correlation matrices beforehand. Variables were selected as predictors with Pearson correlations < 0.7 (see SI_2_correlations_23vars_3.xlsx in Magnin, 2023). Based on prior knowledge and archaeological significance,[1] a set of candidate models was developed for each class of the response variable, including a null model with no predictors (Burnham et al., 2011). This procedure of constructing candidate models with combinations of predictor variables informed by ethnographic, archaeological, and historical knowledge, while evaluating their correlation, is an alternative to the "data-dredging" or "fishing" approach. Rather than constructing models with only particular combinations of predictors, this method compares models with

[1] In the biological literature, the term "biological significance" is used, as opposed to "statistical significance." We use "archaeological significance" here.

all possible combinations of the predictors of interest (the all-subsets approach). However, constructing models with only particular combinations of predictors is not recommended (Burnham & Anderson, 2002, but see Symonds & Moussalli, 2011).

In the model selection process, the model with the lowest AICc value is considered the "best" approximation model among the candidate models (Burnham & Anderson, 2002). Models with a difference in AICc (ΔAICc) of less than two are considered equally explanatory. The degree to which the best model fits the data is obtained through the Akaike weight (w), which describes the probability that a model is the best, given the other models under consideration (Burnham et al., 2011), and the pseudo-R^2, which determines the variability explained by the model relative to the total variation of the data. When the best models included the categorical variable "geomorphology," Tukey's multiple comparisons were performed to assess significant differences between categories. The analyzed models are described in Table 10.3.

The "best model" for a class of evidence is the one that adjusts most efficiently to the observed locational variability among the considered models. This adjustment is defined as substantial, moderate, weak, or very weak (see below).

10.2 Results and Discussion

A "best" model was obtained for nearly all classes except the hunting ground class, whose distribution is indistinguishable from the control sample. The selected models are presented in bold in Table 10.3. Plots of the odds ratios for the explanatory variables in the selected models, as well as a statistical report, can be found in Magnin (2023), "Informe_Análisis_R__1_.html." Table 10.4 shows that the models for the following classes explain between 58 and 24% of the variability observed within each class: lithic workshops, chenque structures, rock art sites, hunting stands, and multiple activities loci. Meanwhile, the models for the following classes explain between 22 and 5% of the observed variability: lithic quarries, transitory campsites, limited activities loci, and occasional knapping.

The pseudo-R^2 values are structural descriptors of spatial patterns at the La Primavera locality (Wandsnider, 1996). All classes except hunting grounds present patterns that are effectively defined by the considered variables and differ significantly from the control sample and other classes (Cohen, 1988). In particular, the lithic workshop, chenque structures, rock art sites, hunting stands, and multiple activity sites present the highest pseudo-R^2 values. The following matters are considered to discuss the obtained results: (1) The patterns observed between "explanatory variables" and "response variables" do not describe causal relationships between the two. This statistical terminology should not be confused with explanations of human behavior. That is, a "functional view of structure" (Wandsnider, 1996, p. 326) is not employed here. Rather, both groups of variables were designed based on relevant ethnographic and archaeological information, as well as geoarchaeological formation processes, in order to investigate the existence of spatial structure in the data

Table 10.3 Explanatory variables (EVs) included in each model

Class of evidence	Candidate models
Rock art (A)	M1 = terrain surface (EV 2, 3) M2 = terrain surface (EV 2, 3) + distance to resources (EV10) M3 = terrain surface (EV 2, 3) + distance to resources (EV10) + distance to roads (EV 13) **M4 = terrain surface (EV 2, 3) + distance to paths (EV 13) + accessibility (EV 7)** M5 = terrain surface (EV 2, 3) + distance to paths (EV 13) + accessibility (EV 7) + visibility (EV 9)
Hunting stands (HS)	**M1 = terrain surface (EV 1, 2, 3) + geoform (EV 4)** M2 = terrain surface (EV 1, 2, 3) + geoform (EV 4) + hillshade (EV 6) M3 = terrain surface (EV 1, 2, 3) + geoform (EV 4) + hillshade (EV 6) + visibility (EV 9) M4 = terrain surface (EV 1, 2, 3) + geoform (EV 4) + hillshade (EV 6) + visibility (EV 9) + accessibility (EV 7) M5 = terrain surface (EV 1, 2, 3) + geoform (EV 4) + hillshade (EV 6) + visibility (EV 9) + accessibility (EV 7) + distance to resources (EV 11)
Quarries (Q)	**M1 = terrain surface (EV 1, 2, 3) + geoform (EV 4)** M2 = terrain surface (EV 1, 2, 3) + geoform (EV 4) + hillshade (EV 6) M3 = terrain surface (EV 1, 2, 3) + geoform (EV 4) + hillshade (EV 6) + distance to roads (EV 13) M4 = terrain surface (EV 1, 2, 3) + geoform (EV 4) + hillshade (EV 6) + distance to roads (EV 13) + accessibility (EV 7) M5 = terrain surface (EV 1, 2, 3) + geoform (EV 4) + hillshade (EV 6) + distance to roads (EV 13) + accessibility (EV 7) + visibility (EV 9)
Hunting ground (HG)	M1 = terrain surface (EV 2, 3) + geoform (EV 4) M2 = terrain surface (EV 2, 3) + geoform (EV 4) + distance to resources (EV 11) M3 = terrain surface (EV 2, 3) + geoform (EV 4) + distance to resources (EV 11) + settlement pattern (EV 16, 21) M4 = terrain surface (EV 2, 3) + geoform (EV 4) + distance to resources (EV 11) + settlement pattern (EV 16, 21) + visibility (EV 9)
Chenques (Ch)	M1 = terrain surface (EV 2, 3) + geoform (EV 4) **M2 = terrain surface (EV 2, 3) + geoform (EV 4) + visibility (EV 9)** M3 = terrain surface (EV 2, 3) + geoform (EV 4) + visibility (EV 9) + settlement pattern (EV 21) M4 = terrain surface (EV 2, 3) + geoform (EV 4) + visibility (EV 9) + settlement pattern (EV 21) + accessibility (EV 7)
Transitory camps (TC)	**M1 = terrain surface (EV 1, 2, 3) + geoform (EV 4)** M2 = terrain surface (EV 1, 2, 3) + geoform (EV 4) + hillshade (EV 5) M3 = terrain surface (EV 1, 2, 3) + geoform (EV 4) + hillshade (EV 5) + distance to resources (EV 10, 11, 12) M4 = terrain surface (EV 1, 2, 3) + geoform (EV 4) + hillshade (EV 5) + distance to resources (EV 10, 11, 12) + visibility (EV 9)

(continued)

Table 10.3 (continued)

Class of evidence	Candidate models
Limited activities site (LAS)	**M1 = terrain surface (EV 1, 2, 3) + geoform (EV 4)** M2 = terrain surface (EV 1, 2, 3) + geoform (EV 4) + distance to resources (EV 10, 11, 12) M3 = terrain surface (EV 1, 2, 3) + geoform (EV 4) + distance to resources (EV 10, 11, 12) + settlement pattern (EV 16, 17)
Multiple activities site (MAL)	M1 = terrain surface (EV 1, 2, 3) + geoform (EV 4) M2 = terrain surface (EV 1, 2, 3) + geoform (EV 4) + hillshade (EV 5) **M3 = terrain surface (EV 1, 2, 3) + geoform (EV 4) + hillshade (EV 5) + distance to resources (EV 10, 12)** M4 = terrain surface (EV 1, 2, 3) + geoform (EV 4) + hillshade (EV 5) + distance to resources (EV 10, 12) + visibility (EV 9) M5 = terrain surface (EV 1, 2, 3) + geoform (EV 4) + hillshade (EV 5) + distance to resources (EV 10, 12) + visibility (EV 9) + settlement pattern (EV 17, 18)
Lithic workshop (LW)	M1 = terrain surface (EV 1, 2, 3) M2 = terrain surface (EV 1, 2, 3) + hillshade (EV 5) M3 = terrain surface (EV 1, 2, 3) + hillshade (EV 5) + distance to resources (EV 12) **M4 = terrain surface (EV 1, 2, 3) + hillshade (EV 5) + distance to resources (EV 12) + visibility (EV 9)**
Occasional knapping (OK)	**M1 = terrain surface (EV 1, 2, 3) + geoform (EV 4)** M2 = terrain surface (EV 1, 2, 3) + geoform (EV 4) + distance to resources (EV 11)

The modeling process excluded some of the explicative variables in the candidate models (EV 8, 14, 15, 19, 20–23). See supplementary data SI_1_data.xlsx, in Magnin (2023) for details. The bold typography indicates the best models

Table 10.4 Coefficients of the best models for each class of evidence, obtained through the selection of models based on AICc

Class of evidence	Best model	Pseudo-R^2
Lithic workshop (LW)	M4	0.58
Chenque structures (Ch)	M2	0.55
Rock art sites (A)	M4	0.40
Hunting stands (HS)	M1	0.29
Multiple activities site/Residential campsite (MAL)	M3	0.24
Quarries (Q)	M1	0.22
Transitory camps (TC)	M1	0.15
Limited activities site (LAL)	M1	0.14
Occasional knapping (OK)	M1	0.05
Hunting ground (HG)	–	–

from a formational perspective that considers both natural and human processes (Van Leusen, 2002; Wandsnider, 1996); (2) the most relevant information to generate new knowledge about the archaeological case under analysis is the existence or lack of this spatial structure, the intensity with which it occurs, the variables included in the best models, and the comparison of these parameters between the different types of site; (3) in the interpretation, the specificity of each functional category (Jochim, 2022) must be considered because it will impact the intensity of the locational patterns. As in all archaeological studies, the selection of different variables will impact the results of the locational analysis. In this regard, to fulfill the recommended spatial analysis procedures (i.e., Gillings et al., 2020; Mehrer & Wescott, 2006), we provide the data and metadata used for the statistical and GIS analyses to enable future restudies (Magnin, 2023).

10.2.1 Locational Patterns at La Primavera

The following paragraphs characterize the best models obtained for each class, presenting the explanatory variables included and ranked by order of significance. The higher value of pseudo-R^2 is for the lithic workshops (LW), indicating a stronger locational pattern. Due to the low frequency of cases ($n = 5$), a correction for small samples was introduced into the statistical procedure, which allowed for their analysis and gave us confidence in the results. In La Primavera, lithic workshops are located in areas of the landscape with low slope angles (EV2) and high altitudes (EV1). These areas have wide visual basins (EV9), are exposed to solar radiation (high insolation values, EV5) and westerly winds (land orientations to the south and west, EV3). As expected for this type of activity, the workshops are located near quarries (EV12), which have a high availability of raw materials for flintknapping. The locations of the workshops, in high and exposed areas, coincide with rocky outcrops that were used as primary quarries for lithic raw materials. This suggests that local geology was a significant factor in the decision to locate workshops. Where rocks suitable for flintknapping are naturally available, people remained for a period of time to manufacture lithic tools. The model obtained for the workshops presents a spatial trend that explains 58% of the intra-class variability (Table 10.4) (Kintigh, 1988). The spatial organization and large quantity of worked materials in the workshops suggest that activities were performed at these sites frequently over time. Unlike the placement of multiple activity loci, hunting stands, and chenques, the location of lithic workshops has not received much attention in archaeological research of Patagonia. This is partly due to the low frequency of this class of evidence, but the statistical analysis we applied allowed us to include it. The information presented here encourages expanding the sample of known LW locations in the study area.

Another class with a clearer spatial structure is the chenque (Ch), as the best-analyzed model explains 55% of the intra-class variability (Table 10.4). They are located in high-elevation areas of the landscape (EV1), which provide wide visual basins (EV9). Regarding the geoforms (EV4), they are found on plateau tops and high plateau slopes. This characterization coincides with descriptions of the location of mortuary stone structures in various bibliographic sources (De Aparicio, 1933–35; Deodat, 1946; Goñi & Barrientos, 2000; Goñi et al., 2004; Musters, [1871] 1997; Viganti, 1934). As mentioned in previous chapters, they date to the Late Holocene (ca. 1150–350 AD) in the northwest area of Santa Cruz (Goñi, 2000–2002; García Guráieb, 2011; García Guráieb et al., 2015).

Regarding the cardinal orientation of the terrain where they are located (EV3), the chenques are predominantly situated on terrain inclined toward the south and west. This particular location in unsheltered areas with difficult access makes us think that the activities carried out in these places, such as the burial of human corpses or the marking of the landscape, were short-term and highly symbolic. The inter-visibility pattern between the structures also suggests this (Magnin, 2013b). Chenques are placed in sectors of high archaeological visibility. They are in places where erosive processes predominate due to wind and pluvial action. These places are away from sedimentary environments and have little to no soil development. The artifactual assemblages associated with the chenques are small, with an average of 6.64 lithic artifacts recorded per structure, indicating short-term activities.

The rock art sites (A) model best explains 40% of the observed internal variability (Table 10.4). The analysis indicates that this class of evidence is located on steep slopes and near permanent water sources (EV2, EV7, and EV10). Rock art is also found in terrain oriented west or north (EV3). These characteristics coincide with natural cave formation conditions involving water springs and wind action. The wind mainly blows from the western quadrant, eroding rocky outcrops (Panza, 2001). The resulting caves and rock shelters provide support for the rock art on their walls and roofs while protecting it from destruction (Gómez & Magnin, 2009). Additionally, rock art is located in areas of high natural accessibility and close to the network of paths between residential campsites. This pattern suggests that rock art painting may be related to residential movement. It also indicates a connection between public rock art performances (Carden & Miotti, 2020) and the visualization of human or animal movement in the landscape.

The best model for the hunting stands (HS) explains 29% of the observed variation. Similar to the chenque, the hunting stands are located in high places (EV1) with gentle slopes (EV2). They are found on plateau tops and slopes more often than on other geoforms. In addition to being in high, exposed areas, a final factor that characterizes hunting stands is their terrain orientation (EV3). Hunting stands are found on terrain surfaces oriented to the east and north. These places are topographically protected from winds blowing from the west and south. Therefore, the stone structures that equip some of the stands could have been used to hide hunters from their prey. This pattern of hunting blind placement has been observed in other places in extra-Andean Patagonia and dated to the last 2500 years BP (Belardi et al., 2017; Miotti et al., 2021).

Multiple activity sites, quarries, transitory camps, and limited activity loci models are less robust than previous classes since their best models only explain between 24 and 14% of the observed variability (Table 10.4). Multiple activity loci (MALs), which are indicators of residential campsites, are located near water sources

(EV10). Fresh water is a critical resource necessary not only for human consumption but also for concentrating other resources, such as fauna and vegetation (Jochim, 1968; Martinić 1995; Vignati, 1934). In terms of terrain, they are located on ground surfaces facing east and north (EV3). These surfaces are protected from the stronger westerly winds in summer. Additionally, they are located in sunny areas of the landscape (EV5) with gentle slopes (EV2) and at low altitudes (EV1). Geomorphologically, they are associated with mass-removal deposits (EV4), which coincide with deep canyons dissected by fluvial action on basaltic plateaus. These deposits occur at the foot of basaltic plateaus where gravity deposits them (Gómez & Magnin, 2009). These flat, sheltered terrains meet the criteria for establishing a residential campsite (Jochim, 1968; Martinić, 1995). The low slope and sheltered conditions of these places favor sediment accumulation and good preservation conditions for buried evidence. This has been verified at the open-air site La Quinta (Chap. 8). The last variable of the model indicates a tendency regarding the proximity to quarries (EV12). As analyzed in Chap. 7, most of the rocks used to manufacture artifacts found at campsites are locally sourced due to the high availability of lithic raw materials. These include andesite, rhyolite, basalt, petrified wood, and other siliceous rocks (Magnin, 2015). However, the most commonly used raw materials in each MAL were not always found at the shortest distance quarry. Obsidian, on the other hand, required provisioning from greater distances (Hermo, 2008; Skarbun et al., 2022). This information indicates that the provisioning distances ranged between 91 m and 4.9 km on average (Magnin, 2015).

The quarries (Q) are associated with terraces (EV4) in low-altitude areas (EV1) and also to steeper slopes (EV2). Consequently, the location of the quarries is unclear. This is likely due to the wide distribution and variety of rocks sources in this area (Hermo et al., 2020a). They are both primary quarries, which are rocky outcrops associated with hilly areas and steep slopes, and secondary quarries, which are located on terraces that form pediment levels in low-lying areas of the landscape.

Transitory campsites (TC) are defined by evidence of multiple activities, similar to residential campsites. However, they have lower artifactual density, which could indicate that they were occupied less frequently and/or by smaller groups of people. As expected for residential areas (EV2), they are located on flat terrain. The main difference in location between residential campsites and MAL is that TC is in high areas of the landscape (EV1) and on north- and west-facing terrain (EV3). Their high position means that they are not protected from low winter temperatures. While the northward orientation provides optimal topographic shelter from cold southern winds, the westward terrain aspect is not protected. Overall, the location of the transitory campsites in high places suggests that these were occupations requiring the use of skin windbreaks or awnings—"kau" in native language Tehuelche and "toldo" in Spanish—probably made of wood and leather, as documented ethnographically (Martinić, 1995; Musters, [1871] 1997).

In the case of limited activity loci (LAL) and occasional knapping (OK) classes, the unexplained variability is too large. Only 14% and 5%, respectively, of the observed variability are explained by the best models. This indicates that their distribution is close to random. This result is consistent with the low functional specificity of these classes of evidence. Limited activity sites tend to be located in areas with low slopes (EV2) and in hills more frequently than in other geomorphic forms (EV4). Occasional knapping is more frequent in plateaus and areas of mass removal (EV4), at low altitudes (EV1), on low slopes (EV2), and facing south (EV3).

Regarding hunting grounds (HG), they cannot be distinguished from control sites, suggesting that the entire landscape was used for hunting. This may indicate that guanacos and rheas, the main game animals in the region, were hunted in large areas of the landscape. Fieldwork records show that traces of these animals, such as guanaco wallows and rhea nests, as well as paths and sightings of both species, are distributed throughout the locality. Moreover, this pattern is expected, considering the variable distances covered in hunts (2.16–9.5 km in communal ñandú hunts using horses, according to Giardina et al., 2021) between stalking and slaughter sites, as represented by isolated finds of stone hunting weapons. This maximum distance is close to the total analyzed area (10 by 10 km).

In summary, locational patterns were observed for the lithic workshop, chenques, rock art, and hunting stand classes, all of which exhibit pronounced tendencies of substantial significance (Cohen, 1988). In contrast, the locational patterning of residential base camps, quarries, transitory camps, and loci of limited activities present moderate significance. These classes exhibit higher locational selectivity because they are associated with repetitive and prolonged use of certain areas (Jochim, 2022). In contrast, occasional knapping showed very weak significance, and the hunting grounds class was statistically indistinguishable from the control sample.

The work performed with GLM provided additional insight beyond that obtained in the previous univariate study. The results indicate the existence of spatial structuring in La Primavera that sheds light on ancient decisions regarding past landscape use. With the exception of hunting grounds, the spatial locations of all classes of archaeological evidence differ significantly from the control sample. Additionally, the different classes exhibit interesting interclass locational variability. Analyzing their coefficients revealed that the activities performed by indigenous people that were more spatially stable were the manufacturing of lithic artifacts in workshops (LW), cave and rock shelter art painting (A), chenque construction for burials and/or monuments (Ch), and stalking prey (HS). Other activities, such as residential long-term camps (MAL), transitory base camps (TC), and other undefined limited activity areas (LAL), were comparatively less stable.

This study has important implications and provides new insights into the archaeological study of the region. The spatial analysis performed revealed itself to be a relevant, independent line of analysis that allows new questions to be asked about the use of the landscape. The locational tendencies and patterns obtained are significant because they involve routine landscape use. According to practice theory, these patterns reveal deep cultural knowledge derived from dwelling (Ingold, 1993). Since time and space result from situated practices and are inseparable, the studied patterns respond to a recurrence of the same spatial practices throughout the hunter-gatherer dwelling process, pointing to its temporal depth. The spatial structuring in the La Primavera locality supports the idea of a consolidated occupation of well-known landscapes within the Deseado Massif. Further studies are needed to address some emerging questions related to the variability detected in the location of residential sites (MAL and TC). Are these differences related to the seasonal occupation of the area? Or, are they the result of changes in the settlement involving a different chronology? When was this dwelling pattern established, and how did it change with the establishment of rural ranches in the nineteenth century? To address these questions, we aim to excavate open-air sites to obtain chronological information and intensify research on the historical period at La Primavera. We are also replicating the formalized methodology to achieve a comparative perspective with other archaeological sites within the annual range of mobile hunter-gatherers (Magnin, 2015). Lastly, we will discuss our most challenging objective.

10.3 The Social Importance of Our Research

The original research presented in this book was part of a doctoral thesis for a PhD in Natural Sciences, obtained in 2010. To achieve this goal, we defined our study area around Cueva Maripe, an intensely studied archaeological site. Our aim was to expand our knowledge of hunter-gatherer activities in this landscape. The cave owes its name to a Tehuelche countryman who used it as a refuge during spring guanaco hunts, or "guanaqueadas" (Echeverría Baleta, 1999). The 1931 census of Santa Cruz Province registered two men and three women with the last name Maripe in Gobernador Moyano, the closest town located 65 km northeast of the cave (Aguerre, 2008). Although the indigenous identity of the material studied was never questioned throughout our research, the archaeological work we performed was disconnected from indigenous communities. The growing visibility of indigenous people in the Santa Cruz province over the last decade has prompted deep reflection on our practice. Our work must help raise awareness of the 16 officially recognized urban and rural indigenous communities throughout the province belonging to the Mapuche, Mapuche-Tehuelche, and Tehuelche peoples (Rodríguez & Horlent, 2016), as well as the various identities that emerge in our society. In many cases, these identities have been hidden and negated for the effective devaluation of Indigenous cultures in the territory that is now Argentina throughout the history of the construction of the national state. This process led to the subjugation of these peoples and their rights.

In pursuit of a more socially engaged perspective consistent with political archaeology (Curtoni & Olivan, 2023), we initiated dialogues and promoted collaborative scientific research with indigenous communities in the province. For instance, we developed university extension projects at the University of La Plata (UNLP), such as "Picaderos, chenques, and paintings: Archaeology of the Santa Cruz Plateaus," developed between 2018 and 2020, and the "Archaeologists in the School" project (RESOL-2021-249-APN-SECPU#ME), sponsored by the Secretariat of University Policies of the Nation (Argentine Ministry of Education). Current collaborative efforts seek to establish constructive dialogue between indigenous peoples and archaeologists, as demonstrated by a symposium to be held in Río Gallegos, Santa Cruz, Argentina, in October 2024, during the XII Jornadas de Arqueología de la

Patagonia. All of these efforts aim to build constructive dialogue between indigenous peoples and archaeologists.

In Argentina's current political and social climate, the scientific work of public universities and the National Council of Scientific and Technical Research is undervalued by society due to ultra-conservative political thought. In this context, positive critique on the social importance of our work is fundamental, as it strengthens the bond between archaeologists and indigenous peoples, allowing them to achieve shared goals. We hope that the results of this work are significant and help raise awareness of the current indigenous existence and their claims. The patterns detected demonstrate repeated locational customs performed in this landscape for thousands of years. The implications of that landscape knowledge must be recognized in today's society.

References

Aguerre, A. M. (1977). A propósito de un nuevo fechado radiocarbónico para la Cueva de las Manos. *Relaciones De La Sociedad Argentina De Antropología, 11*, 129–142.

Aguerre, A. M. (1987). Investigaciones arqueológicas en el área de La Martita. Departamentode Magallanes, provincia de Santa Cruz. In *Primeras Jornadas de Arqueología de La Patagonia. Serie Humanidades 2* (pp. 11–26). Gobierno de la provincia de Chubut.

Aguerre, A. M. (2000). *Las vidas de Pati en la toldería tehuelche del Río Pinturas y después.* Universidad de Buenos Aires, Facultad de Filosofía y Letras.

Aguerre, A. M. (2008). *Genealogía de familias tehuelches-araucanas de la Patagonia central y meridional argentina.* Editorial de la Facultad de Filosofía y Letras, Universidad de Buenos Aires.

Albeck, M. E. (1995–1996). Utilización de la liquenometría como indicador cronológico en las estructuras agrícolas prehispánicas de Coctaca. *Shincal, 5*, 67–89.

Alberdi, M. T., Miotti, L., &Prado, J. L. (2001). Hippidion saldiasi Roth. 1899 (Equidae. Perissodactyla) at the Piedra Museo Site (Santa Cruz. Argentina): Its implication for the regional economy and environmental reconstruction. *Journal of Archaeological Science, 28*, 411–419.

Albert, B., & Le Tourneau, F. (2007). Ethnogeography and resource use among the Yanomami. *Current Anthropology, 48*(6), 584–592.

Aldenderfer, M. (1998). *Montane foragers: Asana and the South-Central andean archaic.* University of Iowa Press.

Allen, K., Green, S. W., & Zubrow, E. (Eds.). (1990). *Interpreting space: GIS and archaeology.* Taylor & Francis.

Allen, M. J. (1991). Analyzing the landscape: A geographical approach to archaeological problems. In A. J. Schofield (Ed.) *Interpreting artefact scatters: Contributions to plough zone archaeology. Oxbow monograph 4* (pp. 39–57). Oxbow Books.

Alonso, F., Gradin, C., Aschero, C., & Aguerre, A. (1984–1985). Algunas consideraciones sobre dataciones radiocarbónicas para el área del Río Pinturas. *Relaciones de la Sociedad Argentina de Antropología, 16*, 275–285.

Altschul, J. (1988). Models and the modeling process. Quantifying the present and predicting the past. In W. J. Judge, & L. Sebastian (Eds.). *Theory method and application of archaeological predictive modeling,* (pp. 61–87). US Bureau of Land Management. Department of Interior. US Government Printing Office.

Álvarez, M. (2004). El uso de materias primas vegetales en la costa norte del canal de Beagle a través del análisis funcional de base microscópica. In T. Civalero, P. Fernández, & G. Guráieb (Eds.), *Contra Viento y Marea. Arqueología de Patagonia,* (pp. 279–294). INAPL-Sociedad Argentina de Antropología.

L. A. Magnin, *The Landscape and Mobility of Hunter-Gatherer Societies*, The Latin American Studies Book Series, https://doi.org/10.1007/978-3-032-08127-8

Ambrústolo, P. (2020). Obsidiana negra en la costa norte de Santa Cruz: Interacciones humanas y aprovisionamiento lítico. *Comechingonia, 24*(3), 83–110. https://doi.org/10.37603/2250.7728.v24.n3.31092

Andrefsky, W. (2005). *Lithics. Macroscopic approaches to análisis. Manuals in archaeology.* Cambridge University Press.

APN Administración de Parques Nacionales. (2018). *Plan de Gestión Parque Nacional Bosques Petrificados de Jaramillo y Monumento Natural Bosques Petrificados.* IF-2018-60876708-APN-DRPA#APNAC. https://sib.gob.ar/archivos/IF-2018-60876708-APN-DRPA-APNAC.pdf. Accessed September 15, 2024.

Aragón, E., & Franco, N. (1997). Características de rocas para la talla por percusión y propiedades petrográficas. *Anales Del Instituto De La Patagonia, 25*, 187–199.

Ariztegui, D., Anselmetti, F. S., Gilli, A., & Waldmann, N. (2008). Late Pleistocene environmental change in Eastem Patagonia and Tierra del Fuego. A limno geological approach. In J. Rabassa (Ed.). *The late cenozoic of patagonia and tierra del Fuego: Developments in quaternary sciences series 11,* (pp. 241–253). Elsevier.

Aschero, C. (1975). *Ensayo para una clasificación morfológica de artefactos líticos aplicada a estudios tipológicos comparativos.* Informe al CONICET. Ms.

Aschero, C. (1983). *Ensayo para una clasificación morfológica de artefactos líticos aplicada a estudios tipológicos comparativos. Revisión.* Informe al CONICET. Ms.

Aschero, C. (1996). ¿Adonde van esos guanacos? In J. Gómez Otero (Ed.). *Arqueología solo patagonia* (pp: 153–162). CENPAT-CONICET.

Aschero, C., Bellelli, C., & Goñi, R. A. (1992–1993). Avances en las Investigaciones Arqueológicas del Parque Nacional Perito Moreno (Provincia de Santa Cruz. Patagonia Argentina). *Cuadernos del Instituto Nacional de Antropología y Pensamiento Latinoamericano, 14*, 143–170.

Aschero, C., Goñi, R., Civallero, M. T., Molinari, R., Espinosa, S., Guraieb, G., & Bellelli, C. (2005). Holocenic park: Arqueología del parque nacional perito moreno. *Anales De Parques Nacionales, 17*, 71–119.

Baetti, C. (2004). Unidades fisionómicas y florísticas de vegetación. In L. González, & P. Rial (Eds.) *Guía Geográfica Interactiva de Santa Cruz.* http://www.inta.gov.ar/santacruz/info/documentos/teledet/GuiaSC/. Accessed June 1, 2010.

Bagolini B. (1968). *Ricerche sulle dimensioni dei manufatti litici preistorici non ritoccati* (Vol. XV, pp. 195–219). Annali Univo Ferrara, n.s.

Bamforth, D. B. (1986). Technological efficiency and tool curation. *American Antiquity, 51*(1), 38–50.

Barberena, R, & Borrero, L. (2010). Geoarqueología y distribuciones subsupericiales de materiales arqueológicos: Localidad Cabo Vírgenes. In L. Borrero, & J. Charlin (Eds.). *La Arqueología de Pali Aike y Cabo Vírgenes* (pp: 103–121). Dunken.

Barbosa, C. E., & Gradín, C. (1986–87). Estudio composicional por difracción de rayos X de los pigmentos provenientes de la excavación del Alero Cárdenas (Provincia de Santa Cruz). *Relaciones de la Sociedad Argentina de Antropología, 17*(1), 143–171.

Bayón, C., & Flegenheimer, N. (2004). Cambio de planes a través del tiempo para el traslado de roca en la pampa bonaerense. *Estudios Atacameños, 28*, 59–70.

Bayón, C., Flegenheimer, N., & Pupio, A. (2006). Planes sociales en el abastecimiento y traslado de roca en la Pampa bonaerense en el Holoceno temprano y tardío. *Relaciones De La Sociedad Argentina De Antropología, 31*, 19–45.

Behrensmeyer, A. K. (1978). Taphonomic and ecologic information from bone weathering. *Paleobiology, 4*(2), 150–162.

Belardi, J. B. (1992). De lo espacial a lo temporal: Explorando distribuciones de artefactos. *Revista De Estudios Regionales CEIDER, 10*, 35–67.

Belardi, J. B. (2005). *Paisajes Arqueológicos: Un estudio comparativo de diferentes ambientes patagónicos.* Archaeopress. Publish of British Archaeological Reports (BAR).

Belardi, J., & Borrero, L. (1999). El paisaje arqueológico de la margen norte del Lago Argentino (provincia de Santa Cruz). *Praehistoria, 3*, 35–64.

Belardi, J. B., Borrero, L. A., Campan, P., Carballo Marina, F., Franco, N., García, M. F., Horwitz, V., Lanata, J. L., Martin, F. M., Muñoz, F. E., Muñoz, A. S., & Savanti, F. (1992). Intensive archaeological survey in the Upper Santa Cruz Basin. Southern Patagonia. *Current Anthropology, 33*(4), 451–454.

Belardi, J., Carallo Marina, F., Madrid, P., Barrientos, G., & Campan, P. (2017). Late Holocene guanaco hunting grounds in southern Patagonia: Blinds, tactics and differential landscape use. *Antiquity, 91*, 718–731.

Belardi, J., & Carballo Marina, F. (2003). Tafonomía regional en la cuenca media del río Coyle (Santa Cruz. Patagonia Argentina). *Intersecciones En Antropología, 4*, 59–97.

Belardi, J. B., Espinosa, S., & Cassiodoro, G. (2005). Un Paisaje de puntas: Las cuencas de los lagos Cardiel y Strobel (Provincia de Santa cruz. Patagonia Argentina). *Werken, 7*, 57–76.

Belardi, J. B., & Goñi, R. (2006). Representaciones rupestres y convergencia poblacional durante momentos tardíos en Santa Cruz (*Patagonia argentina*). El caso de la meseta del Strobel. In D. Fiore, & M. M. Podestá (Eds.), *Tramas en la Piedra. Producción y usos del arte rupestre,* (pp 85–94). Sociedad Argentina de Antropología. Asociación Amigos del Instituto Nacional de Antropología. World Archaeological Congress.

Belardi, J. B., & Goñi, R. A. (2002). Distribución espacial de motivos rupestres en la cuenca del Lago Cardiel (Patagonia Argentina). *Boletín SIARB, 16*, 29–38.

Belardi, J. B., Tiberi, P., Stern, C., & Súnico, A. (2006). Al este de Cerro Pampa: Ampliación del área de disponibilidad de obsidiana de la Pampa del Asador (Provincia de Santa Cruz). *Intersecciones En Antropología, 7*, 27–36.

Bellelli, C., Pereyra, F., Fernández, P., Scheinsohn, V., Carballido, M., & Podestá, M. (2000). Aproximación geoarqueológica del sector sur de la Comarca Andina del Paralelo 42° (Cholila. Chubut). *Cuaternario y Ciencias Ambientales, 1*, 15–21.

Berón, M. (1994). El recurso y el método: Estrategias de movilidad y asentamiento en la subregión Pampa Seca. *Arqueología, 4*, 213–234.

Berón, M. (2003). El Sitio Chenque I. Un cementerio de cazadores-recolectores en la pampa seca. *Atekna "En la tierra", 1*, 242–272.

Berón, M. (2006). Base regional de recursos minerales en el occidente Pampeano. Procedencias y estrategias de aprovisionamiento. *Relaciones De La Sociedad Argentina De Antropología, 31*, 47–88.

Berón, M., Baffi, E., Molinari, R., Barrientos, G., Aranda, C., & Luna, L. (2000). Estructuras funerarias de momentos tardíos en Pampa-Patagonia. El "Chenque" de Lihué Calel. In J. B. Belardi, F. Carballo Marina, & S. Espinosa (Eds.), *Desde el País de los Gigantes. Perspectivas arqueológicas en Patagonia* (pp: 141–160). Universidad Nacional de la Patagonia Austral.

Bettinger, R. (1980). Explanatory-predictive models of hunter-gatherer behavior. *Advances in Archaeological Theory and Method, 3*, 189–255.

Binford, L. (1962). Archaeology as anthropology. *American Antiquity, 28*(2), 217–225.

Binford, L. (1977). Introducción. In L. Binford (Comp.), *For theory building in archaeology: Essays of faunal remains spatial analysis and systemic modelling* (pp. 1–10). Academic Press.

Binford, L. (1979). Organization and formation processes: Looking at curated technologies. *Journal of Anthropological Research, 35*(3), 255–273.

Binford, L. (1980). Willow smoke and dogs' tails: Hunter-gatherer settlement systems and archaeological site formation. *American Antiquity, 45*(1), 4–20.

Binford, L. (1981). *Bones: Ancient men and modern myths*. Academic Press.

Binford, L. (1982). The archaeology of place. *Journal of Anthropological Archaeology, 1*(1), 5–31.

Binford, L. (1988). *In pursuit of the past: Decoding the archaeological record*. Thames and Hudson.

Binford, L. R. (1983). *In pursuit of the past. Decoding the archaeological record*. Thames and Hudson.

Bird, J. (1938). Antiquity and migrations of the early inhabitants of Patagonia. *The Geographical Review, 28*, 250–275.

Blanco, R. (2015). *El arte rupestre en los Macizos del Deseado y Somuncurá: la producción de grabados y pinturas entre cazadores-recolectores del Holoceno Medio*. Doctoral Thesis.

Facultad de Ciencias Naturales y Museo—Universidad Nacional de La Plata. https://sedici.unlp.edu.ar/handle/10915/44509

Blanco, R., & Magnin, L. (2023). Los grabados de Seccional Cerro Horqueta (Parque Nacional Bosques Petrificados de Jaramillo), nuevos aportes al conocimiento regional. *Libro de Resúmenes del XXI CNAA* (pp. 87–88). Facultad de Humanidades, Universidad Nacional del Noreste.

Blankholm, H. (1991). *Instrasite spatial analysis in theory and practice.* Aarhus University Press.

Bonomo, M. (2006). Un acercamiento a la dimensión simbólica de la cultura material en la región pampeana. *Relaciones De La Sociedad Argentina De Antropología, 31*, 89–116.

Borrazzo, K. (2004). *Hacia una tafonomía lítica: el análisis tafonómico y tecnológico de los conjuntos artefactuales líticos de superficie provenientes de los loci San Genaro 3 y 4 (Bahía San Sebastián-Tierra del Fuego. Argentina).* Doctoral Thesis. Facultad de Filosofía y Letras. Universidad Nacional de Buenos Aires. https://www.academia.edu/234476/Hacia_una_tafonom%C3%ADa_l%C3%ADtica_Towards_a_lithic_taphonomy_undergraduate_thesis?hb-sb-sw=75082789. Accessed September 15, 2024.

Borrazzo, K. (2005). Al sur de Baguales: Cambio en el uso tecnológico y uso del espacio. In *Libro de resúmenes de las Sextas Jornadas de Arqueología de la Patagonia* (pp. 23–24). Punta Arenas.

Borrero, J., Franco, N., Lanata, J. L., & Belardi, J. B. (1991). Distribuciones arqueológicas y tafonómicas en la margen norte del Lago Argentino (Santa Cruz. Argentina). In *Boletín 4 del Museo Regional de la Araucanía* (pp. 23–31). Sociedad Chilena de Arqueología.

Borrero, L. (1985). *La economía prehistórica de los habitantes del norte de la Isla Grande de Tierra del Fuego.* Doctoral Thesis. Facultad de Filosofía y Letras—Universidad Nacional de Buenos Aires. http://repositorio.filo.uba.ar/handle/filodigital/1661. Accessed September 15, 2024.

Borrero, L. (1989). Replanteo de la arqueología patagónica. *Interciencia, 14*(3), 127–135.

Borrero, L. (1994/95). Arqueología de la Patagonia. *Palimpsesto. Revista de Arqueología, 4*, 9–69.

Borrero, L. (1995). Historia reciente de la arqueología patagónica. *Runa, 22*, 51–176.

Borrero, L. (2001). Regional taphonomy: Background noise and the integrity of the archaeological record. In L. Kuznar (Ed.) *Contributions to archaeological method and theory, ethnoarchaeological series 4* (pp. 243–254). International Monographs in Prehistory.

Borrero, L. A. (1995). Historia reciente de la arqueología patagónica. *RUNA, Archivo Para Las Ciencias Del Hombre, 22*(1), 151–176. https://doi.org/10.34096/runa.v22i1.1322

Borrero, L., & Lanata, J. (1992). Introducción. In L. A. Borrero, & J. L. Lanata (Eds.), *Análisis espacial en la arqueología patagónica* (pp. 7–8). Ayllu.

Borrero, L. A., Lanata, J. L., & Ventura, B. (1992). Distribuciones de hallazgos aislados en Piedra del Aguila. In J. L. Borrero, & J. L. Lanata, (Eds.), *Análisis espacial en la arqueología patagónica* (pp. 9–20). Ayllu.

Borrero, L., & Manzi, L. (2007). Arqueología supra-regional y biogeografía en la Patagonia meridional. en F. M. Morello, A. Martinic, A. Prieto y G. Bahamonde (Eds.), *Arqueología de Fuego-Patagonia. Levantando piedras, desenterrando huesos... y develando arcanos* (pp. 163–172). Punta Arenas.

Borrero, L. A., & Nami, H. (1996). Algunas hipótesis y propuestas de trabajo para una arqueología regional. *Praehistoria, 2*, 35–43.

Borrero, L. A., Zárate, M., Miotti, L., & Massone, M. (1998). The Pleistocene-Holocene transition and human occupations in the Southern Cone. *Quaternary International, 49–50*, 191–199.

Borromei, A. M. (2003). Palynology at Piedra Museo locality. Santa Cruz province. Argentina. In Miotti L., Salemme, M., & Flegenheimer, N. (Eds.), *Ancient evidence for paleo South Americans: From where the south winds blow* (pp. 113–119). Centre for the Study of First Americans-Texas A & M University Press.

Bozzuto, D. (2007). Utilización del recurso topográfico en la localidad Cerro Casa de Piedra (Provincia de Santa Cruz. Argentina). In F. Morello, M. Martinic, A. Prieto, & G., Bahamonde (Eds.). *Arqueología de Fuego-Patagonia. Levantando piedras. desenterrando huesos... y develando arcanos* (pp. 219–234). CEQUA.

Brook, G. A., Franco, N. V., Ambrústolo, P., Mancini, M. V., & Fernández, P. (2015). Evidence of the earliest humans in the Southern Deseado Massif (Patagonia, Argentina), Mylodontidae, and changes in water availability. *Quaternary International, 363*, 107–125.

Burroni, D., Donahue, R., Pollard, A., & Mussi, M. (2002). The surface alteration features of flint artifacts as a record of environmental processes. *Journal of Archaeological Science, 29*, 1277–1287.

Butzer, K. (1982). *Arqueología. Una ecología del hombre: Método y teoría para un enfoque contextual.* Bellaterra.

Cabrera, A. (1971). Fitogeografía de la República Argentina. *Boletín De La Sociedad Argentina De Botánica, 14*, 1–42.

Cabrera, A., & Willink, A. (1980). *Biogeografía de América Latina. Monografía 13. Serie de Biología.* Secretaría General de la Organización de los Estados Americanos.

Camilli, E. (1989). The occupational history of sites and the interpretation of prehistoric technological systems: an example from Cedar Mesa. Utah. In R. Torrence (Ed.) *Time energy and stone tools* (pp: 17–26). Cambridge University Press.

Carballo Marina, F. (2006). *La Cuenca superior del río Santa Cruz: las poblaciones humanas y el uso del espacio.* Doctoral Thesis. Facultad de Ciencias Naturales, Universidad Nacional de La Plata.

Carballo Marina, F., Belardi, J., Espinosa, S., & Ercolano, B. (2000–2002). Tecnología y movilidad en la cuenca media del río Coyle. Santa Cruz. *Cuadernos del Instituto Nacional de Antropología y Pensamiento* Latinoamericano*, 19*, 89–107.

Carballo Marina, F., & Sáenz, J. L. (1992). Diseño de un plan para estudiar la distribución de sitios en la Provincia de Santa Cruz. In L. A. Borrero, & J. L. Lanata (Comp.), *Análisis espacial en la arqueología patagónica* (pp. 125–148). Ayllu.

Carden, N. (2004). Valga la redundancia. Arte rupestre e información en el Macizo del Deseado. In T. Civalero, P. Fernández, & G. Guráieb (Eds.), *Contra Viento y Marea. Arqueología de Patagonia* (pp. 613–623). INAPL-Sociedad Argentina de Antropología.

Carden, N. (2008). Territories among hunter-gatherers and the ritual dimension of landscapes: The central Patagonian plateau. Argentina. *Before Farming, 1*, 1–19.

Carden, N. (2009). *Imágenes a través del tiempo. Arte rupestre y construcción social del paisaje en la Meseta Central de Santa Cruz.* Sociedad Argentina de Antropología.

Carden, N., Blanco, R., Poiré, D., Genazzini, C., Magnín, L., & García, P. (2014). Análisis de pigmentos del Macizo del Deseado: El abastecimiento de materias primas y la producción de pinturas rupestres en Cueva Maripe (Santa Cruz, Argentina). *Relaciones De La Sociedad Argentina De Antropología, 39*(2), 483–508.

Carden, N., Magnin, L., & Miotti, L. (2009). Distribución de figuras animales y dinámica poblacional: Un estudio comparativo en Patagonia (provincia de Santa Cruz). In M. Sepúlveda, L. Briones, & J. Chacama, (Eds.) *Crónicas sobre la Piedra. Arte rupestre de las Américas* (pp. 153–174). Ediciones Universidad de Tarapacá.

Carden, N., & Miotti, L. (2020). Unraveling rock art palimpsests through superimpositions: The definition of painting episodes in Los Toldos (southern Patagonia) as a baseline for chronology. *Journal of Archaeological Science: Reports, 30*, Article 102265.

Cardich, A. (1973). Secuencia arqueológica y cronología radiocarbónica de la Cueva 3 de Los Toldos (Santa Cruz. Argentina). *Relaciones De La Sociedad Argentina De Antropología, 7*, 87–122.

Cardich, A. (1977). Las culturas pleistocénicas y post-pleistocénicas de Los Toldos y un bosquejo de la prehistoria de Sudamérica. *Obras Del Centenario Del Museo De La Plata, 2*(149), 172.

Cardich, A. (1978). Un motivo sobresaliente en las pinturas de El Ceibo. *Revista Relaciones De La Sociedad Argentina De Antropología, 13*, 163–182.

Cardich, A. (1984). Paleoambientes y la más antigua presencia del hombre. *Las culturas de América en la época del descubrimiento* (pp. 13–36). Cultura Hispánica.

Cardich, A. (1985). Una fecha radiocarbónica más de la Cueva 3 de los Toldos (Santa Cruz. Argentina). *Relaciones De La Sociedad Argentina De Antropología N. S., 16*, 269–273.

Cardich, A. (1987). Arqueología de Los Toldos y El Ceibo (provincia de Santa Cruz. Argentina). *Estudios Atacameños, 8*, 95–113.

Cardich, A., Cardich, L., & Hadjuk, A. (1973). Secuencia arqueológica y cronología radiocarbónica de la Cueva 3 de Los Toldos (Santa Cruz. Argentina). *Relaciones de la Sociedad Argentina de Antropología, VII*, 87–122.

Cardich, A., Mansur-Franchomme, M., Giesso, M., & Durán, V. (1982). Arqueología de las Cuevas de El Ceibo (Provincia de Santa Cruz. Argentina). *Relaciones Sociedad Argentina de Antropología, 14*(2), 173–209.

Cardich, A., & Miotti, L. (1983). Recursos faunísticos en la economía de los cazadores recolectores de Los Toldos (provincia de Santa Cruz. Argentina). *Relaciones De La Sociedad Argentina De Antropología, 15*, 147–157.

Cardich, A., & Paunero, R. (1991–1992). Arqueología de la Cueva II de Los Toldos (Santa Cruz. Argentina). In *Anales de Arqueología y Etnología* (pp. 46–47, 49–71). Universidad de Cuyo. Mendoza.

Cardich, A., & Paunero, R. (1992). Arqueología de la Cueva II de Los Toldos (Santa Cruz. Argentina). In *Anales de Arqueología y Etnología. Universidad de Cuyo. Mendoza, 46*(47), 49–71.

Casamiquela, R. (2000). Temas patagónicos de interés arqueológico VI. Análisis etnográfico de la morfología del toldo tehuelche y sus derivaciones etnológicas (hacia una 'retroetnología'). *Intersecciones En Antropología, 1*, 3–33.

Cassiodoro, G. (2004). La tecnología cerámica en cazadores recolectores de la provincia de Santa Cruz. In *Libro de resúmenes del XV Congreso Nacional de Arqueología Argentina* (p. 92). Universidad Nacional de Río Cuarto.

Cassiodoro, G. (2008). La tecnología cerámica en cazadores recolectores del noroeste de la Provincia de Santa Cruz. In A. Austral, & M. Tamagnini (Comps.), *Problemáticas de la Arqueología Contemporánea. Tomo II* (pp: 227–237). Universidad Nacional de Río Cuarto.

Cassiodoro, G. (2011). *Movilidad y uso del espacio de cazadores-recolectores del Holoceno tardío. Estudio de la variabilidad del registro tecnológico en distintos ambientes del noroeste de la provincia de Santa Cruz (Argentina).* South American Archaeology Series 13, BAR International Series 2259. Archaeopress. https://doi.org/10.30861/9781407308302

Cassiodoro, G., Aragone, A., & Re, A. (2004) Más allá de los chenques… Registro arqueológico de sitios a cielo abierto en la cuenca de los Lagos Salitroso y Posadas Pueyrredón. In T. Civalero, P. Fernández, & G. Guráieb (Eds.), *Contra Viento y Marea. Arqueología de Patagonia* (pp. 325–338). INAPL-Sociedad Argentina de Antropología.

Cassiodoro, G., Flores Coni, J., Agnolin, A., & Goñi, R. (2020). Caracterización de puntas de proyectil apedunculadas. Un aporte al poblamiento del centro-oeste de la provincia de Santa Cruz (Argentina). *Revista del Museo de La Plata, 5*(1), 126–141. https://doi.org/10.24215/25456377e099

Castro, A., & Moreno, E. (2000). Noticia sobre enterratorios humanos en la costa norte de Santa Cruz—Patagonia Argentina. *Anales Instituto Patagonia. Serie Ciencias Humanas, 28*, 225–231.

Castro, A., Moreno, E., Andolfo, M., & Zubimendi, M. (2001). Distribución espacial de sitios en la localidad de Punta Medanosa. Santa Cruz (Argentina). *Relaciones De La Sociedad Argentina De Antropología, 26*, 303–321.

Cattáneo, R. (1999). Organización de la tecnología en la Patagonia Centro Meridional: El caso de la localidad arqueológica Piedra Museo. Provincia de Santa Cruz. In C. Diez Marín (Ed.), *Actas del XII Congreso Nacional de Arqueología Argentina III* (pp. 16–24). Universidad Nacional de La Plata.

Cattáneo, R. (2000). El paisaje y la distribución de recursos líticos en el Nesocratón del Deseado. In L. Miotti, R. Paunero, M. Salemme, & G. Cattáneo (Eds.), *Guía de campo de la visita a las localidades arqueológicas. Taller Internacional del INQUA: La colonización del Sur de América durante la Transición Pleistoceno/Holoceno* (pp. 26–35). INQUA.

Cattáneo, R. (2002). *Una aproximación a la organización de la tecnología lítica entre los cazadores-recolectores del Holoceno medio/Pleistoceno final en la Patagonia austral (Argentina).* Doctoral

Thesis. Facultad de Ciencias Naturales y Museo-Universidad Nacional de La Plata. La Plata. http://sedici.unlp.edu.ar/handle/10915/4613

Cattáneo, R. (2005). Tecnología lítica en la localidad arqueológica Piedra Museo (Santa Cruz. Argentina). *Relaciones De La Sociedad Argentina De Antropología, 30*, 79–103.

Cattáneo, R., Di Lello, C., & Gómez, J. C. (2006). Cuantificación y análisis de la distribución de rocas útiles para la manufactura de instrumentos a través del uso de sistemas de información geográfica (SIG) en el área de Piedra Museo. Santa Cruz. Argentina. In A. Austral, & M. Tamagnini (Eds.), *Problemáticas de la Arqueología contemporánea. Tomo I* (pp. 165–175). Universidad Nacional de Río Cuarto.

Chaile, C., Goñi, R., & Cassiodoro, G. (2019). La cerámica del fin del mundo: el uso de contenedores cerámicos en el centro-oeste de Santa Cruz (Argentina). In A. Laguens, M. Bonnin, & B. Marconetto (Comp.), *Libro de Resúmenes XX Congreso Nacional de Arqueología Argentina* (pp. 1515–1519). IDACOR-UNC.

Chapa, T., Pereira, J., Madrigal, A., & Mayoral, V. (1998). *La necrópolis ibérica de los Castellones de Céal (Hinojares. Jaén)*. Colección Arqueología. Junta de Andalucía. Universidad de Jaén.

Chapa, T., Vallejo, I., Belén, M., Martínez-Navarrete, M. I., Ceprián, B., Rodero, A., & Pereira, J. (2009). El trabajo de los escultores ibéricos: Un ejemplo de Porcuna (Jaén). *Trabajos De Prehistoria, 66*(1), 161–173.

Ciampagna, M. L. (2014). *Estudio de la interacción entre grupos cazadores recolectores de Patagonia y las plantas silvestres: El caso de la costa norte de Santa Cruz durante el Holoceno medio y tardío*. Doctoral Thesis. Universidad Nacional de La Plata. La Plata. http://sedici.unlp.edu.ar/handle/10915/45420. Accessed September 15, 2024.

Cione, A., Tonni, E., & Soilbelzon, L. (2009). Did humans cause the late pleistocene-early holocene mammalian extinctions in South America in a context of shrinking open areas? In G. Haynes (Ed.), *American megafaunal extinctions at the end of the Pleistocene* (pp. 125–144). Springer.

Cirigliano, N., Montenegro, T., Franco, N., Brook, G., & Cherkinsky, A. (2019). Hallazgos cerámicos y su cronología en el área comprendida entre el extremo sur del Macizo del Deseado y la cuenca del río Santa Cruz (Patagonia sur, Argentina). *Magallania, 47*(2), 132–149.

Civalero, M. T. (1999). Obsidiana en Santa Cruz. Una problemática a resolver. In J. B. Belardi, P. M. Fernández, R. A. Goñi, A. G. Guráieb, & M. De Nigris (Eds.), *Soplando en el viento...Actas de las terceras Jornadas de Arqueología de la Patagonia* (pp. 155–164). Instituto Nacional de Antropología y Pensamiento Latinoamericano. Universidad Nacional del Comahue.

Claraz, J. (1988) [1865–66]. *Diario de viaje de exploración al Chubut. 1865–1866*. Rodolfo Casamiquela. Marymar.

Clarke, D. L. (1972). Models and paradigms in contemporary archaeology. In D. Clarke (Ed.), *Models in archaeology* (pp: 1–60). Methuen and Co LTD.

Clarke, D. L. (1977). Spatial information in archaeology. In D. Clarke (Ed.), *Spatial archaeology* (pp. 1–32). Academic Press.

Clarke, K., Parks, B., & Crane, M. (2002). Modeling human-environmental systems. In K. Clarke, B. Parks, O. Bradley, & M. Crane (Eds). *Geographic information systems and environmental modeling* (pp. 160–187). Upper Saddle River.

Cohen, J. (1988). *Statistical power analysis for the behavioral sciences*. Routledge.

Conolly, J., & Lake, M. (2006). *Geographical information systems in archaeology*. University Press.

Corbella, H., & Lara, L. (2008). Late cenozoic quaternary volcanism in patagonia and Tierra del Fuego. In J. Rabassa (Ed.), *The late cenozoic of patagonia and Tierra del Fuego* (pp. 437–483). Elsevier.

Correa, M. N. (1971). Flora patagónica. *Boletín De La Sociedad Argentina De Botánica, 14*, 1–2.

Crabtree, D. (1972). *An introduction to flintworking. Occasional papers nº 28*. Idaho State University Museum.

Criado Boado, F. (1993). Límites y posibilidades de la arqueología del paisaje. *Revista De Prehistoria y Arqueología De La Universidad De Sevilla, 2*, 9–55.

Criado Boado, F. (1999). Del terreno al espacio: Planteamientos y perspectivas para la Arqueología del Paisaje. *CAPA, 6*, 1–58.

Crivelli Montero, E., Fernández, M., Franco, N., Pardiñas, U., & Tavella, Z. (1990–1992). Prospecciones arqueológicas en el partido de Tapalqué (provincia de Buenos Aires). *Paleoetnológica, 6*, 31–46.

Crivelli Montero, E., Fernández, M., Franco, N., Pardiñas, U., & Tavella, Z. (1992). Prospecciones arqueológicas en el partido de Tapalqué (provincia de Buenos Aires). *Paleoetnológica, 6*, 31–46.

Crivelli Montero, E., & Fernández, M. M. (2004). Demografía, movilidad y tecnología bifacial en sitios de la cuenca del río Limay, In *Contra Viento y Marea. Arqueología de Patagonia* (pp. 89–103). Instituto Nacional de Antropología y Pensamiento Latinoamericano, Sociedad Argentina de Antropología.

Crivelli Montero, E., Pardiñas, U., & Fernández, M. (1996). Introducción, procesamiento y almacenamiento de macro-vegetales en la Cueva Epullán Grande (provincia del Neuquén). In J. Gómez Otero (Ed.) *Arqueología. Sólo Patagonia. Ponencias de las II Jornadas de Arqueología de la Patagonia* (pp. 49–57). Centro Nacional Patagónico.

Cueto, M. (2015) *Análisis de los procesos de uso de artefactos líticos en sociedades cazadoras-recolectoras. Ocupaciones correspondientes a la transición Pleistoceno/Holoceno, Meseta Central de Santa Cruz.* BAR International Series.

Cueto, M., & Frank, A. (2008). Tratamiento térmico en la producción de artefactos líticos: Estudios experimentales. In A. Austral & M. Tamagnini (Eds.), *Problemáticas de la Arqueología Contemporánea II* (pp. 375–383). Universidad Nacional de Río Cuarto.

Curtoni, R. (2007). *Arqueología y paisaje en el área centro-este de la provincia de La Pampa: la especialidad humana y la formación de territorios.* Doctoral Thesis. Facultad de Ciencias Naturales y Museo, Universidad Nacional de La Plata. http://sedici.unlp.edu.ar/handle/10915/4347

Curtoni, R., & Oliván, A. (2023). Arqueología como política. De la res cogitans a la decolonialidad de la práctica. *Revista Latino-Americana de Arqueología Histórica, 17*(2), 25–42.

Dalla Bona, L., & Larcombe, L. (1996). Modelling prehistoric land use in Northern Ontario. In H. Maschner (Ed.), *New methods. Old problems. Geographic information systems in modern archaeological research.* Occasional Paper No. 23 (pp. 252–271). Center of Archaeological Investigations—Southern Illinois University at Carbondale.

De Aparicio, F. (1933/1935). Viaje preliminar de exploración en el territorio de Santa Cruz. *Publicaciones del Museo Antropológico y Etnográfico. Serie A, 3*, 71–92.

De Giusto, J., Di Persia, C., & Pezzi, E. (1980). Nesocratón del deseado. *Geología Regional Argentina, 2*, 1389–1462.

De Nigris, M., & Mengoni Goñalons, G. (2004). El guanaco como fuente de carne. In M. T. Civalero, P. Fernández, G. Guraieb (Eds.), *Contra Viento y Marea. Arqueología de Patagonia* (pp. 469–476). Instituto Nacional de Antropología y Pensamiento Latinoamericano. Sociedad Argentina de Antropología.

De Porras, M. E. (2010). *Dinámica de la vegetación de la Meseta Central de Santa Cruz durante los últimos 11.000 años a partir del análisis polínico: Forzantes bióticos y abióticos.* Doctoral Thesis. Facultad de Ciencias Exactas y Naturales, Universidad Nacional de Mar Del Plata, Mar del Plata.

De Porras, M. E., Mancini, M. V., & Prieto, A. R. (2008). Cambios de la vegetación del cañadón de Los Toldos (Santa Cruz) durante el Holoceno medio-tardío en relación con las variaciones climáticas y las ocupaciones humanas. In F. Morello, M. Martinic, A. Prieto, & G. Bahamonde (Eds.), *Arqueología de Fuego-Patagonia. Levantando piedras, desenterrando huesos… y develando arcanos* (pp. 53–60). CEQUA.

De Porras, M. E., Mancini, M. V., & Prieto, A. R. (2009). Vegetation changes and human occupation in the Patagonian steppe, Argentina, during the late Holocene. *Vegetation History Archaeobotany, 18*(3), 235–244. https://doi.org/10.1007/s00334-008-0200-8

Deodat, L. (1946). ¿"Chenke" o "Coshom"? Una pequeña cuestión lexicológica de la Patagonia argentina. *Argentina Austral, 15*(180), 4–13.

Dewar, R., & McBride, K. (1992). Remnant settlement patterns. In J. Rossignol, & L. Wandsnider (Eds.), *Space time and archaeological landscapes* (pp. 227–255). Plenum Press.

Dincauze, D. (2000). *Environmental archaeology. Principles and practice.* Cambridge University Press.

Dobres, M. A., & Hoffman, C. (1994). Social agency and the dynamics of prehistoric technology. *Journal of Archaeological Method and Theory, 1*(3), 211–258.

Dunnell, R. (1980). Evolutionary theory and archaeology. In M. Schiffer (Ed.), *Advances in archaeological method and theory* (pp. 35–99). Academic Press.

Dunnell, R. C., & Dancey, W. S. (1983). The siteless survey: A regional scale data collection strategy. In M. Schiffer (Ed.), *Advances in archaeological method and theory* (pp. 267–287). Academic Press.

Durán, V. (1986/1987). Estudio tecno-tipológico de los raspadores del sitio El Verano. Cueva 1. Patagonia Centro meridional. Santa Cruz. Argentina. In *Anales de Arqueología y Etnología* (pp. 41–42:129–163).

Ebert. J., & Kohler, T. (1988). The theoretical basis of archaeological predictive modeling and a consideration of appropriate data-collection methods. In W. J. Judge, & L. Sebastian (Eds.), *Quantifying the present and predicting the past: Theory. Method and application of archaeological predictive modeling* (pp. 97–172). US Bureau of Land Management. Department of Interior. US Government Printing Office.

Echeverría Baleta, M. (1982). *Toponimia indígena de Santa Cruz.* Fundación Banco Almafuerte.

Echeverría Baleta, M. (1999). *Toponimia indígena de Santa Cruz.* Cumacú.

Engelmark, R., & Larsson, T. (2005). Rock art and environment: Towards increased contextual understanding. In M. Santos Estévez, & A. Troncoso Meléndez (Coord.), *Trabajos de Arqueoloxía e Patrimonio. Reflexiones sobre Arte Rupestre. paisaje. forma y contenido (TAPA 33)* (pp. 113–122). Instituto de Estudos Galegos Padre Sarmiento.

ERSDAC. (2003). *Aster reference Guide Version 1.0.* Earth Remote Sensing Data Analysis Center. https://unit.aist.go.jp/igg/rs-rg/ASTERSciWeb_AIST/en/documnts/users_guide/part1/pdf/Part1_4E.pdf

Espinosa S. (2000). Los conjuntos artefactuales líticos de la estepa y el bosque en el Parque Nacional Perito Moreno (Santa Cruz. Argentina). In J. Belardi, F. Carballo Marina, & S. Espinosa (Eds). *Desde el País de los Gigantes. Perspectivas arqueológicas en Patagonia* (pp. 357–367). Universidad Nacional de la Patagonia Austral.

Espinosa, S., & Goñi, R. (1999). ¡Viven!: una fuente de obsidiana en la provincia de Santa Cruz. In J. B. Belardi, P. M. Fernández, R. A. Goñi, A. G. Guráieb, & M. De Nigris (Eds.), *Soplando en el viento... Actas de las Terceras Jornadas de Arqueología de la Patagonia* (pp. 177–188). Instituto Nacional de Antropología y Pensamiento Latinoamericano y Universidad Nacional del Comahue.

Fábrega Álvarez, P., & Parcero Oubiña, C. (2007). Proposals for an archaeological analysis of movement and pathways. *Archeologia e Calcolatori, 18*, 121–140.

Fanning, P., & Holdaway, S. (2001). Stone artifact scatters in Western NSW. Australia: Geomorphic controls on artifact size and distribution. *Geoarchaeology: An International Journal, 16*(6), 667–686.

Farrand, W. R. (1985). Rockshelter and cave sediments. In J. Stein, & W. Farrand (Eds.), *Archaeological sediments in context. Peopling of the Americas. Volume 1* (pp. 21–40). Center for the Study of Early Man. Institute for Quaternary Studies. University of Maine.

Favier Dubois, C. (1997). Indicadores de tasas de depositación relativas en el registro arqueológico. Su aplicación en sitios de Fuego Patagonia. *Arqueología, 7*, 51–75.

Favier Dubois, C. (2001a). *Análisis geoarqueológico de los procesos de formación del registro, en sitios arqueológicos de Fuego-Patagonia.* Doctoral Thesis, Facultad de Ciencias Exactas, Universidad Nacional de Buenos Aires, Buenos Aires. https://bibliotecadigital.exactas.uba.ar/download/tesis/tesis_n3404_FavierDubois.pdf

Favier Dubois, C. (2001b). *Análisis geoarqueológico de los procesos de formación del registro, en sitios arqueológicos de Fuego-Patagonia.* Doctoral Thesis. Facultad de Ciencias Exactas. Universidad Nacional de Buenos Aires. Buenos Aires. http://repositorio.filo.uba.ar/handle/filodigital/9887

Favier Dubois, C. (2004). Fluctuaciones climáticas referibles al período cálido Medieval en Fuego-Patagonia: indicadores indirectos y el aporte de modelos climáticos. In M. T. Civalero, P. M. Fernández, & A. G. Guráieb (Eds.). *Contra Viento y Marea. Arqueología de Patagonia* (pp. 545–556). INAPL-SAA.

Fernández Cacho, S. (2006). *Patrimonio arqueológico y políticas territoriales en Andalucía.* Doctoral Thesis. Departamento de Prehistoria y Antropología. Universidad de Sevilla. http://hdl.handle.net/11441/57784. Accessed September 15, 2024.

Fernández, M. (2013). *Los paleoambientes de Patagonia meridional, Tierra del Fuego e Isla de los Estados en los tiempos de las primeras ocupaciones humanas. Estudio basado en el análisis de diatomeas.* Doctoral Thesis. Universidad Nacional de La Plata. https://sedici.unlp.edu.ar/handle/10915/34395

Ferring, C. (1986). Rates of fluvial sedimentation: Implications for archaeological variability. *Geoarchaeology: An International Journal, 1*(3), 259–274.

Fidalgo, F. (1973). Consideraciones sobre los bajos situados al norte de la provincia de Santa Cruz. *Actas del 5° Congreso Geológico Argentino. Tomo V* (pp. 123–137).

Figuerero Torres, M. J., Pereyra, F., Curato, L., & Chiara, M. (2004). Modelo SIG de dinámica del paisaje arqueológico en Patagonia Meridional (Cuenca del Lago Posadas). In *Libro de resúmenes del XV Congreso Nacional de Arqueología Argentina* (p. 188). Universidad Nacional de Río Cuarto.

Fisher, P., Farrelly, C., Maddocks, A., & Ruggles, C. (1997). Spatial analysis of visible areas from the bronze age cairns of mull. *Journal of Archaeological Science, 24*(7), 581–592.

Fisher, P., Wood, J., & Cheng, T. (2004). Where is Helvellyn? Fuzziness of multi-scale landscape morphometry. *Transactions of the Institute of British Geographers, 29*, 106–128.

Flegenheimer, N., & Bayón, C. (2008). Abastecimiento de calizas silicificadas en las ocupaciones tempranas en Pampa Bonaerense. In A. Austra, & M. Tamagnini (Comps.), *Problemáticas de la Arqueología Contemporánea. Tomo II* (pp. 479–481). Universidad Nacional de Río Cuarto.

Flores Coni, J. (2019a). Tecnología de parapetos en Patagonia meridional: El caso de la meseta del Strobel. *Relaciones de la Sociedad Argentina de Antropología, 43*(1), 131–153. http://sedici.unlp.edu.ar/handle/10915/89986

Flores Coni, J. (2019b). *Poblamiento humano y uso del espacio en la meseta del Strobel (provincia de Santa Cruz). Un análisis sobre la variabilidad tecnológica durante el Holoceno.* Doctoral Thesis. Facultad de Filosofía y Letras. Universidad de Buenos Aires. http://hdl.handle.net/11336/82893

Flores Coni, J. (2019c). Tecnología de parapetos en Patagonia meridional: El caso de la meseta del Strobel. *Relaciones De La Sociedad Argentina De Antropología, 44*(1), 131–153.

Flores Coni, J., Dellepiane, J., Cassiodoro, G., Goñi, R., & Agnolin, A. (2021). Technological strategies and guaranteed return: Hunting blinds and patagonian plateaus. In J. B. Belardi, D. L. Bozzuto, P. M. Fernández, E. A. Moreno, & G. A. Neme (Eds.), *Ancient hunting strategies in southern South America* (pp. 293–312). Springer.

Foley, R. (1981). A model of regional archaeological structure. *Proceedings of the Prehistoric Society, 47*, 1–17. https://doi.org/10.1017/S0079497X00008823

Franco, N. (2002). ¿Es posible diferenciar los conjuntos líticos atribuidos a la exploración de un espacio de los correspondientes a otras etapas del poblamiento? El caso del extremo sur de Patagonia. *Werken, 3*, 119–132.

Franco, N., & Borrero, L. A. (1996). El estrés temporal y los artefactos líticos. La cuenca superior del río Santa Cruz. In J. Gómez Otero (Ed.). *Arqueología. Sólo Patagonia*, (pp. 341–348). CENPAT-CONICET.

Franco, N., & Borrero, L. A. (1999). Metodología de análisis de la estructura regional de recursos líticos. In C. Aschero, A. Korstanje, & P., Vuoto (Eds.). *En los tres reinos: Prácticas de recolección en el cono sur de América* (pp. 27–37). Magna Publicaciones.

Franco, N., & Carballo Marina, F. (1993). Variabilidad en raederas en Lago Argentino (Santa Cruz. Argentina). *Arqueología, 3*, 213–232.

Franco, N., Castro, A., Cardillo, M., & Charlin, J. (2009). La importancia de las variables morfológicas, métricas y de microdesgaste para evaluar las diferencias en diseños de puntas de proyectil

bifaciales pedunculadas. Un ejemplo del sur de Patagonia Continental. *Magallania, 37*(1), 99–112.
Franco, N. V., Ambrústolo, P., Acevedo, A., Cirigliano, N. A., & Vommaro, M. (2013). Prospecciones en el sur del Macizo del Deseado (Provincia de Santa Cruz). Los casos de La Gruta y Viuda Quenzana. In F. J. Zangrando, R. Barberena, A. F. Gil, G. A. Neme, M.A. Giardina, L. Luna, C. Otaola, S. L. Paulides, L. M. Salgán, & A. M. Tivoli (Eds.), *Tendencias teórico metodológicas y casos de estudio en la arqueología de la Patagonia* (pp. 371–378). Museo de Historia Natural de San Rafael. Altuna Impresores.
Franco, N. V., Ambrústolo, P., Martucci, M., Brook, G. A., Mancini, M. V., & Cirigliano, N. A. (2010a). Early human occupation in the Southern part of the Deseado Massif (Patagonia, Argentina). *Current Research in the Pleistocene, 27*, 13–16.
Franco, N. V., Brook, G. A., Ambrústolo, P., Iglesias, C., Stern, C., Glascock, M. D., & Vetrisano, L. (2023). Recientes datos de elementos traza de una nueva fuente primaria de obsidiana negra en el noreste de la provincia de Santa Cruz, Patagonia, Argentina. *Magallania*, 51, 1–14. https://doi.org/10.22352/MAGALLANIA202351004
Franco, N. V., Brook, G. A., Cirigliano, N. A., Stern, C., & Vetrisano, L. (2017). 17 de Marzo (Santa Cruz, Argentina): A new distal source of Pampa del Asador type black obsidian and its implications for understanding hunter-gatherer behavior in Patagonia. *Journal of Archaeological Science Reports, 12*, 232–243.
Franco, N. V., Martucci, M., Ambrústolo, P., Brook, G., Mancini, M. V., & Cirigliano, N. A. (2010b). Ocupaciones humanas correspondientes a la transición Pleistoceno-Holoceno al sur del Macizo del Deseado: El área de La Gruta (provincia de Santa Cruz, Argentina). *Relaciones De La Sociedad Argentina De Antropología, 35*, 301–308.
Franco, N. V., Vetrisano, L., Gilio, B. L., Cirigliano, N. A., & Bianchi, P. E. (2021). Hunting blinds in the Southern end of the Deseado Massif: Collective hunting strategies during the late Holocene. In J. B. Belardi, D. L. Bozzuto, P. M. Fernández, E. A. Moreno, & Neme, G. A. (Eds.), *Ancient hunting strategies in Southern South America* (pp. 313–341). Springer. http://hdl.handle.net/11336/184966
Franco, N. V., Vetrisano, L. J., Mancini, M. V., & Brook, G. (2020). Nueva información referida a la transición Pleistoceno-Holoceno y al Holoceno temprano en el extremo sur del Macizo del Deseado (Patagonia, Argentina). *Revista del Museo de La Plata, 5*(1), 109–125. https://doi.org/10.24215/25456377e098
Frank, A. (2011). *Tratamiento térmico y manejo del fuego en sociedades cazadoras-recolectoras de la Meseta Central de Santa Cruz.* Doctoral Thesis. Facultad de Ciencias Naturales y Museo. Universidad Nacional de La Plata. La Plata. http://sedici.unlp.edu.ar/handle/10915/5318
Gaffney, V., Stančič, Z., & Watson, H. (1996). Moving from catchments to cognition: Tentative steps toward a larger archaeological context for GIS. In M. Aldenderfer, & H. Maschner (Eds.) *Anthropology space and geographic information systems* (pp.132–154). Oxford University Press.
Gamble, C. (1982). Interaction and alliance in paleolithic society. *Man, 17*(1), 92–107.
Gamble, C. (1990). *El Poblamiento Paleolítico de Europa.* Crítica.
García Añino, E. (2018). *Estrategias de consumo de grandes mamíferos a lo largo del Holoceno entre los cazadores-recolectores de la Meseta Central de Santa Cruz. El caso del guanaco en el sitio Cueva Maripe.* Doctoral Thesis. Facultad de Ciencias Naturales y Museo. Universidad Nacional de La Plata. La Plata. https://doi.org/10.35537/10915/71185
García Guraieb, S., Goñi, R., & Tessone, A. (2015). Paleodemography of Late Holocene hunter-gatherers from Patagonia (Santa Cruz, Argentina): An approach using multiple archaeological and bioarchaeological indicators. *Quaternary International, 356*, 147–158. https://doi.org/10.1016/j.quaint.2014.09.054
García Guráieb, S., González, P., & Bernal, V. (2007). Estructura de sexo y edad de la muestra de restos humanos del Holoceno Tardío del Lago Salitroso (Santa Cruz. Argentina). In F. Morello, M., Martinic, A. Prieto, & G. Bahamonde (Eds.), *Arqueología de Fuego-Patagonia. Levantando piedras. desenterrando huesos… y develando arcanos* (pp. 367–374). CEQUA.

García Guráieb, S. G. (2011). *Bioarqueología de cazadoresrecolectores del Holoceno tardío de la Cuenca del Lago Salitroso (Santa Cruz): aspectos paleodemográficos y paleopatológicos.* Tesis Doctoral. Facultad de Filosofía y Letras, Universidad de Buenos Aires. http://repositorio.filo.uba.ar/handle/filodigital/1655

García, L., & Pérez de Micou, C. (1980). Aproximación a un análisis funcional de parapetos pertenecientes al Complejo Patagoniense en la Meseta de Somuncurá. Provincia De Río Negro. *Sapiens, 4*, 139–144.

García Sanjuán, L. (2005). *Introducción al Reconocimiento y Análisis Arqueológico del Territorio.* Ariel.

Garcia Sanjuán, L. (2006), Funerary ideology and social inequality in the Late Prehistory of the Iberian South-West (c. 3300–850 cal BC)". In P. del Río, & L. García Sanjuán (Eds.). *Social inequality in Iberian late prehistory* 1525 (pp. 149–170). British Archaeological Reports International Series. Archaeopress.

García Sanjuán, L., Metcalfe-Wood, S., Rivera Jimenez, T., & Wheatley, D. (2006). Análisis de pautas de visibilidad en la distribución de monumentos megalíticos de Sierra Morena Occidental. In I. Grau Mirá (Ed.), *La aplicación de los SIG en la arqueología del paisaje* (pp. 181–200). Universidad de Alicante.

Giardina, M., Hermo, D., Miotti, L., & Vazquez, M. (2000). Resolución e Integridad Arqueológica del Componente Inferior de Piedra Museo. In *Guía de Campo de la Visita a las Localidades Arqueológicas: La Colonización del Sur de América Durante la Transición Pleistoceno/Holoceno* (pp. 81–83). INQUA.

Gillings, M., Hacıgüzeller, P., & Lock, G. (2020). Archaeology and spatial analysis. In *Archaeological spatial analysis: A methodological guide.* Routledge.

Goebel, T., Waters, M., & O'Rourke, D. (2008). The late pleistocene dispersal of modern humans in the Americas. *Science, 14*, 319(5869), 1497–1502.

Gómez, J. C., & Magnin, L. (2008). Cartografía geomorfológica aplicada a un sector de interés arqueológico en el Macizo del Deseado. Santa Cruz (Patagonia Argentina). *Investigaciones Geográficas. Boletín del Instituto de Geografía. UNAM, 65*, 22–37. https://doi.org/10.14350/rig.17972

Gómez Otero, J. (1986–87). Posición estratigráfica particular de puntas de los períodos IV y V de Bird en el alero Potrok-Aike (Santa Cruz). In *Primeras Jornadas de Arqueología de la Patagonia* (pp. 125–130). Dirección de Cultura de la Provincia del Chubut.

Goñi, R. (2000). Arqueología de momentos históricos fuera de los centros de conquista y colonización: Un análisis de caso en el sur de la Patagonia. In J. Belardi. F. Carballo Marina, & S. Espinosa (Eds). *Desde el País de los Gigantes. Perspectivas arqueológicas en Patagonia Tomo I* (pp. 283–293). Universidad Nacional de la Patagonia Austral.

Goñi, R. (2000–2002). Fechados radiocarbónicos y registro arqueológico en la cuenca de los lagos Salitroso/Posadas (Santa Cruz). *Cuadernos del Instituto Nacional de Antropología y Pensamiento Latinoamericano, 19*, 666–669.

Goñi, R. A., Belardi, J. B., Re, A., Nuevo Delaunay, A., Molinari, R. L., & Ferraro, L. (2007). Los grabados de la meseta del lago Strobel (*Patagonia argentina*) desde una perspectiva regional. In R. Hostnig, M. Strecker, & J. Guffroy (Eds.), *Actas del Primer Simposio Nacional de Arte Rupestre* (pp. 427–438). Instituto Francés de Estudios Andinos-Instituto de Investigación para el Desarrollo-Embajada de Alemania en el Perú. Actes and Memoires de l'Institut Francais d'Etudes Andines 12.

Goñi, R., & Barrientos, G. (2000). Estudio de chenques del Lago Salitroso. In J. B. Belardi, F. Carballo Marina, & S. Espinosa (Eds.), *Desde el País de los Gigantes. Perspectivas arqueológicas en Patagonia, Tomo I* (pp. 161–175). Universidad Nacional de la Patagonia Austral.

Goñi, R., & Barrientos, G. (2004). Poblamiento tardío y movilidad en la cuenca del Lago Salitroso. In M. T. Civalero, P. Fernández, & G. Guraieb (Eds.), *Contra Viento y Marea. Arqueología de Patagonia* (pp. 313–324). Instituto Nacional de Antropología y Pensamiento Latinoamericano, Sociedad Argentina de Antropología.

Goñi, R., Barrientos, G., Figuerero, M., Mengoni, G., Mena, F., Lucero, V., & Reyes, O. (2004). Distribución espacial de entierros en la cordillera de Patagonia centro-meridional (Lago Salitroso-Paso Roballos, Argentina. Entrada Baker-Chacabuco, Chile). *Chungara, 36*, 1101–1107. https://doi.org/10.4067/S0717-73562004000400045

Goñi, R., Belardi, J. B., Cassiodoro, G., Rindel, D., García Guraieb, S., & Bourlot, T. (2014). Registro arqueológico y cronología de las ocupaciones cazadoras recolectoras en la cuenca del lago Cardiel. In R. Goñi, J. B. Belardi, G. Cassiodoro, & A. Re (Eds.). *Arqueología de las cuencas de los lagos Cardiel y Strobel. Poblamiento humano y paleoambientes en Patagonia* (pp. 41–66). Aspha.

Goñi, R., Belardi, J. B., Espinosa, S., & Savanti, F. (2004). Más vale tarde que nunca: Cronología de las ocupaciones cazadoras—Recolectoras en la cuenca del lago Cardiel (Santa Cruz, Argentina). In *Contra viento y marea. Arqueología de la Patagonia*, (pp. 237–248). INAPL-SAA.

Goñi, R. G., Barrientos, M., Figuerero, G., Mengoni, F., Mena, V., Lucero, M., & Reyes, O. (2004). Distribución espacial de entierros en la cordillera de Patagonia centro-meridional (Lago Salitroso-Paso Roballos, Argentina. Entrada Baker-Chacabuco, Chile). *Chungara, 36*, 1101–1107.

González Dubox, R. (2024). *Arte, comunicación y paisaje en sociedades de cazadores-recolectores del sur de la Meseta Central de Santa Cruz*. Doctoral Thesis. Facultad de Ciencias Naturales y Museo. Universidad Nacional de La Plata. http://sedici.unlp.edu.ar/handle/10915/172112

González Pérez, C., Del Río Pose, J., Bóveda López, M., & Criado Boado, F. (1997). Tecnologías orientadas a objetos en la gestión de recursos culturales. *Boletín Del Instituto Andaluz Del Patrimonio Histórico, 21*, 137–140.

Gradín, C. (1959–60). Petroglifos de la meseta del lago Strobel (Provincia de Santa Cruz. Argentina). *Acta Praehistorica, 3–4*, 23–143.

Gradín, C. (1959–60). Tres informaciones referentes a la meseta del Lago Strobel (Prov. De Santa Cruz. Argentina). *Acta Praehistorica, 3–4*, 144–149.

Gradín, C. (1962). Tres informaciones referentes a la meseta del Lago Strobel. (Provincia de Santa Cruz, Argentina). *Acta Praehistorica, 3–4*, 144–149.

Gradín, C. (1976a). Parapetos de piedra y grabados rupestres en la meseta del lago Buenos Aires. In *Actas y Memorias del IV Congreso Nacional de Arqueología Argentina, tomo III (1–4)* (pp. 315–337). Museo de Historia Natural de San Rafael.

Gradin, C. (1976b). Parapetos de piedra y grabados rupestres de la meseta del lago Buenos Aires. In *Actas y Memorias IV Congreso Nacional de Arqueología Argentina* (pp. 315–337). Revista del Museo de Historia Natural de San Rafael.

Gradín, C. (1976c). Parapetos de piedra y grabados rupestres en la meseta del Lago Buenos Aires. *Revista Museo Historia Natural de San Rafael. Tomo I*, 315–337.

Gradín, C. (1976db). Parapetos de piedra y grabados rupestres en la meseta del lago Buenos Aires. *Revista Museo Historia Natural De San Rafael, 1*, 315–337.

Gradín, C., & Aguerre, A. (1992). Nuevo aporte al conocimiento de la dinámica poblacional en la cuenca del Río Pinturas (Provincia de Santa Cruz. República Argentina). In L. A. Borrero, & J. L. Lanata (Eds.). A*nálisis Espacial en la Arqueología Patagónica* (pp. 83–120). Ayllu.

Gradin, C., & Aguerre, A. (1994). *Contribución a la Arqueología del Río Pinturas, provincia de Santa Cruz*. Ayllu.

Gradin, C., Aschero, C., & Aguerre, A. (1976). Investigaciones arqueológicas en la Cueva de Las Manos, Estancia Alto Río Pinturas (Santa Cruz). *Relaciones De La Sociedad Argentina De Antropología, 10*, 201–250.

Gradín, C., Aschero, C., & Aguerre, A. (1987). Primeros niveles culturales en el Área Río Pinturas (provincia de Santa Cruz, Argentina). *Estudios Atacameños, 8*, 115–136.

Gradín, C. J. (1980). Secuencias radiocarbónicas del sur de la Patagonia Argentina. *Relaciones De La Sociedad Argentina De Antropología, 14*, 177–194.

GRASS Development Team. 2016. (2017, march 15). Geographic Resources Analysis Support System (GRASS) Software, Version 7.0.4RC1. Open Source Geospatial Foundation. http://grass.osgeo.org

Green, S. (1990). Approaching archaeological space: An introduction to the volume. In K. Allen, S. Green, & E. Zubrow (Eds.), *Interpreting space: GIS and archaeology* (pp. 3–8). Taylor & Francis.

Guichón, F., & Re, A. (2014). Un lugar muy particular. Caza, convergencia de poblaciones y circulación de información en la meseta del Strobel. In R. Goñi, J. B. Belardi, G. Cassiodoro, & A. Re (Eds). *Arqueología de las cuencas de los lagos Cardiel y Strobel. Poblamiento humano y paleoambientes en Patagonia* (pp. 155–186). Aspha.

Guráieb, A. G. (1999). Análisis de la diversidad en los conjuntos líticos de Cerro de los Indios (Lago Posadas. Santa Cruz). *Relaciones De La Sociedad Argentina De Antropología, 24*, 293–306.

Gusinde, M. (1982–1989). *Los Indios de Tierra del Fuego. Tomo 1*. CAEA.

Haberzettl, T., Anselmetti, F., Bowen, S., Fey, M., Mayr, C., Zolitschka, B., Ariztegui, D., Maíz, B., Ohlendorf, C., Kastner, S., Lucke, A., Schabitz, F., & Wille, M. (2009). Late Pleistocene dust deposition in the Patagonian steppe. Extending and refining the paleoenvironmental and tephrochronological record from Laguna Potherk Aike back to 55 ka. *Quaternary Science Reviews, 28*, 2927–2939.

Haloua, P. (1997). L*a vegetación patagónica en el yacimiento El Huemul—Koluel Kaike*. In O. Padín, & R. Portal (Coord.). Total Austral.YPF.

Harris, G. (2008). *Guía de aves y mamíferos de la costa patagónica*. El Ateneo.

Hartley, R. J. (1992). *Rock art on the Northern Colorado Plateau: Variability in content and context*. Avebury.

Hasenstab, R. (1996). Settlement as adaptation: Variability in iroquois village site selection as inferred through GIS. In H. Maschner (Ed.). *New methods. Old problems. Geographic information systems in modern archaeological research* (pp. 223–241). Center of Archaeological Investigations. Southern Illinois University at Carbondale.

Hawkes, C. (1954). Archaeological theory and method: Some suggestions from the Old World. *American Anthropologist, 56*, 155–168.

Hermo, D. (2005). Fuentes de aprovisionamiento de xilópalo en el Monumento Natural Bosque Petrificado, provincia de Santa cruz. In *Entre Pasados y Presentes. Trabajo de las VI Jornadas de Jóvenes investigadores en Ciencias Antropológicas* (pp. 235–246). Instituto Nacional de Antropología y Pensamiento Latinoamericano.

Hermo, D. (2008a). *Los cambios en la circulación de las materias primas líticas en ambientes mesetarios de Patagonia. Una aproximación para la construcción de los paisajes arqueológicos de las sociedades cazadoras-recolectoras*. Doctoral Thesis. Facultad de Ciencias Naturales y Museo. Universidad Nacional de La Plata. La Plata. https://www.academia.edu/4943318/Tesis_Hermo_2008_Tomos_Iy_II

Hermo, D. (2008b). *Los cambios en la circulación de las materias primas líticas en ambientes mesetarios de Patagonia. Una aproximación para la construcción de los paisajes arqueológicos de las sociedades cazadoras-recolectoras*. Doctoral Thesis. Facultad de Ciencias Naturales y Museo. Universidad Nacional de La Plata. La Plata. http://sedici.unlp.edu.ar/handle/10915/4414

Hermo, D. (2009). Estructura de los recursos líticos y paisajes arqueológicos en el Nesocratón del Deseado (Santa Cruz. Argentina). *Arqueología Suramericana/Arqueología Sul-americana, 5*(2), 178–203.

Hermo, D. (2014). Diseño instrumental y disponibilidad de materias primas. Buscando relaciones en la Meseta Central de Santa Cruz, Argentina. In P. Escola, & S. Hocsman (Eds.), *Artefactos, movilidad y funcionalidad de sitios: Problemas y perspectivas, BAR International Series 2628* (pp. 106–115). Archaeopress.

Hermo, D., Magnin, L., Moreira, P., & Medel, S. (2015). Variabilidad y distribución de fuentes de materias primas líticas del Macizo del Deseado (Santa Cruz, Argentina). *Intersecciones en Antropología. Volumen especial Materias primas líticas en Patagonia, 2*, 87–100. https://www.redalyc.org/articulo.oa?id=179546078008

Hermo, D., Marchionni, L., Mosquera, B., & Álvarez, V. (2021). Prospecciones en la localidad Alma Gaucha (Macizo del Deseado, Santa Cruz). Implicancias para la arqueología de las Cabeceras del arroyo El Pescado. In *Libro de resúmenes de las XI Jornadas de Arqueología de la Patagonia* (p. 150). Universidad Austral de Chile.

Hermo, D., & Miotti, L. (2003). El efecto zoom para relacionar los desechos líticos de un sitio y los paisajes arqueológicos de cazadores-recolectores. *Revista Intersecciones En Antropología, 4*, 99–108.

Hermo, D., & Miotti, L. (2011). La obsidiana en el Nesocratón del Deseado (Santa Cruz, Argentina): extractos de una oscura biografía. In D. Hermo, & L. Miotti (Coords.), *Biografías de Paisajes y Seres: Visiones desde la Arqueología Sudamericana* (pp.111–131). Encuentro Grupo Editor.

Hermo, D., & Vázquez, M. (1999). ¡Cuánto que caminamos! Primeros resultados de las prospecciones en Cerro Vanguardia y Monumento Natural Bosques Petrificados. In C. Diez Marín (Ed.), *Actas del XII Congreso Nacional de Arqueología Argentina. Tomo III* (pp. 475–483). UNLP.

Hernández, M., González, N., & Hernández, L. (2008). Late cenozoic geohidrology of extra-andean patagonia. Argentina. In J. Rabassa (Ed.). *The late cenozoic of patagonia and Tierra del Fuego: Developments in quaternary sciences series* 11, (pp. 497–509). Elsevier.

Higgs, E., & Vita Finzi, C. (1972a). Prehistoric economies: A territorial approach. In E. Higgs (Ed.), *Papers in economic prehistory* (pp. 27–36). Cambridge University Press.

Hodder, I., & Orton, C. (1990). *Análisis Espacial en Arqueología*. Crítica.

Ingold, T. (1986). *The appropriation of nature: Essays of human ecology and social relations*. Manchester University Press.

Ingold, T. (1993). The temporality of the landscape. *World Archaeology, 25*(2), 152–174.

Iñiguez, A. M., & Gradín, C. (1977). Análisis mineralógico por difracción de rayos X de muestras de pinturas de Cueva de las Manos. Ea. Alto río Pinturas. *Relaciones De La Sociedad Argentina De Antropología, 11*, 121–128.

Jochim, M. (1976). *Hunter-Gatherer subsistence and settlement*. Academic Press.

Jochim, M. (2022). Dots on the map: Issues in the archaeological analysis of site locations. *Journal of Archaeological Method and Theory*. https://doi.org/10.1007/s10816-022-09580-8

Johnson, J. (1979). Archaic biface manufacture production failures. A chronicle of the misbegotten. *Lithic Technology, 10*, 25–35.

Kelly, R. (1992). Mobility/Sedentism: Concepts. Archaeological measures and effects. *Annual Review of Anthropology, 21*, 43–66.

Kintigh, K. (1988). Measuring archaeological diversity by comparison with simulated assemblages. *American Antiquity, 49*(1), 44–54.

Kofalt, M. R., & Mascó, M. (2004). Flora. In L. González, & P. Rial (Eds.) *Guía Geográfica Interactiva de Santa Cruz*. INTA. http://www.inta.gov.ar/santacruz/info/documentos/teledet/GuiaSC/. Acceso Junio 15, 2010.

Kohler, T. (1988). Predictive locational modeling: History and current practice. In W. J. Judge, & L. Sebastian (Eds). *Quantifying the present and predicting the past: Theory. Method and application of archaeological predictive modeling* (pp. 19–59). US Bureau of Land Management. Department of Interior. US Government Printing Office.

Kohler, T., & Parker, S. (1986). Predictive models for archaeological resource location. In M. B. Schiffer (Ed.), *Advances in archaeological method and theory 9* (pp. 397–452). Academic Press.

Kornfeld, M., Vasil'ev, S., Miotti, L. (Eds.) (2007). *On shelter's ledge: Histories. Theories and methods of rockshelter research*. B.A.R. International Series.

Krist, F., & Brown, D. (1994). GIS modeling of Palaeo-Indian period caribou migrations and viewsheds in northeastern Lower Michigan. *Photogrammetric Engineering and Remote Sensing, 60*, 1129–1137.

Kuhn, S. (1994). A formal approach to the design and assembly of mobile toolkits. *American Antiquity, 59*(3), 426–442.

Kuna, M. (2000). Comments on archaeological prediction. In Lock. G. (Ed.), B*eyond the map: Archaeology and spatial technologies* (pp. 180–197). IOS Press.

Kvamme, K. (1988). A simple graphic approach and poor man's clustering technique for investigating surface lithic scatter types. *Plains Anthropologist, 33*, 385–394.

Kvamme, K. (1990). The fundamental principles and practice of predictive archaeological modelling. In A. Voorrips (Ed.), *Mathematics and information science in archaeology: A flexible framework* (pp. 257–295). Holos-Verlag.
Kvamme, K. (1995). *Archaeological spatial analysis using gis: methods and issues presented in the course information systems and geographical networks in archaeology: GIS and INTERNET*. University of Siena and the Italian Council of Research.
Kvamme, K. (2006). Integrating multidimensional geophysical data. *Archaeological Prospection, 13*, 57–72.
Lake, M., & Woodman, P. (2003). Visibility studies in archaeology: A review and case study. *Environment and Planning B: Planning and Design, 30*, 689–707.
Lake, M., Woodman, P., & Mithen, S. (1998). Tailoring GIS software for archaeological applications: An example concerning viewshed analysis. *Journal of Archaeological Science, 25*, 27–38.
Lanata, J. (1996). La diversidad instrumental en el norte de Península Mitre. Tierra del Fuego. *Arqueología, 6*, 159–197.
Leanza, J. (1958). Geología regional. In De Aparicio, F., & Difrieri, H. (Dirs.), *La Argentina. Suma de Geografía 1* (pp. 217–349). Peuser.
Llobera, M. (1996). Exploring the topography of mind: GIS. social space and archaeology. *Antiquity, 70*, 612–622.
Llobera, M. (2003). Extending GIS-based visual analysis: The concept of visualscapes. *International Journal of Geographical Information Science, 17*(1), 25–48.
Lock, G., & Harris, T. (2000). Beyond the map: Archaeology and spatial technologies. In G. Lock (Ed.), *Beyond the map: Archaeology and spatial technologies* (pp. 1–25). IOS Press.
Lyman, R. L. (1994). *Vertebrate taphonomy. Cambridge manuals in archaeology*. Cambridge University Press. https://doi.org/10.1017/CBO9781139878302
Lynch, V. (2016). *Estudio comparativo de la producción y uso de artefactos líticos en el Macizo del Deseado (Santa Cruz, Argentina)*. BAR International Series.
Magnin, L. (2009a). Paisajes digitales. Aproximación teórica. materiales y técnicas. In M. Salemme, F. Santiago, M. Álvarez, E. Piana, M. Vázquez, & M. Mansur (Comps.), *Arqueología de la Patagonia. Una mirada desde el último confín I* (pp. 233–248). Utopías.
Magnin, L. (2009b). Búsqueda de patrones en el emplazamiento de enterratorios humanos en el Macizo del Deseado. Aplicación del análisis de la visibilidad (SIG). In M. Salemme, F. Santiago, M. Álvarez, E. Piana, M. Vázquez, & M. Mansur (Comps.). *Arqueología de la Patagonia. Una mirada desde el último confín II*, (pp: 1045–1060). Utopías.
Magnin, L. (2010). *Distribuciones arqueológicas en la Meseta Central de Santa Cruz. Implicancias para los estudios de uso del espacio y movilidad de sociedades cazadoras recolectoras*. Doctoral Thesis. Facultad de Ciencias Naturales y Museo. Universidad Nacional de La Plata. La Plata. http://sedici.unlp.edu.ar/handle/10915/24773
Magnin, L. (2013a). Análisis de visibilidad aplicado a estructuras rocosas en la localidad La Primavera, Santa Cruz, *Revista Arqueología, 19*(2), 325–346. https://sites.google.com/site/reviarqueol3/home/arqueologa-19-2-2013/magnin
Magnin, L. (2013b). Incorporación de sistemas de Información Geográfica a los estudios arqueológicos de Cazadores Recolectores del Sector Norte del Macizo Central en Santa Cruz, Argentina. In M. J. Figuerero Torres, & A. Izeta (Eds.). *El uso de Sistemas de Información Geográfica (SIG) en la arqueología sudamerican*a. Archae*opress. BAR Archaeological Series 2497, South American Archaeology Series, 18*, 9–30.
Magnin, L. (2013c). Localidades arqueológicas del norte de Santa Cruz y vías de movilidad entre ellas. Los SIG en la integración de paisajes arqueológicos regionales. In A. F. Zangrando, R. Barberena, A. Gil, G. Neme, M. Giardina, L. Luna, C. Otaola, S. Paulides, L. Salgán, & A. Tivoli (Comps.), *Tendencias teórico-metodológicas y casos de estudio en la arqueología de la Patagonia* (pp. 435–433). Museo de Historia Natural de San Rafael.
Magnin, L. (2013d). Las variables locacionales en arqueología de cazadores recolectores. La Primavera (Santa Cruz). *Cuadernos de Antropología, 9*, 207–226. http://sedici.unlp.edu.ar/handle/10915/102948

Magnin, L. (2015a) Hunter-gatherer provisioning strategies in a landscape with abundant lithic resources (La Primavera, Santa Cruz, Argentina). *Quaternary International, 375*, 55–71.https://www.researchgate.net/publication/273176481_Hunter-gatherer_provisioning_strategies_in_a_landscape_with_abundant_lithic_resources_La_Primavera_Santa_Cruz_Argentina

Magnin, L. (2015ba). Hunter-gatherer provisioning strategies in a landscape with abundant lithic resources (La Primavera, Santa Cruz, Argentina). *Quaternary International, 375*, 55–71. https://doi.org/10.1016/j.quaint.2014.11.064

Magnin, L. (2016). Avances en un modelo predictivo para sitios arqueológicos en el Macizo del Deseado (Santa Cruz, Argentina). In *Actas del XIX Congreso Nacional de Arqueología Argentina, San Miguel de Tucumán, Serie Monográfica y didáctica* (Vol. 54, pp. 3152–3156).

Magnin, L. (2023). *Locational Analysis at La Primavera: Data (23 explicative variables, 10 archaeological classes, 1 control data set).* Consejo Nacional de Investigaciones Científicas y Técnicas. http://hdl.handle.net/11336/197748

Magnin, L., & Blanco, R. (2006). Análisis espacial de artefactos líticos dentro de un conjunto de recolección superficial. Sitio arqueológico La Playita. Cañadón de La Primavera. Santa Cruz. In *VII Jornadas de Jóvenes Investigadores en Ciencias Antropológicas* (pp. 46–47). Instituto Nacional de Antropología y Pensamiento Latinoamericano.

Magnin, L., & Blanco, R. (2019). Primera noticia de arte rupestre en el Parque Nacional Bosques Petrificados de Jaramillo (Santa Cruz, Argentina), *Revista Relaciones, 54*(2), 341–348. http://sedici.unlp.edu.ar/handle/10915/90166

Magnin, L., Hermo, D., Lynch, V., & Mosquera, B. (2017). Mapping triangular unstemmed points across the Deseado Massif (Santa Cruz, Argentina), In J. Alberti et al. (Eds.), *Resúmenes 11th International Symposium on Knappable Materials,* (p. 203). IMHICIHU. https://www.imhicihu-conicet.gob.ar/wp-content/uploads/2020/11/11th-ISKMFromToolstone.pdf. Acceso: 1 de agosto, 2024

Magnin, L., Hermo, D., & Weitzel, C. (2015). Estrategias de caza en la localidad arqueológica de La Primavera, Santa Cruz (Argentina). Análisis de visibilidad y accesibilidad mediante SIG. In H. González, & M. Sepúlveda (comp), *Libro de las Actas del XIX Congreso Nacional de Arqueología Chilena* (pp. 63–67). Ediciones Universidad de Tarapacá—Sociedad Chilena de Arqueología.

Magnin, L., Miotti, L., & Hermo, D. (2018). Perspectivas del estudio arqueológico del paisaje en el Parque Nacional Bosques Petrificados de Jaramillo (Santa Cruz, Argentina). *Revista del Museo de La Plata, 3*(1), 132–147. https://publicaciones.fcnym.unlp.edu.ar/rmlp/article/view/2206/1523

Magnin, L., & Moreira, G. (2024). *Mamíferos y aves de la Meseta Central de Santa Cruz (Argentina).* Consejo Nacional de Investigaciones Científicas y Técnicas. http://hdl.handle.net/11336/237481. Access September 15, 2024.

Magnin, L., Rodríguez Ceja, M., & Blanco, R. (2023). Primeros datos de cronología absoluta para la arqueología del Parque Nacional Bosques Petrificados de Jaramillo (Santa Cruz, Argentina). In *Libro de resúmenes del II Conferencia Latinoamericana de Radiocarbono (CLARa 2)*, (p. 66). UNAM.

Mancini, M. (1998). Vegetational changes during the Holocene in Extra-Andean Patagonia, Santa Cruz Province, Argentina. *Palaeogeography, Palaeoclimatology, Palaeoecology, 138*, 207–219.

Mancini, M. V., Franco, N. V., & Brook, G. (2013). Palaeoenvironment and early human occupation of southernmost South America (South Patagonia, Argentina). *Quaternary International, 299*, 13–22.

Mancini, M. V., Páez, M. M., & Prieto, A. R. (2002). Cambios paleoambientales durante los últimos 7000 14C años en el ecotono bosque-estepa, 47–48°S, Santa Cruz, Argentina. *Ameghiniana, 39*(2), 151–162.

Mancini, M. V., Páez, M. M., & Prieto, A. R. (2005). Mid-Holocene paeonviroments and human occupation in southern South America. *Quaternary International, 132*(1), 47–59.

Mansur, E. (1999). Análisis funcional de instrumental lítico: problemas de formación y deformación de rastros de uso. In C. Diez Marín (Ed.). *Actas del XII Congreso Nacional de Arqueología Argentina I* (pp. 335–366). UNLP.

Manzi, L. (1999). Diseño exploratorio acerca del uso del espacio por grupos cazadores-recolectores pedestres en la isla grande de Tierra del Fuego. In J. B. Belardi., P. M. Fernández, R. A. Goñi, A. G. Guráieb, & M. De Nigris (Eds.), *Soplando en el Viento. Actas de las Terceras Jornadas de Arqueología de la Patagonia* (pp. 419–438). Universidad Nacional del Comahue e Instituto Nacional de Antropología y Pensamiento Latinoamericano.

Manzi, L., & Spikins, P. (2008). El fuego en las altas latitudes: Los Selk'nam de Tierra del Fuego como referente etnográfico para el Mesolítico Europeo. *Complutum, 19*(1), 79–96.

Marchionni, L. (2013). *Comparación de las distintas historias tafonómicas en conjuntos zooarqueológicos provenientes de la Meseta Central de la provincia de Santa Cruz*. Doctoral Thesis. Facultad de Ciencias Naturales y Museo-Universidad Nacional de La Plata. https://doi.org/10.35537/10915/27075

Marchionni, L., Hermo, D., Magnin, L., Miott, L., & García Añino, E. (2020). Actualistic taphonomy in the Northeast of the Santa Cruz Plateau (Argentina). Advances and archaeological implications. In *Actualistic taphonomy in South America. Topics in geobiology* (Vol. 48, pp. 169–191). Springer.

Martíquez, G. (Dir.) (1997). *El gran libro de la Patagonia*. Planeta.

Martínez, G. (1999). *Tecnología. subsistencia y asentamiento en el curso medio del Río Quequén Grande. Un enfoque arqueológico*. Tesis doctoral inédita. Facultad de Ciencias Naturales y Museo. Universidad Nacional de La Plata. La Plata.

Martinez, G. (2003a). Late Holocene human occupation of the Quequen Grande River valley bottom: Settlement systems and an example of a built environment in the Argentine Pampas. *Before Farming, 4*(1), 11–27.

Martínez, J. (2003b). *Ocupaciones humanas tempranas y tecnología de caza en la microrregión de Antofagasta de la Sierra (10000–7000 AP)*. Doctoral Thesis. Facultad de Ciencias Naturales e Instituto Miguel Lillo. Universidad Nacional de Tucumán. San Miguel de Tucumán.

Martinić, M. (1995a). *Los Aónikenk. Historia y cultura*. Universidad de Magallanes.

Martinić, M. (1995b). *Los Aónikenk. Historia y cultura*. Ediciones Universidad de Magallanes.

Maschner. H. (1996a). *New methods. Old problems. Geographic information systems in modern archaeological research. Occasional paper No. 23* (pp. 1–21). Center of Archaeological Investigations. Southern Illinois University at Carbondale.

Maschner, H. (1996b). Geographic information systems in archaeology. In H. Maschner (Ed.). *New methods. Old problems. Geographic information systems in modern archaeological research. Occasional paper No. 23* (pp. 1–21). Center of Archaeological Investigations. Southern Illinois University at Carbondale.

Matteucci, S. D., & Scheinsohn, V. (2004). Procesamiento de imágenes. SIG y modelos ecológicos aplicados a la arqueología. *GeoFocus, 4*, 93–109.

Mazzoni, E. (2001). Distribución espacial y caracterización geomorfológica de "bajos sin salida" de la Patagonia Austral extracordillerana. *Anales Del Instituto De La Patagonia-Serie Ciencias Naturales, 29*, 5–24.

Mazzoni, E. (2007). *Geomorfología y evolución geomorfológico de los escoriales volcánicos y sus mallines asociados en diferentes ambientes de Patagonia extra-andina*. Doctoral Thesis. Departamento de Geografía. Universidad Nacional del Sur. Bahía Blanca.

Mazzoni, E., & Rabassa, J. (2007). Volcanic landscapes of patagonia: A geomorphological map of Piedra del Aguila Volcanic Plateau; Province of Neuquén, Argentina. *Journal of Maps, 3*(1), 311–322.

Mazzoni, E., & Rabassa, J. (Ed.). (2018). Volcanic landscapes and associated wetlands of lowland patagonia. In *The Latin American studies book series*. Springer. https://doi.org/10.1007/978-3-319-71921-4_1

Mazzoni, E., & Vázquez, M. (1999). Aplicación de un S.I.G. para la delimitación de unidades de paisaje naturales y determinación de áreas con mallines. In *Actas Congreso Nacional de Geología—60º Semana de Geografía* (pp. 463–475). Sociedad Argentina de Estudios Geográficos.

Mazzoni, E., & Vázquez, M. (2004). Consecuencias de la posición geográfica de la provincia de Santa Cruz. In L. González, & P. Rial (Eds.), *Guía Geográfica Interactiva de Santa Cruz*. INTA. http://www.inta.gov.ar/santacruz/info/documentos/teledet/GuiaSC/. Accessed Jun 15, 2010.

Mazzoni, E., Vázquez, M., Ferrari, S., Belardi, J. B., Albrieu, C., Carballo Marina, F., Espinosa, S., Larrosa, J., & Gismondi, D. (1999a). Consideraciones acerca del estudio de una de las geoformas típicas del paisaje patagónico: Las cuencas endorreicas o "bajos sin salida." *Espacios, 16*, 45–55.

Mazzoni, E., Vázquez, M., Ferrari, S., Belardi, J. B., Albrieu, C., Carballo Marina, F., Espinosa, S., Larrosa, J., & Gismondi, D. (1999b). Consideraciones acerca del estudio de una de las geoformas típicas del paisaje patagónico: Las cuencas endorreicas o "bajos sin salida." *Espacios, 16*, 45–50.

Mazzoni, E., Vázquez, M., & Ruiz Posse, E. (2000). Dinámica Geomorfológico de depósitos eólicos asociados a "Bajos sin salida" de la Patagonia Austral. In *IV Jornadas Nacionales de Geografía Física* (pp. 165–176). Universidad Nacional de Mar del Plata.

Mehrer, M., & Wescott, K. (2006). *GIS and archaeological site location modeling*. Taylor and Francis.

Melendi, E. (2023). *Controles litogénicos en el desarrollo de mallines en Patagonia Extra-andina (Provincia De Santa Cruz) como fuente de abastecimiento de agua en zonas áridas*. Doctoral Thesis, Facultad de Ciencias Naturales y Museo, Universidad Nacional de La Plata. La Plata. http://sedici.unlp.edu.ar/handle/10915/162032

Mena, F., & Lucero, V. (2004). En torno a las últimas poblaciones indígenas de la cordillera centro-Patagónica: estudio comparado de tres valles en Aisén oriental (Chile). In M. Civalero, P. Fernández, & A. Guráieb (Eds.), *Contra viento y marea. Arqueología de Patagonia* (pp. 643–657). Instituto Nacional de Antropología y Pensamiento Latinoamericano.

Mena, F., & Reyes, O. (2001). Montículos y cuevas funerarias en Patagonia. Una visión desde Cueva Baño Nuevo-1. XI Región. *Chungará, 33*(1), 21–30.

Menghín, O. (1952). Fundamentos cronológicos de la prehistoria de Patagonia. *RUNA, V*(1–2), 5–22.

Miotti, L. (1989). *Zooarqueología de la Meseta Central y costa de la provincia de Santa Cruz: Un enfoque de las estrategias adaptativas aborígenes y los paleoambientes*. Doctoral Thesis. Facultad de Ciencias Naturales y Museo. Universidad Nacional de La Plata. La Plata. http://sedici.unlp.edu.ar/handle/10915/4884

Miotti, L. (1991). Manifestaciones rupestres de Santa Cruz: La localidad Arqueológica Piedra Museo. In M. Podestá, M. Hernández Llosas, & S. Renard de Coquet (Eds.), *El Arte Rupestre en la Arqueología contemporánea* (pp. 132–139). FECIC.

Miotti, L. (1992). Paleoindian occupation at Piedra Museo Locality. Patagonian Region. Argentina. *Current Research in the Pleistocene, 9*, 30–32.

Miotti, L. (1995). Piedra Museo locality: A special place in the new world. *Current Research in the Pleistocene, 12*, 37–40.

Miotti, L. (1996a). Los cazadores recolectores de la Patagonia. Horizontes perdidos. *Revista de la Fundación Museo de La Plata, 2*(8), 33–39.

Miotti, L. (1996b). Piedra Museo (Santa Cruz): Nuevos datos para el debate de la ocupación Pleistocénica en Patagonia. In J. Gómez Otero (Ed.), *Arqueología, sólo Patagonia* (pp. 27–38). CENPAT-CONICET.

Miotti, L. (1997). Piedra Museo. Una ventana abierta al pasado de los cazadores-recolectores de Patagonia y el proceso de colonización hace 13.000 años. *Revista La Ventana, 14*, 9–14.

Miotti, L. (2000). Localidad Arqueológica Piedra Museo. In L. Miotti, R. Paunero, M. Salemme, & R. Cattáneo (Eds.). *Guía de Campo del Taller Internacional: La colonización del Sur de América durante la transición Pleistoceno/Holoceno* (pp. 55–56). INQUA.

Miotti, L. (2001). La ocupación humana en el Holoceno medio. In *Libro de resúmenes del XIV Congreso Nacional de Arqueología Argentina* (pp. 254–255). Universidad Nacional de Rosario.

Miotti, L. (2003). Patagonia: A paradox for building images of the first Americans during the Pleistocene/Holocene Transition. *Quaternary International, 109*(110), 147–173.

Miotti, L. (2006a). Paisajes domésticos y sagrados desde la arqueología de cazadores recolectores del Macizo del Deseado. Provincia de Santa Cruz. *Cazadores Recolectores Del Cono Sur. Revista De Arqueología, 1*, 13–42.

Miotti, L. (2006b). La fachada atlántica, como puerta de ingreso alternativa de la colonización humana de América del Sur durante la transición Pleistoceno-Holoceno. In J. C. Jiménez, & S. González (Eds.), *II Simposio Internacional El Hombre Temprano en América* (pp. 155–188). Consejo Nacional para la Cultura y las Artes (CONACULTA)-Instituto Nacional de Antropología e Historia (INAH).

Miotti, L. (2008). Household and sacred landscapes among Holocene hunter-gatherers of Patagonias Central Plateau. *Before Farming, 3*, 5–44. https://doi.org/10.3828/bfarm.2008.3.1

Miotti, L., & Carden, N. (2006). Rumbo a la nada. La experiencia en la práctica arqueológica de un lugar de Patagonia. In A. Austral, & M. Tamagnini (Comps.), *Problemáticas de la Arqueología contemporánea. XV Congreso Nacional de Arqueología Argentina Tomo I* (pp. 309–312). Universidad Nacional de Río Cuarto.

Miotti, L., & Carden, N. (2007). The relationships between rock art and archaeofaunas in the Central Patagonian Plateau. In M. Gutierrez, L. Miotti, M. Salemme, G. Mengoni, & G. Barrientos (Eds.), *Taphonomy and archaeology in Argentina* (pp. 203–217). BAR International Series.

Miotti, L., Carden, N., & Canosa, M. J. (1999). Paisajes arqueológicos de Cazadores-recolectores. Arte rupestre y Lagunas. In C. Diez Marín (Ed.), *Actas del XII Congreso Nacional de Arqueología Argentina. Tomo III* (pp. 54–63). Universidad Nacional de La Plata.

Miotti, L., Carden, N., & Magnin, L. (2007a). Tendencia central. dispersión y orientación en el paisaje regional. Estudios de distribución del arte rupestre en la Meseta Central de Santa Cruz. In F. Morello, M. Martinic, & A. Prieto, G. Bahamonde (Eds.), *Arqueología de Fuego-Patagonia. Levantando piedras. desenterrando huesos... y develando arcanos* (pp. 601–612). CEQUA.

Miotti, L., & Cattáneo, R. (1997). Bifacial technology at 13.000 years ago in Southern Patagonia. *Current Research in the Pleistocene, 14*, 65–68.

Miotti, L., & Cattáneo, R. (2003). Variation in the strategies of lithic production and faunal exploitation during the Pleistocene/Holocene transition at Piedra Museo and surrounding region. In L. Miotti, M. Salemme, & N. Flehengeimer (Eds.), *Ancient evidence for paleo South American: Form where the south winds blow* (pp. 105–111). Texas A&M University Press.

Miotti, L., & Hermo, D. (2000). Ambientes y paisajes del Nesocratón del Deseado. In L Miotti, R. Paunero, M. Salemme, & R. Cattáneo (Eds.), *Guía de Campo del Taller Internacional: La colonización del Sur de América durante la transición Pleistoceno/Holoceno* (pp. 14–26). INQUA.

Miotti, L., Hermo, D., Magnin, L., Carden, N., Marchionni, L., Alcaraz, A., Mosquera, B., Terranova, E., & Salemme, M. (2007). Resolución e Integridad Arqueológica de la Cueva Maripe (Santa Cruz. Argentina). In F. Morello, M. Martinic, A. Prieto, & G. Bahamonde (Eds.), *Arqueología de Fuego-Patagonia. Levantando piedras. desenterrando huesos... y develando arcanos* (pp. 555–569). CEQUA.

Miotti, L., Hermo, D., Salemme, M., Magnin, L., & Marchionni, L. (2009). Cueva Maripe y su excavación. Implicancias en los estudios regionales del Macizo del Deseado. In A. Austral y M. Tamagnini (Eds.), *Problemáticas de la arqueología contemporánea* (pp. 1031–1035). Universidad Nacional de Río Cuarto.

Miotti, L., Hermo, D., & Vázquez, M. (1997). *Evaluación de recursos culturales en el Monumento Natural Bosque Petrificado de Jaramillo. Santa Cruz. Argentina. Informe técnico desarrollado para la Administración de Parques Nacionales.* Ms.

Miotti, L., Magnin, L., & Terranova, E. (2022). Challenges for the twenty-first century: The patrimonialization of Piedra Museo. In L. Miotti, M. Salemme, & D. Hermo (Eds.), *Archaeology of Piedra Museo Locality* (pp. 423–447). The Latin American Studies Book Series. Springer. https://doi.org/10.1007/978-3-030-92503-1_15

Miotti, L., & Marchionni, L. (2009). Procesando Huesos: entre la Arqueología y la Etnografía. In M. Salemme, F. Santiago, M. Álvarez, E. Piana, M. Vázquez, & E. Mansur (Eds.), *Arqueología de la Patagonia. Una mirada desde el último confín* (pp. 787–798). Utopías.

Miotti, L., Marchionni, L., Hermo, D., Terranova, E., Magnin, L., Lynch, V., Mosquera, B., Vargas Gariglio, J., & Carden, N. (2021). *Changes and continuities of hunting practices from the late Pleistocene to the late Holocene among nomadic societies of the Patagonian plateaus.* In B. Belardi, D. Bozzuto, P. M. Fernández, E. Moren, & G. Neme (Eds.), *Ancient Hunting Strategies in Southern South America* (pp. 259–291). Springer The Latin American Studies Book Series.

Miotti, L., & Podgorny, I. (1995). Una flecha en mi sopa. La convivencia con los restos arqueológicos en la región del río Deseado, pcia. de Santa Cruz, República Argentina. *Cuadernos Del Instituto Nacional De Pensamiento Latinoamericano, 16*, 343–356.

Miotti, L., & Salemme, M. (1999). Biodiversity, taxonomic richness and specialists-generalista during Late Pleistocene/Early Holocene times in Pampa and Patagonia (Argentina, Southern South America). *Quaternary International, 53–54*, 53–68. https://doi.org/10.1016/S1040-6182(98)00007-X

Miotti, L., & Salemme, M. (2003). When Patagonia was colonized: People mobility at high latitudes during Pleistocene/Holocene transition. *Quaternary International, 109*, 95–111.

Miotti, L., & Salemme, M. (2004a). Poblamiento. movilidad y territorios entre las sociedades cazadoras-recolectoras de Patagonia: Cambios desde la transición Pleistoceno/Holoceno al Holoceno medio. *Complutum, 15*, 177–206.

Miotti, L., & Salemme, M. (2004b). Hunting and butchering events in the late Pleistocene and early Holocene in Piedra Museo (Patagonia. Southernmost South America). In R. Bonnichsen (Ed.), *Paleoamerican prehistory: Colonization models. Biological populations and human adaptations* (pp. 141–151). Center for the Study of the First Americans. University of Texas A&M.

Miotti, L., Salemme, M., Hermo, D., Vázquez, M., Carden, N., Giardina, M., & Magnin, L. (2005). Aguada del Cuero: Un nuevo escalón en la arqueología de la Meseta Central de Santa Cruz. In E. Berberián (Ed.), *Libro de Actas del XIII Congreso Nacional de Arqueología Argentina* (pp. 55–62). Universidad Nacional de Córdoba.

Miotti, L., Salemme, M., & Rabassa, J. (2003). Radiocarbon chronology at Piedra Museo Locality. In L. Miotti, M. Salemme, & N. Flegeheimer (Eds.), *Ancient evidence for Paleo South Americans, From Where the South Winds Blow,* (pp. 99-104). Center for the Study of First Americans; A&M University Press.

Miotti, L., Terranova, E., Blanco, R., Marchionni, L., Hermo, D., & Magnin, L. (2014). Entre basaltos y lagunas: las estructuras de piedra de la meseta de Somuncurá. Apuntes para la reflexión de los patrones de movilidad de los cazadores-recolectores. In *Libro de Resúmenes IX Jornadas de Arqueología de la Patagonia* (p. 76). CIEP.

Miotti, L., Vázquez, M., & Hermo, D. (1999a). Piedra Museo un Yamnagoo Pleistocénico en la Colonización de la Meseta de Santa Cruz. El estudio de la Arqueofauna. In J. B. Belardi., P. M. Fernández, R. A. Goñi, A. G. Guráieb, & M. De Nigris (Eds.), *Soplando en el Viento. Actas II Jornadas de Arqueología de la Patagonia* (pp. 113–135). Universidad Nacional del Comahue. Facultad de Humanidades.

Miotti, L., Vázquez, M., & Hermo, D. (1999b). Piedra Museo un Yamnagoo Pleistocénico en la Colonización de la Meseta de Santa Cruz. El estudio de la Arqueofauna. In J. B. Belardi, P. M. Fernández, R. A. Goñi, A. G. Guráieb, M. De Nigris (Eds.). *Soplando en el Viento. Actas II Jornadas de Arqueología de la Patagonia* (pp. 113–135). Universidad Nacional del Comahue, Instituto de Pensamiento Latinoamericano.

Mlekuz, D. (2004). Listening to landscape: Modelling past soundscapes in GIS. *Internet Archaeology, 16*. http://intarch.ac.uk/journal/issue16/6/index.html. Accessed September 15, 2024.

Molinari, R., & Espinosa, S. (1999). Brilla tu diamante "loco"… In J. B. Belardi, P. M. Fernández, R. A. Goñi, A. G. Guráieb, & M. De Nigris (Eds.) *Soplando en el Viento. Actas II Jornadas de Arqueología de la Patagonia* (pp. 189–198). Universidad Nacional del Comahue, Instituto de Pensamiento Latinoamericano.

Molinari, R., & Ferraro, L. (2000). Estancia Menelik: Implicancias para el estudio de recursos culturales en la zona de amortiguación del Parque Nacional Perito Moreno. In M. T. Civalero, P. Fernández, & G. Guráieb (Eds.), *Contra Viento y Marea. Arqueología Sólo Patagonia* (pp. 625–634). Instituto Nacional de Antropología y Pensamiento Latinoamericano.

Molinari, R., & Ferraro, L. (2004). Estancia Menelik: implicancias para el estudio de recursos culturales en la zona de amortiguación del Parque Nacional Perito Moreno. In M. T. Civalero, P. Fernández, G. Guráieb (Eds.). *Contra Viento y Marea. Arqueología Sólo Patagonia*, (pp. 625–634). Buenos Aires: Instituto Nacional de Antropología y Pensamiento Latinoamericano.

Mondragón, C. (2011). A weft of nexus: changing notions of space and geographical identity in Vanuatu, Oceania. In P. W. Kirby (Ed.), *Boundless worlds: An anthropological approach to movement* (pp. 115–134). Berghahn Books. http://www.jstor.org/stable/j.ctt9qchtv.9

Morales, M., Barberena, R., Belardi, J. B., Borrero, L. A., Cortegoso, V., Durán, D., Guercie, A., Goñi, R., Gil, A., Neme, G., Yacobaccio, H., & Zárate, M. (2009). Reviewing human–environment interactions in arid regions of southern South America during the past 3000 years. *Palaeogeography. Palaeoclimatology. Palaeoecology, 281*(3–4), 283–295.

Moreira, P., Echeveste, H., Fernández, R., Hartman, L., Orestes, J., & Schalamuk, I. (2009). Depositional age of Jurassic epithermal gold-silver ore in the Deseado Massif, Patagonia, Argentina, based on Manantial Espejo and La Josefina prospects. *Neues Jahrbuch Für Geologie und Paläontologie, 253*(1), 25–40.

Moreno, F. (1997 [1879]). *Viaje a la Patagonia Austral*. Elefante Blanco.

Mosquera, B. (2016). *Geoarqueología de los Zanjones Blanco y Rojo, Macizo del Deseado, provincia de Santa Cruz*. Doctoral Thesis. Facultad de Ciencias Naturales y Museo. Universidad Nacional de La Plata. La Plata. https://doi.org/10.35537/10915/62539

Mosquera, B. (2018). Análisis de la información radiocarbónica de sitios arqueológicos del Macizo del Deseado, provincia de Santa Cruz, Argentina. *InterSecciones en Antropología, 19*(1), 25–36. https://doi.org/10.37176/iea.19.1.2018

Mosquera, B., & Mancini, M. V. (2021). Paleoenvironmental analysis of Deseado Massif wet meadow. Implications to the Holocene occupations of Argentinean Patagonia. *The Holocene, 31*. https://doi.org/10.1177/09596836211025969

Musters, G. (1997 [1871]). *Vida entre los patagones*. Elefante Blanco.

Nami, H. (1987). Cueva del Medio: Perspectivas arqueológicas para Patagonia Austral. *Anales Del Instituto De La Patagonia, 17*, 73–106.

Nami, H. (1992). El subsistema tecnológico de la confección de instrumentos líticos y la explotación de los recursos del ambiente: Una nueva vía de aproximación. *Shincal, 2*, 33–53.

Naranjo, J. A., & Stern, C. R. (1998). Holocene explosive activity of Hudson volcano, southern Andes. *Bulletin of Volcanology, 59*, 291–306.

Navas, J. (1987). Los vertebrados exóticos introducidos en la Argentina. *Revista del Museo Argentino de Ciencias Naturales "Bernardino Rivadavia", 14*(2), 7–38.

Negre, J., Briz i Godino, I., Pal, N., & Álvarez, M. (2018). Estrategias de asentamiento Y movilidad en sociedades cazadoras- recolectoras-pescadoras del sector meridional de la costa atlántica Fueguina. *Arqueología, 1*, 24, 101–122.

Nelson, M. (1991). The study of technological organization. In M. Schiffer (Ed.), *Archaeological method and theory* (pp. 57–100). Arizona Press.

Nielsen, A. E. (1994). Como es arriba es abajo: evaluación crítica de las posibilidades del análisis de micro artefactos para la inferencia arqueológica. *Arqueología*, 9–41.

Oliva, G., Rial, P., González, L., & Mazzoni, E. (1995). Evaluación del estado actual de la desertificación en la transecta Santa Cruz (Capítulo IV.1.). In H. Eiden, G. Mensching, & H. J. Georgen (Eds.), *Lucha contra la desertificación en la Patagonia. Cooperación Técnica Argentino-Alem*ana (pp. 65–85). Proyecto INTA-GTZ.

Oliva, G., Rial, P., González, L., & Rial, P. (2004). *Áreas ecológicas. Guía Geográfica Interactiva de Santa Cruz*. In L. González, & P. Rial (Eds.), INTA.

Oria, J. (2008) *Patrones de movilidad pre europeos en el norte de Tierra del Fuego. Una aproximación geo arqueológica*. Doctoral Thesis. Facultad de Ciencias Naturales y Museo. Universidad Nacional de La Plata. La Plata. http://sedici.unlp.edu.ar/handle/10915/34272

Orquera. L. A., & Piana, E. (1986). *Normas para la descripción de objetos arqueológicos de piedra tallada. Contribución científica 1*. Publicación Especial CADIC.

Orton, C. (2000). *Sampling in archaeology*. Cambridge University Press.

Outes, F. (1905a). *La edad de piedra en Patagonia.* Anales del Museo Nacional de Buenos Aires XII.
Outes. F. (1905b). *La edad de piedra en Patagonia.* Imprenta de Juan A. Alsina.
Páez, M. M., Prieto, A., & Manzini, M. (1999). Fossil pollen from Los Toldos locality: A record of Late-Glacial transition in the Extra-Andean Patagonia. *Quaternary International, 53–54*, 69–76.
Pallo, M. C. (2012). El estrés invernal como generador de áreas marginales en el extremo sur de Patagonia Continental durante el Holoceno tardío. *Comechingonia Virtual, 6*, 86–114.
Panza, J. L. (2001). Hoja Geológica 4769-IV. Monumento natural bosques petrificados. Provincia de Santa Cruz. In *Dirección Nacional del Servicio Geológico. Boletín 258.* Servicio Geológico Minero Argentino.
Panza, J., Zubia, M., Genini, A., & Godeas, M. (1994). *Hoja Geológica 4969-II, Tres Cerros, provincia de Santa Cruz. Dirección Nacional del Servicio Geológico, Boletín 213* (pp. 1–103). Servicio Geológico Minero Argentino.
Parcero, C. (2002). *La Construcción del Paisaje Social en la Edad del Hierro del Noroeste Ibérico. Colección de Monografías Ortegalia 1.* Fundação Federico Maciñeira.
Paunero, R. (1994a). El sitio Cueva 1 de la Localidad Arqueológica Cerro Tres Tetas (Estancia San Rafael. Provincia de Santa Cruz). *Anales De Arqueología y Etnología, 48*(49), 73–90.
Paunero, R. (2000). Localidad arqueológica Cerro Tres Tetas. In L. Miotti, R. Paunero, M. Salemme, & R. Cattáneo (Eds.), *La colonización de América durante la transición Pleistoceno/Holoceno". Guía de Campo de la visita a las localidades arqueológicas* (pp. 90–100). INQUA.
Paunero, R. (2003a). The Cerro Tres Tetas Locality (C3T) in Central Plateau of Santa Cruz. Argentina. In L. Miotti, M. Salemme, & N. Flegenheimer (Eds.), *Ancient evidence for Paleo South Americans. From Where the South Winds Blow* (pp. 133–141). Center for the Study of First Americans. A&M University Press.
Paunero, R. (2003b). The presence of a Pleistocenic colonizing culture in La Maria Archaeological locality: Casa del Minero I. In L. Miotti, M. Salemme, & N. Flegenheimer (Eds.), *Ancient evidence for Paleo South Americans. From Where the South Winds Blow* (pp. 127–132). Center for the Study of First Americans. A&M University Press.
Paunero, R. (2009). La colonización humana de la meseta central de Santa Cruz durante el Pleistoceno final: indicadores arqueológicos, referentes estratigráficos y nuevas evidencias. In M. Álvarez, E. Piana, M. Vázquez, E. Mansur, M. Salemme, & F. Santiago (Eds.), *Arqueología de Patagonia: Una mirada desde el último confín*, (pp. 85–100). Utopías.
Paunero, R., Frank, A., Skarbun, F., Rosales, G., Cueto, M., Zapata, G., Paunero, M., Lunazzi, N., & Del Giorgio, M. (2007). Investigaciones arqueológicas en el sitio Casa del Minero 1. Estancia la María. Meseta Central de Santa Cruz. In F. Morello, M. Martinic, A. Prieto, & G. Bahamonde (Eds.), *Arqueología de Fuego-Patagonia. Levantando piedras, desenterrando huesos... y develando arcanos* (pp. 577–588). CEQUA.
Pérez de Micou, C., Bellelli, C., & Aschero, C. (1992). Vestigios minerales y vegetales en la determinación del territorio de explotación de un sitio. In L. A. Borrero, & J. L. Lanata (Eds.), *Análisis espacial en la arqueología patagónica* (pp. 53–82). Ayllu.
Pérez de Micou, C., & Ratto, N. (2007). Las plantas silvestres como recurso en regiones áridas (Patagonia y Puna). In T. Civalero. P. Fernández, & G. Guraieb (Eds.), *Contra viento y marea. Arqueología de Patagonia* (pp. 167–176). Instituto Nacional de Antropología y Pensamiento Latinoamericano.
Perez, I., Postillone, M., Rindel, D., Gobbo, D., Gonzalez, P., et al. (2016). Peopling time, spatial occupation and demography of Late Pleistocene-Holocene human population from Patagonia. *Quaternary International, 425*, 214–223. https://doi.org/10.1016/j.quaint.2016.05.004
Pigafetta, A. (2001). *Primer Viaje Alrededor del Mundo.* Elefante Blanco.
Politis, G. (2006). The different dimensions of mobility among the Nukak foragers of the Colombian Amazon. In F. Sellet, R. Greaves, & P. L. Yu (Eds.), *Archaeology and ethnoarchaeology of mobility* (pp. 23–43). University of Florida Press.

Politis, G. (2007). *Nukak. Ethnoarchaeology of an Amazonian People*. University College London Institute of Archaeology Publications.

Politis, G., & Borrero, L. (2024). *The Archaeology of the Pampas and Patagonia*. Cambridge University Press.

Politis, G., & Saunders, N. (2002). Archaeological correlates of ideological activity: Food taboos and spirit-animals in an amazonian hunter-gatherer society. In P. Miracle & N. Milner (Eds.), *Consuming passions and patterns of consumption* (pp. 113–130). Mc Donald Institute Monographs.

Prates, L., Politis, G. G., & Perez, S. I. (2020). Rapid radiation of humans in South America after the last glacial maximum: A radiocarbon-based study. *PLoS ONE, 15*(7), Article e0236023. https://doi.org/10.1371/journal.pone.0236023

R CoreTeam. (2016). *R: A language and environment for statistical computing*. R Foundation for Statistical Computing. http://www.R-project.org

Ramos, V. (1999). Las Provincias Geológicas del Territorio Argentino. *Geología Argentina. Instituto de Geología y Recursos Minerales. Anales, 29*(3), 41–96.

Ratto, N. (1994). Funcionalidad vs. adscripción cultural: Cabezales líticos de la margen norte del Estrecho de Magallanes. In J. L. Lanata, & L. A. Borrero (Eds.), *Arqueología de cazadores-recolectores. Límites. Casos y aperturas* (pp. 105–120). Programa de Estudios Prehistóricos. CONICET.

Ravesloot, J., & Waters, M. (2004). Geoarchaeology and Archaeological site patterning of the Middle Gila River. Arizona. *Journal of Field Archaeology, 29*, 203–214.

Re, A., Goñi, R., Flores Coni, J., Guichón, F., Dellepiane, J., & Umaño, M. (2017). Arqueología de la meseta del Strobel Argentina (Patagonia meridional): 15 años después. *Relaciones de la Sociedad de Antropología, 42*, 133–158. https://www.saantropologia.com.ar/wp-content/uploads/2017/09/06-Re.pdf. Acceso: 1 de agosto, 2024.

Re, A., Nuevo Delaunay, A., & Ferraro, L. (2005). Grabados en la meseta del lago Strobel (Provincia de Santa Cruz): El sitio Laguna del Faldeo Verde. *Relaciones De La Sociedad Argentina De Antropología, 30*, 245–256.

Redford, K. H., & Eisenberg, J. F. (1992). *Mammals of the neotropics: The Southern Cone* (Vol. 2). The University of Chicago Press.

Renfrew, C., & Bahn, P. (1993). *Archaeology. Theories. Methods and practice*. Thames & Hudson.

Reyes, O., Méndez, C., Velásquez, H., & Trejo, V. (2006). Distribuciones espaciales y contextos arqueológicos de cazadores recolectores esteparios en Alto río Cisnes (XI Región de Aisén). *Magallania, 34*(2), 75–90.

Rick, J. (1976). Downslope movement and archaeological intrasite spatial analysis. *American Antiquity, 41*, 133–144. https://doi.org/10.2307/279164

Rindel, D., Cassiodoro, G., & Aragone, A. (2007). La utilización de mesetas altas durante el Holoceno tardío: El sitio Cerro Pampa 2 Ojo de Agua (Santa Cruz). In F. Morello, M. Martinic, A. Prieto, & G. Bahamonde (Eds.), *Arqueología de Fuego-Patagonia. Levantando piedras, desenterrando huesos...y develando arcanos* (pp. 649–662). CEQUA.

Roberts, B. (1987). Landscape archaeology. In J. M. Wagstaff (Ed.), *Landscape and Culture. Geographical and archaeological perspectives* (pp. 77–95). Basil Blackwell.

Rodríguez, M. E., & Horlent, L. (2016). *Tehuelches y selk'nam (Santa Cruz y Tierra del Fuego): No desaparecimos*. Ministerio de Educación y Deportes. http://socialesyescuela.com.ar/items/show/476

Roper, D. (1979). The method and theory of site catchment analysis: A review. *Advances in Archaeological Method and Theory, 2*, 119–140.

Rossignol, J. (1992). Concepts, methods and theory building: A landscape approach. In J. Rossignol, & L. Wandsnider (Eds.), *Space. Time and archaeological landscapes* (pp. 3–16). Plenum Press.

Rossignol, J., & Wandsnider, L. (Eds.). (1992). *Space. Time and archaeological landscapes*. Plenum Press.

Roumec, A., Zubimendi, M. A., & Ciampagna, M. L. (2020). *Estudio de la cerámica de la costa norte de Santa Cruz: Análisis iniciales de distribución y caracterización tecnomorfológica* (pp. 17–34). Fundación de Historia Natural Félix de Azara.

Sackett, J. R. (1985). Style and ethnicity in the Kalahari: A reply to wiessner. *American Antiquity, 50*(1), 154–159. https://doi.org/10.2307/280642

Salemme, M., & Miott, L. (1987). Zooarchaeology and palaeoenvironmental: Some examples from Patagonian and Pampean regions (Argentina). *Quaternary of South America and Antarctic Peninsula, 5*, 33–57.

Salemme, M., & Miotti, L. (2008). Archaeological hunter-gatherer landscapes since the latest Pleistocene in Fuego-Patagonia. In J. Rabassa (Ed.), *The late cenozoic of Patagonia and Tierra del Fuego* (pp. 437–483). Elsevier.

Salemme, M., Santiago, F., Miotti, L., & Magnin, L. (2006). Prospección no invasiva en sitios a cielo abierto y en cuevas (Patagonia Austral y Tierra del Fuego). *Libro de Resúmenes del III Congreso Geomorfologia y Cuaternario* (pp. 11–13). Universidad Nacional de Córdoba.

Santiago, F. (2009). *La ocupación humana del norte de Tierra del Fuego durante el Holoceno medio y tardío. Su vinculación con el paisaje.* Doctoral Thesis Facultad de Ciencias Sociales. Universidad Nacional del Centro de la Provincia de Buenos Aires. https://www.ridaa.unicen.edu.ar/xmlui/handle/123456789/2903

Sauvet, G. (2019). The lifeworld of hunter-gatherers and the concepts of territory. *Quaternary International, 503*, 191–199.

Savage, S. (1990). GIS in archaeological research. In K. Allen, S. W. Green, & E. B. Zubrow (Eds.), *Interpreting space: GIS and archaeology* (pp. 22–32). Taylor and Francis.

Scheinsohn, V. (2001a). 2001 Odisea del espacio. Paisajes y distribuciones artefactuales en Arqueología. Resultados y propuestas. *Relaciones De La Sociedad Argentina De Antropología, 24*, 285–302.

Scheinsohn, V. (2001b). Odisea del espacio. Paisajes y distribuciones artefactuales en Arqueología. Resultados y propuestas. *Relaciones De La Sociedad Argentina De Antropología, 26*, 285–302.

Scheinsohn, V., & Mateucci, S. (2004). Obtención de un modelo regional de distribuciones arqueológicas mediante SIG. In M. Tamagnini & O. Mendonza (Eds.), *Libro de resúmenes del XV Congreso Nacional de Arqueología Argentina* (p. 188). Universidad Nacional de Río Cuarto.

Schiffer, M. (1983). Towards the identification of formation processes. *American Antiquity, 48*, 675–706.

Schofield, A. J. (1991). Interpreting artefact scatters: An introduction. In A. J. Schofield (Ed.), *Interpreting artefact scatters: Contributions to ploughzone archaeology* (pp. 3–8). Oxbow Books.

Shennan, S. (1992). *Arqueología cuantitativa.* Crítica.

Shott, M. (1986). Technological organization and settlement mobility: An ethnographic examination. *Journal of Anthropological Research, 42*(1), 15–51.

Silvertsen, B. (1980). A site activity model for Hill and butchering activities at hunter-gatherer sites. *Journal of Field Archaeology, 7*, 423–442.

Skarbun, F. (2011). *La organización tecnológica en grupos cazadores recolectores desde las ocupaciones del pleistoceno final al Holoceno Tardío, en la Meseta Central de Santa Cruz, Patagonia).* Archaeopress. Publish of British Archaeological Reports (BAR).

Skarbun, F., Cueto, M., Frank, A., & Paunero, R. (2022). Tecnología lítica de las primeras sociedades del extremo sur de Sudamérica September 2021. *Latin American Antiquity, 33*(3), 1–21. https://doi.org/10.1017/laq.2021.50

Soto, J., & Vázquez, M. (2000). Las condiciones climáticas de la Provincia de Santa Cruz. In A. García, & E. Mazzoni (Dir). *El gran libro de la provincia de Santa Cruz. Tomo 1.* Milenio-Alfa Centro Literario.

SRTM (2013). *Shuttle Radar Topography Mission (SRTM) Global.* Distributed by OpenTopography.https://doi.org/10.5069/G9445JDF.Accessed:2024-09-24

Sruoga, P., Busteros, A., Giacosa, R., Martínez, H., Kleiman, L., Japas, S., Maloberti, A., & Gayone, M. R. (2008). Análisis litofacial y estructural del complejo volcánico Bahía Laura en el área El

Dorado-Montserrat. Provincia de Santa Cruz. *Revista de la Asociación Geológica Argentina, 63*(4), 653–664.

Stančič, Z., & Kvamme, K. (1999). Settlement patterns modeling through Boolean overlays of social and environmental variables. In J. A. Barceló. I. Briz, & A. Vila (Eds). *New Techniques for Old Times. CAA 98: Proceedings of the 26th Conference* (pp. 231–237). BAR International Series.

Stančič, Z., & Veljanovski, T. (2000). Understanding Roman settlement patterns through multivariate statistics and predictive modelling. In G. Lock (Ed.), *Beyond the map: Archaeology and spatial technologies* (pp. 147–198). IOS Press.

Steele, J., & Politis, G. (2009b). AMS 14 C dating of early human occupation of southern South America. *Journal of Archaeological Science, 36*(2), 419–429. https://doi.org/10.1016/j.jas.2008.09.02

Stein, J. K., Deo, J. N., & Phillips, L. S. (2003). Big sites—Short time: Accumulation rates in archaeological sites. *Journal of Archaeological Science, 30*, 297–316.

Stern, C. (1999). Black obsidian from Central South Patagonia: Chemical characteristics. Sources and regional distribution of artifacts. In J. B. Belardi, P. M. Fernández, R. A., Goñi, A. G., Guráieb, & M. De Nigris (Eds.), *Soplando en el viento... Actas de las III Jornadas de Arqueología de la Patagonia* (pp. 221–234). Instituto Nacional de Antropología y Pensamiento Latinoamericano y Universidad Nacional del Comahue.

Stern, C. (2004). Obsidian in Southern Patagonia: Review of the current information. In T. Civalero. P. Fernández, & G. Guraieb (Eds.), *Contra viento y marea. Arqueología de Patagonia* (pp. 167–176). Instituto Nacional de Antropología y Pensamiento Latinoamericano.

Stine, S., & Stine, M. (1990). A record from Lake Cardiel of climate change in Southern South America. *Nature, 345*, 705–708.

Surovell, T., & Waguespack, N. (2007). Folsom hearth-centered use of space at Barger Gulch. Locality B. In R. S. Brunswig, & B. Pitblado (Eds.), *Emerging frontiers in colorado paleoindian archaeology* (pp. 219–259). University of Colorado Press.

Taborton, D. (2002). *Terrain análisis using digital elevation models* (TauDEM). http://www.engenieering.usu.edu/dtrab

Tavarone, E. (2004). *Análisis de la factibilidad de reintroducción del guanaco (Lama guanicoe) en el Parque Nacional Quebrada del Condorito.* Master degree Thesis. Programa de Maestría en Manejo de Vida Silvestre-Centro de Zoología Aplicada. Facultad de Ciencias Exactas. Físicas y Naturales. Universidad Nacional de Córdoba.

Thomas, D. (1975a). Nonsite sampling in archaeology: Up the creek without a site? In W. James (Ed.), *Sampling in archaeology* (pp. 61–83). The University of Arizona Press.

Thomas, D. (1975b). Nonsite sampling in archaeology: Up the creek without a site? In J. W. Meuller (Ed.), *Archaeological sampling* (pp. 61–81). University of Arizona Press.

Tilley, C. (1994). *A phenomenology of landscape: Places. Paths and monuments.* Berg Publisher.

Tilley, C., & Bennet, W. (2001). An archaeology of supernatural places: The case of West Penwith. *Journal of the Royal Anthropological Institute, 7*(2), 335–362.

Tobler, W. (1993). *Three presentations on geographical analysis and modelling.* National Centre for Geographic Information and Analysis, Santa Barbara, Technical report 93–1. https://www.researchgate.net/publication/2460331_Three_Presentations_On_Geographical_Analysis_And_Modeling. Accessed September 15, 2024.

Torrence, R. (1983). Time budgeting and hunter-gatherer technology. In G. Bailey (Ed.), *Hunter-gatherer economy in prehistory* (pp. 11–22). Cambridge University Press.

Travaini, A., Bustamante, J., Rodríguez, A., Zapata, Z., Procopio, D., Pedrana, J., Martínez, & Peck, R. (2007). An integrated framework to map animal distributions in large and remote regions. *Biodiversity Research, 13*, 289–298.

Trigger, B. (1992). *Historia del pensamiento arqueológico.* Crítica.

USGS. (2005). United States geological survey, NASA shuttle radar topography mission, global 1 arc second V003. http://srtm.usgs.gov

Valiza Dvis, C. (2020). *Zooarqueología de cazadores-recolectores que habitaron la Meseta Central de Santa Cruz durante la transición Pleistoceno-Holoceno.* Doctoral Thesis. Facultad de Ciencias Naturales y Museo. Universidad Nacional de La Plata. https://naturalis.fcnym.unlp.edu.ar/handle/628872547/28248

Van Leusen, M. (1996). GIS and locational modeling in dutch archaeology: A review of current approaches. In H. Maschner (Ed.), *New methods. OldProblems. Geographic information systems in modern archaeological research. Occasional paper 23*, (pp. 177–197). Centre of Archaeological Investigations. Southern Illinois. University at Carbondale.

Van Leusen, M. (2002a). *Pattern to process. Methodological investigation and interpretation of spatial patterns in archaeological landscapes.* Doctoral Thesis, Rijksuniversiteit Groningen. https://pure.rug.nl/ws/portalfiles/portal/7002343/thesis.pdf

Van Leusen, M. (2002b). *Line-of-sight and cost-surface analysis using GIS Pattern to process. Methodological investigation and interpretation of spatial patterns in archaeological landscapes.* Doctoral Thesis. Rijksuniversiteit Groningen, Groningen. https://research.rug.nl/files/7002329/ch6.pdf. Accessed September 15, 2024.

VanPool, C., & VanPool, T. (1999). The scientific nature of postprocessualism. *American Antiquity Society for American Archaeology, 64*(1), 33–53.

Vázquez, M., Mazzoni, E., Ruiz Posse, E., & Baetti, C. (2004). Aplicación de un Sistema de Información Geográfica para el estudio de pequeñas cuencas endorreicas del sur de la Patagonia. In *Jornadas nacionales de geografía física: Actas* (p. 205). http://redbiblio.unne.edu.ar/opac/cgi-bin/pgopac.cgi?VDOC=41.121501. Accessed June 15, 2010.

Vázquez, M., Mazzoni, E., Ruiz Posse, E., & Baetti, C. (2004). Aplicación de un S.I.G. para el estudio de Pequeñas Cuencas Endorreicas del Sur de la Patagonia. In *Jornadas de Geografía Física Argentina* (pp. 205–213). Universidad Católica de Santa Fé.

Vázquez, A. (2012). Erupción del volcán Hudson y crisis ganadera en la meseta central santacrucenia, Patagonia argentina, https://www.researchgate.net/publication/254411820_ERUPCION_DEL_VOLCAN_HUDSON_Y_CRISIS_GANADERA_EN_LA_MESETA_CENTRAL_SANTACRUCENA_PATAGONIA_ARGENTINA. Accessed September 15, 2024.

Vecchi, R. (2011). Bolas de boleadora del curso inferior del río Salado: Análisis de procedencia de materias primas. In J. G. Martínez, & D. L. Bozzuto (Comp.), *Armas prehispánicas: Múltiples enfoques para su estudio en Sudamérica* (pp. 195–204), Fundación de Historia Natural Félix Azara.

Velásquez, H., Méndez, C., Reyes, O., Trejo, V., Sanhueza, L., Quiroz, D., & Jackson, D. (2007). Late open air residential camps at upper Río Cisnes (Aisén Región): Appeleg 1(CIS 009). *Magallania, 35*, 121–132.

Vicent, J. (1991). Fundamentos teórico-metodológicos para un programa de investigación arqueogeográfica. In P. Lopez García (Ed.) *El cambio cultural del IV al II milenios a. C. en la comarca noroeste de Murcia* (pp. 31–117). Consejo Superior de Investigaciones Científicas.

Vignati, M. (1934). Resultados de una excursión por la margen sur del río Santa Cruz. *Notas preliminares del Museo de La Plata. Tomo II* (pp. 78–151).

Wandsnider, L. A. (1996). Describing and comparing archaeological spatial structures. *Journal of Archaeological Method and Theory, 3*(4), 319–384. http://digitalcommons.unl.edu/anthropologyfacpub/70

Wandsnider, L., & Camilli, E. (1992). Character of surface archaeological deposits and its influence on survey accuracy. *Journal of Field Archaeology, 19*(2), 169–188.

Warren, R. (1990a). Predictive modelling of archaeological site location: A case study in the Midwest. In K. Allen, S. Green, & E. Zubrow (Eds.), *Interpreting space: GIS and archaeology* (pp. 201–215). Taylor and Francis.

Warren, R. (1990b). Predictive modelling in archaeology: A primer. In K. Allen, S. Green, & E. Zubrow (Eds.), *Interpreting space: GIS and archaeology* (pp. 90–107). Taylor and Francis.

Waters, M. (1992). Geoarchaeological foundations. *Principles of Geoarchaeology* (pp. 15–114). The University of Arizona Press.

Waters, M., & Kuehn, D. (1996). The geoarchaeology of place: The effect of geological processes on the preservation and interpretation of the archaeological record. *American Antiquity, 61*(3), 483–497.

Weitzel, C., & Colombo, M. (2006). ¿Qué hacemos con los fragmentos? Un experimento en fractura de artefactos líticos tallados. *La Zaranda De Ideas Revista De Jóvenes Investigadores En Arqueología, 2*, 19–33.

Wescott, K. (2000). Introduction. In K. Wescott, & R. Brandon (Eds.), *Practical applications of GIS for archaeologists. A predictive modeling toolkit* (pp. 1–4). Taylor and Francis Inc.

Wescott, K. L., & Brandon, R. (Eds.). (2000). *Practical applications of GIS for archaeologists: A predictive modeling kit*. London.

Wheatley, D. (1993). Going over old ground: GIS archaeological theory and the act of perception. In J. Andresen, T. Madsen, & I. Scollar (Eds.), *Computing the past: Computer applications and quantitative methods in archaeology* (pp. 133–138). Aharhus University Press.

Wheatley, D. (1995). Cumulative viewshed analysis: A GIS-based method for investigating intervisibility and its archaeological application. In G. Lock, & Z. Stančič (Eds.). *GIS and archaeology: A European perspective, 1* (pp.70–185). Taylor and Francis.

Wheatley, D. (1996). The use of GIS to understand regional variation in earlier neolithic wessex. In H. Maschner (Ed.), *New methods. Old problems. Geographic information systems in modern archaeological research. Occasional paper No. 23* (pp. 75–103). Southern Illinois University at Carbondale.

Wheatley, D., & Gillings, M. (2002). *Spatial technology and archaeology*. Taylor and Francis.

Winterbottom, S. J., & Long, D. (2006). From abstract to rich virtual environments: Landscape contexts in Kilmartin. Scotland. *Journal of Archaeological Science, 33*, 1356–1367.

Woodman, P. (2000a). Beyond significant patterning towards past intentions: The location of Orcadian chambered tombs. In C. Buck, V. Cummings, C. Henley, S. Mills, & S.Trick (Eds.), *U.K. chapter of computer applications and quantitative methods in archaeology, Oxford, British archaeological reports, 2000* (pp. 91–105). Archaeopress.

Woodman, P. (2000b). A predictive model for Mesolithic site location on Islay using logistic regression and GIS. In S. J. Mithen (Ed.). *Hunter-gatherer landscape archaeology: The Southern hebrides mesolithic project 1988–98. The McDonald Institute for Archaeological Research. Cambridge. Vol. 2. Archaeological Fieldwork on Colonsay. Computer modelling. Experimental archaeology and final interpretations* (pp. 445–464). McDonald Institute Monographs.

Woodman, P. (2000c). Beyond significant patterning towards past intentions: The location of Orcadian chambered tombs. In *Proceedings of the UK Chapter of Computer Applications and Quantitative Methods in Archaeology*. In C. Buck, V. Cummings, C. Henley, S. Mills, Trick (Eds.), *Reports international series 844* (pp. 91–105). Archaeopress.

Woodman, P., & Woodward, M. (2002). The use and abuse of statistical methods in archaeological site location modelling. In D. Wheatley, G. Earl, & S. Poppy (Eds.), *Contemporary themes in archaeological computing* (pp. 22–27). Oxbow Books.

Yellen, J. (1976). Settlement patterns of the !Kung: An archaeological perspective. In R. Lee & I. DeVore (Eds.), *Kalahari·hunter-gatherers: Studies of the !Kung and their neighbors* (pp. 47–72). Harvard University Press.

Zárate, M., Blasi, A., & Rabassa, J. (2000). Geoarqueología de la Localidad Piedra Museo. In L. Miotti, R. Paunero, M. Salemme, & R. Cattáneo (Eds.), *Taller Internacional del INQUA La colonización de América durante la transición Pleistoceno/Holoceno. Guía de Campo de la visita a las localidades arqueológicas* (pp. 56–60). INQUA.

Zilio, L. (2013). Chenques en Patagonia Centro-meridional: Análisis de los patrones de distribución espacio-temporales. *Comechingonia, 17*(2), 237–254. Recuperado en 19 de agosto de 2024, de http://www.scielo.org.ar/scielo.php?script=sci_arttext&pid=S1851-00272013000200010&lng=es&tlng=es

Zilio, L., Hammond, H., & Castro, A. S. (2017). Levantamiento planimétrico y análisis liquenométrico en el sitio Campo de Chenques, costa norte de Santa Cruz (Patagonia Argentina). *Chungara, 49*, 65–80.

Zvelebil, M. (2003). Enculturation of mesolithic landscapes mesolithic on the move. In L. Larsson, H. Kindgren., K. Knutsson, D. Loeffer, & A. Akerlund (Eds.), *Papers Presented at the Sixth International Conference on the Mesolithic in Europe, Stockholm 2000* (pp. 65–73). Oxbow.
Zvelebil, M., Green, S., & Macklin, M. (1992). Archaeological landscapes. Lithic scatters and human behaviour. In J. Rossignol, & L. Wandsnider (Eds.), *Space. Time and archaeological landscapes* (pp.193–226). Plenum Press.

MIX
Papier aus verantwortungsvollen Quellen
Paper from responsible sources
FSC® C105338

If you have any concerns about our products,
you can contact us on
ProductSafety@springernature.com

In case Publisher is established outside the EU,
the EU authorized representative is:
Springer Nature Customer Service Center GmbH
Europaplatz 3, 69115 Heidelberg, Germany

Printed by Libri Plureos GmbH
in Hamburg, Germany